Disaster in the Early Modern World

How did early modern societies think about disasters, such as earthquakes or floods? How did they represent disaster, and how did they intervene to mitigate its destructive effects? This collection showcases the breadth of new work on the period ca. 1300–1750.

Covering topics that range from new thinking about risk and securitisation to the protection of dikes from shipworm, and with a geography that extends from Europe to Spanish America, the volume places early modern disaster studies squarely at the intersection of intellectual, cultural and socio-economic history. This period witnessed fresh speculation on nature, the diffusion of disaster narratives and imagery and unprecedented attempts to control the physical world.

The book will be essential to specialists and students of environmental history and disaster, as well as general readers who seek to discover how pre-industrial societies addressed some of the same foundational issues we grapple with today.

Ovanes Akopyan is a Marie Skłodowska-Curie fellow at Ca' Foscari University of Venice.

David Rosenthal is a research fellow at the University of Exeter and co-director of Hidden Cities apps.

Routledge Studies in Renaissance and Early Modern Worlds of Knowledge

Series Editors: Harald E. Braun[1] and
Emily Michelson[2]

[1]*University of Liverpool, UK*
[2]*University of St Andrews, UK*

This series explores Renaissance and Early Modern Worlds of Knowledge (c.1400–c.1700) in Europe, the Americas, Asia and Africa. The volumes published in this series study the individuals, communities and networks involved in making and communicating knowledge during the first age of globalisation. Authors investigate the perceptions, practices and modes of behaviour which shaped Renaissance and Early Modern intellectual endeavour and examine the ways in which they reverberated in the political, cultural, social and economic spheres.

The series is interdisciplinary, comparative and global in its outlook. We welcome submissions from new as well as existing fields of Renaissance Studies, including the history of literature (including neo-Latin, European and non-European languages), science and medicine, religion, architecture, environmental and economic history, the history of the book, art history, intellectual history and the history of music. We are particularly interested in proposals that straddle disciplines and are innovative in terms of approach and methodology.

The series includes monographs, shorter works and edited collections of essays. The Society for Renaissance Studies (http://www.rensoc.org.uk) provides an expert editorial board, mentoring, extensive editing and support for contributors to the series, ensuring high standards of peer-reviewed scholarship. We welcome proposals from early career researchers as well as more established colleagues.

For more information about this series, please visit: https://www.routledge.com/Routledge-Studies-in-Renaissance-and-Early-Modern-Worlds-of-Knowledge/book-series/ASHSER4043

Disaster in the Early Modern World

Examinations, Representations, Interventions

Edited by Ovanes Akopyan and David Rosenthal

Routledge
Taylor & Francis Group

NEW YORK AND LONDON

First published 2024
by Routledge
605 Third Avenue, New York, NY 10158

and by Routledge
4 Park Square, Milton Park, Abingdon, Oxon, OX14 4RN

Routledge is an imprint of the Taylor & Francis Group, an informa business

ISBN: 978-0-367-46597-1 (hbk)
ISBN: 978-1-032-58019-7 (pbk)
ISBN: 978-1-003-02982-3 (ebk)

DOI: 10.4324/9781003029823

Typeset in Sabon
by MPS Limited, Dehradun

Contents

Figures

Table

Introduction

Ovanes Akopyan and David Rosenthal

1 November 1755 became one of the grimmest days in the history of Portugal.[1] A massive earthquake, followed by a tsunami, destroyed most of Lisbon and claimed thousands of lives. It significantly affected the state's political and social stability and disrupted its colonial ambitions. What is more, the 'great Lisbon earthquake' heavily impacted territories outside Portugal, causing extensive damage and numerous casualties in Spain and northwest Africa. It is no wonder, therefore, that such an event did not go unnoticed among contemporaries. News about the catastrophe's devastating effects quickly spread across Europe, prompting reflections on the part of many prominent scholars and philosophers.[2] For some, who were attempting to establish the natural conditions behind disastrous occurrences, the Lisbon earthquake provided a wealth of information about the course and consequences of the calamities. Among those was Immanuel Kant, who over the course of three months in 1756 published three essays on the origins of earthly tremors in a weekly paper *Wöchentliche Königsbergische Frag- und Anzeigungs-Nachrichten*.[3] For others, such as Voltaire, the disaster reinforced the question of theodicy and led to another round of intense discussions on where to draw a line between the idea of a loving and almighty God and the human suffering occasioned by destructive manifestations of nature, such as the one that hit Lisbon. Against the adherents of philosophical optimism, for whom people live in the best of all possible worlds that, in spite of its seeming imperfections, is driven towards the Good, Voltaire composed two texts. In the *Poème sur le désastre de Lisbonne*, published as early as March 1756, he firmly dismissed the notion that any world shaken by such events as the Lisbon earthquake could be conceived and governed by a benevolent deity.[4] Sometime later followed his renowned satirical novel *Candide*, where under the guise of professor Pangloss, who 'taught metaphysico-theologico-cosmocodology', Voltaire mocked what he saw as the Christian optimistic bias and fatalistic acceptance of evil.[5] This bold appeal to get rid of the 'superstitious' understanding of nature, along with a Europe-wide consciousness

DOI: 10.4324/9781003029823-1

about the events that was achieved through new channels of information, permitted some scholars to label the Lisbon earthquake as the first disaster of modernity.[6]

Attractive as it may seem, this straightforward binary of rational modernity and an irrational and traditional past is clearly both a simplistic and outdated construct. First, a closer look at the general public's reactions to various cataclysms, which have become more frequent and extreme in the light of the current climate crisis, reveals a more complex picture than is usually thought. It would be naïve to assume that the pervasive infiltration of media and technology into our everyday lives has led to a universal awareness, or acceptance, of scientifically expounded causes of natural disasters. On the contrary, 'irrational' vision remains to some extent popular, at times promoted by the media. Moreover, from an anthropological point of view, the belief that catastrophes have a hidden providential character and therefore urge humankind to transform in one way or another seems to be deeply rooted in human imagination; and it might be beyond the power of science to change the way many people reflect on such matters.[7]

Second, contrary to an opinion articulated in some publications, Kant's contribution could hardly be regarded as the first proper, natural philosophical attempt to investigate the origins of earthquakes. Already in his *Natural Questions* the famous Stoic philosopher Seneca had warned against ascribing the causes of earthly tremors to the intervention of a vengeful deity and argued for a purely natural explanation.[8] Considering that, unlike many other ancient works on philosophy, the *Naturales quaestiones* did not fall into oblivion after the collapse of the Roman Empire and continued to be read during the Middle Ages, rationalising the disaster of earthquakes was by no means an Enlightenment invention.[9] Furthermore, Kant's contention that tremors might be caused by a combination of the conflagration of certain minerals (such as sulphur) on the Earth's surface and water compressed in extensive caverns beneath it does not in fact constitute a new argument and is largely reminiscent of the solutions suggested by his predecessors.[10]

Instead of searching in vain for a replacement for modernity's 'first disaster', this collection of essays focuses on the gradual emergence of a systematic discourse of disasters, both in theory and practice, which resulted in a significant change in the way humans engaged with nature. In other words, this volume is not about how specific events, be it the Lisbon earthquake or any other remarkable calamity, had a transformative and defining effect on human perception of catastrophes, but rather on the intellectual, socio-political and economic structures behind such thinking and their developments over time. Thus, against a backdrop of environmental history, which sees the relationship between human societies and the environment as a complex and interactive process, we would argue that speculation on 'nature', and ideas and practices aimed at mitigating and coping with its extremes, increasingly coalesced, and that in the early modern period we can see a major acceleration in this process. On the one hand, early modernity witnessed fresh natural philosophical approaches to disaster and hazard, in a complicated relationship with religious or providential explanatory paradigms, as well as the widespread diffusion of information and commentary on disaster through print. Simultaneously, the so-called Little Ice Age and its subsequent effects meant that attempts to control the physical world reached an unprecedented level.[11] In conjunction with the aim of averting damage and mitigating community distress, such efforts were widely exploited to economic and political ends. As a result, the combination of these two fundamental components – new ways of thinking about disaster and intensified environmental interventions – singles out early modernity as a period of particular importance in the development of human interactions with the natural world.

In recent years, a number of key studies have firmly established a conceptual framework that places disaster studies at the intersection of intellectual, cultural and socio-economic history. In *Natural Disasters, Cultural Responses*, for example, Christian Mauch and Christian Pfister gathered case studies to illustrate how the social, cultural, political and economic patterns that determined human responses to catastrophe changed over time and differed according to region. Their cross-chronological and transcultural approach has been echoed in subsequent publications, including *Historical Disasters in Context, Historical Disaster Experiences* and, more recently, *Disasters in History*, which looks to show that 'biophysical shocks and hazards' produce a diversity of outcomes linked to factors such as resilience and adaptability.[12] While the early modern period has also received its share of scholarly attention, to date most studies have addressed a specific topic, such as disaster accounts in a single geographical area or related to a particular type of disaster, or have focused on a specific form of engagement with nature.[13] The aim of this collection is instead to showcase the breadth and diversity of fresh work on early modern disaster, while building on the wide-ranging interdisciplinarity of the more general disaster studies cited above. In line with this goal, the volume is structured according to the main responses to disaster: examinations, representations and interventions, with complementary themes emerging across the three sections.

The case studies gathered in Part 1, 'Examinations', reveal that early modern scholars did not avoid certain patterns of environmental reasoning that had persisted since antiquity and are often reproduced even today. However, there were several distinct features, which make it possible to distinguish an 'early modern disaster'. In the section's opening essay, on the basis of rich linguistic and historical material, Gerrit J. Schenk unveils the changing attitudes to the future roughly between the fourteenth and early sixteenth centuries and what it could bear. By tracing the terminological shift that occurred during this period, he demonstrates how standard notions of *fortuna* and *providentia Dei*, which were usually applied when speaking of cataclysms, came to be complemented with two neologisms, *disastro* and *resicum/rischio*. Following the methodology of *Begriffsgeschichte*, Schenk fairly contends that the coinage of the two terms was not merely accidental but in effect echoed the rise of two new patterns of interpreting calamities. While *dis-astro* suggests that a certain occurrence is influenced by celestial constellations or stars (*astri*) and thus, through uniting the supernatural and the terrestrial worlds, means calamity is induced from the sphere of the divine, *resicum/rischio* originated as a consequence of social and economic developments. First emerging in commercial contracts of Italian merchants, who tried to insure their goods against possible losses at sea, 'risk' represented not only a crucial step towards what Schenk calls 'securitisation' of financial assets and institutionalisation of welfare and insurance policies purported to help people cope with disastrous manifestations of nature; it also pointed to another change of fundamental importance – the ways in which early modernity viewed and confronted fate. It appears that the question of how to calculate personal and financial 'risks', which arose within mercantile circles and was particularly pertinent to long-distance trade, formed an intrinsic element of large-scale discussions on predestination, fatalism and fortune, permeating all layers of early modern thought, from theology and philosophy to the visual arts.[14] It is no surprise that disasters were a natural and integral part of that overarching discourse.

A second feature of early modernity in relation to disaster is the growing number of natural philosophical interpretations of their origins. An avalanche of ancient texts rediscovered in the Renaissance, such as Lucretius' *De rerum natura* or the Platonic corpus, as well as a constantly growing body of experimental evidence amassed throughout Europe and the rest of the world, slowly began to corrode the supremacy of Aristotle's teachings on nature. It would certainly be an exaggeration to claim that Aristotelianism lost its dominance in the early modern period, but the challenges posed to its authority by competing philosophies and new discoveries contributed to diversifying solutions. This meant that the *Meteorologia* and the Stagirite's other works were increasingly subjected to scrutiny and came to be considered as one among a variety of explanations of natural effects.[15] The three essays of Ovanes Akopyan, Lydia Barnett and Sara Miglietti exemplify the changing interpretative perspectives and demonstrate how they related to other scholarly tendencies of the time. By focusing on Giannozzo Manetti's account of the two dreadful earthquakes that hit Naples and southern Italy in December 1456, Ovanes Akopyan suggests that the *De terremotu* belongs to a well-established tradition of the 'politics of disaster' and should be read in a wider context of Italian Renaissance political and moral philosophy. As the essay shows, Manetti's principal aim was to explore whether there existed any connection between the political virtues of the ruler and disasters occurring under their reign. Considering that the *De terremotu* was commissioned by Manetti's patron, Alfonso of Aragon, it is little surprise that Manetti did not respond to the question in the affirmative. Though Manetti's position might appear conformist, it reveals a specific and extremely influential line of thinking about disasters. Although there were certainly more politicised calamities in early modern thought – plagues and floods have been seen as particularly emblematic in this regard – earthquakes were in fact no exception; and Manetti's *De terremotu* was among the most significant texts to view earthquakes in relation to political theory, ethics and discussions on fatalism. At the same time, it illustrates how Aristotelian natural philosophy, which was increasingly seen as just one possible framework for interpreting the effects of nature, albeit still the most authoritative and plausible, was adaptable to early modern scholars' theological and political needs.

The two following essays bring the reader to a period when fresh reflections on disaster and post-Reformation moralisations of nature were progressively converging. By attempting to ground natural knowledge on a literal reading of the Scriptures, natural theology, also termed physico-theology or Mosaic physics, offered a way of reconciling the new sciences with Christian teachings.[16] It postulated that God is indeed a benevolent being that pursues a certain plan. As various manifestations of that plan are scattered across the globe in the form of natural artefacts, it is the job of a natural theologian to uncover their significance within the universal design. Perhaps the most emblematic instance of this kind was the early modern fascination with fossils, which were commonly recognised as a revealing natural argument for the diluvial interpretation of the Earth's history.[17] Given that contemporary communicative means fostered intellectual exchange among even the most distant parts of the *res publica literaria*, it is no surprise that the physico-theological movement attracted numerous followers across the whole of Europe and from all Christian denominations. With particular emphasis on disaster narratives, Lydia Barnett's and Sara Miglietti's articles reveal the multi-faceted and complex character of physico-theological reasoning.

Barnett argues that while modern representations of the biblical Flood usually focus on the happy and heroic preservation of non-human animal species by Noah and his family, its early modern version was far more ambivalent. Along with devoting considerable effort to determining the size and carrying capacity of the Ark, Christian Europeans were equally preoccupied with the millions of animals who were not taken on board the Ark and had died in the Flood. These non-saved animals presented early modern Christians with a moral and theological conundrum, resulting in visual representations of the onset of the Flood with monkeys, cows, lions and other animals being overcome by floodwaters that they had done nothing to deserve or to provoke. This essay shows how the figure of the innocent animal at the time of Noah's Flood worked as a site for articulating the spiritual and environmental distinction between human and non-human animals. Animals killed by the Flood served as emblems of humanity's moral responsibility towards non-human animals, but they also functioned as a means of distinguishing between types of creatures who did and did not possess geologic agency. As sinners, humans possessed the unique capacity to provoke catastrophic changes in nature on a planetary scale, as represented by the Universal Deluge. Other animals, who could not sin or act as spiritual agents, therefore lacked this potent and deeply ambivalent capacity to instigate environmental harm.

As Sara Miglietti demonstrates, one of the most famous proponents of physico-theology, the Swiss naturalist Johann Jakob Scheuchzer, was similarly preoccupied with the issue of human geological agency. In a paper he sent to the Royal Society of London in the winter of 1707/1708, he imagined the catastrophic consequences that climate warming in the Alps would have for all of Europe. Arguing against those who thought that a warmer climate would transform the bleak Swiss highlands into fertile gardens, Scheuchzer called attention to the ecological function of glaciers as reservoirs of Europe's waters and stabilisers of Europe's climate. By examining Scheuchzer's thought experiment against the backdrop of early modern ideologies of environmental improvement that described man as God's helper in perfecting the earth, Miglietti argues that in the *De ignis seu caloris certa portione Helvetiae adsignata* ('On the specific amount of fire or heat assigned to Switzerland'), as well as in his later *Physica sacra*, Scheuchzer advocated for greater humility and wisdom in judging and modifying the earth, God's already perfect creation. Concurrently, Miglietti's close reading of the *De portione* suggests that in his 'visions' of a new devastating Flood, Scheuchzer drew heavily upon not only, or primarily, the biblical account but far more upon Seneca's *Natural Questions*. This detail once again confirms the extent to which early modern scholars remained indebted to the classical tradition and continued to borrow, both in style and content, from their ancient predecessors.

Through the example of a university dissertation published in Halle in 1709 under the name of a certain Georg Remus, William Barton addresses the third problem associated with early modern disasters and concerned with the accumulation and circulation of knowledge about them. Written on the occasion of a particularly cold winter, currently known as the 'Great Frost', and defended under (and, most likely, co-authored with) Christian Wolff, Remus' *Consideratio physico-mathematica hiemis proxime praeterlapsae* ('A Mathematical and Physical Study of Last Winter') was among the most meticulous and sophisticated contemporary treatments of the Frost. What is more, it became the principal source for the Royal Society's account of the Frost, composed by the English clergyman and natural theologian William Derham. While the first part of Derham's work focused on the conditions in Britain and was supplemented with information from numerous letters from his British colleagues, he moved on to the rest of Europe in the second section, where Remus' text (given to Derham by the leading English exponent of physico-theology, antiquarian and geologist John Woodward) is analysed in detail. Thus, besides offering a detailed study of the dissertation, Barton's chapter sets Remus'/Wolff's piece in the context of early eighteenth-century circulation of 'scientific' knowledge and showcases the mechanisms behind its adaptation.

These three pillars – a new socio-political environment that allowed for previously unfeasible ways to tackle disasters' effects; an ever-increasing number of interpretations of calamities' origins, many of which were based on first-hand evidence and/or went beyond the Aristotelian natural philosophical framework; and an extensive network of early modern *literati*, who, be they just curious individuals or members of recently established scholarly institutions, contributed to an effective circulation of knowledge – defined manifold reflections on disasters. Along with the continuous influence of the classical tradition and the development of new artistic tastes, these were the factors determining representations of calamities. With its four essays, the volume's second part sheds light on how some of these representations were executed.

Martin Korenjak traces the history of the scholarly observation and representation of avalanches. Although quite common in mountainous regions, the avalanche's genesis was largely unexplored for a long time, and even today there remain many misconceptions about its cause. As Korenjak indicates, descriptions of avalanches from classical and medieval times were strikingly fragmented, and they first became the object of a more systematic study only in the sixteenth century in the region where the phenomenon occurred on a regular basis – the Old Swiss Confederacy. Usually reported by eyewitnesses, for whom they were frequent events, avalanche accounts never associated these phenomena with divine intervention, nor were they explained in terms of Aristotelian or early modern physics. As a result, and in contrast to what one might anticipate, representations of these disasters were to a great extent practical, and their impressive and horrifying character played little or no role in early modern descriptions.

Although dealing with two states remote from one another, the following two essays investigate how artistic solutions aimed to help local communities overcome fear and anxiety in the face of devastating occurrences, either by invoking divine intercession or by promoting their ruler's image as someone capable of controlling nature's extremes. By looking at public responses to the catastrophic events that afflicted the Spanish-American territories in the early modern period, Milena Viceconte explores how religious imagery, developed in the European context of the Spanish Empire, found its way to the 'New World' and began to intermingle with indigenous culture. Through a meticulous analysis of a substantial body of primary material, including visual sources, archival documents, religious books, travel journals and chronicles, this chapter establishes a common visual language of catastrophe pertinent to the whole of the Empire. Felicia Else, in turn, focuses on early modern Florence's well-grounded fear of high floodwaters. Her study examines a variety of paintings, bronze medals and gilt reliefs to reveal the strategies artists adopted in representing Ducal and Grand Ducal authority over waterways and celebrating the accomplishments of Cosimo I and his heirs. Real and imagined, idealised and practical, these works were prominent markers of the legacies of water control that Medici Dukes and Grand Dukes promoted for themselves and their dynastic line, one of the many ways early modern societies reacted to the hazards of their environment and the fear of natural disasters.

Not only does Pamela Long's chapter reconstruct the formation of a specific literary tradition of flood descriptions that emerged in Renaissance Rome as a consequence of numerous disasters hitting the city over the course of the fifteenth and sixteenth centuries, it serves as a connection to the volume's third part, devoted to 'Interventions'. In the span of approximately a hundred years, between 1476 and 1598, Rome experienced a series of catastrophic floods that caused multiple drownings of humans and animals, regularly destroyed much of the city and its infrastructure, and led to hunger and sometimes even starvation. It is little wonder, therefore, that the constant threat of flooding gave rise to an array of writings: poems, official reports, learned treatises and practical tracts. As Long argues, these works not only described particular floods and discussed their possible origins, usually with extensive references to classical sources, but also sought to offer solutions that would allow to prevent such periodic catastrophes in future.

The Roman texts give the reader a glimpse of the shifting dynamics between society and state in relation to disaster responses or – to use modern parlance – crisis management. The essays gathered in Part 3 investigate this question further. On the one hand, the early modern state accumulated enough power and resources to intervene in cases of disaster.

Following the definition coined by the twentieth-century legal thinker Carl Schmitt, sovereignty should be regarded as coterminous with the authority to declare a state of emergency (*Ausnahmezustand*), and a more complex and sophisticated structure of the state apparatus was instrumental in transforming the organisation and self-representation of power in (early) modernity.[18] In this context, Michel Foucault and modern biopolitical discourse have famously concentrated attention on governmental responses to plagues occurring in the late seventeenth century as a milestone in the evolution of the modern state.[19]

At the same time, there transpired significant changes in 'bottom-up' reactions to disaster, that is, in the ways in which ordinary members of society confronted the politics of disaster and evaluated governmental efficiency in the light of their experiences with various kinds of emergencies. Although as in the Middle Ages, crisis management continued to be in large part exemplified by the ruler or saint reckoned to be capable of, among other things, producing miracles, curing the citizens' diseases or diverting nature's negative influences,[20] early modern discourse added a new dimension. Preoccupied with new political and social norms, such as the introduction of insurance and other means aimed at protecting the community and individuals alike, the idea that the state ought to act as effectively and aggressively as possible to prevent disaster from happening became more commonplace. As a result, it developed the technologies and resources to try to reshape nature at its own discretion. This view of the state seems to have planted the seeds for an emerging theory of social contract between populace and government, which was expected to preserve at any cost the existing social order and offer protection in the face of disasters. Thus, Part 3 reconstructs a broad spectrum of approaches to the nexus of socio-political, economic and natural events as perceived by both power holders and the general populace.

Emanuela Ferretti investigates the connections between Leonardo da Vinci and hydraulic works on the Arno, in particular the failed war scheme to divert the river in 1503–1504. Through observations on this river, the artist developed a number of studies on flooding, examining bridges, dams, channels, canals and levees connected with the Arno. Leonardo's activity for Cesare Borgia (1502) and then for Pier Soderini (1503–1504), in the context of plans to divert the Arno near Pisa, points up the relationship among hydraulic engineering, observations on nature and the measurement of the terrain. Drawings and writings in Leonardo's corpus, such as the well-known Windsor drawings (ca. 1503), which describe aspects of Leonardo's project for the 'gran canale', can be related to his studies on the Arno to protect Florence from flooding. This chapter thus highlights the relationship between control, transport, regulation and measurement – both in peacetime and in wartime – to prevent flooding or, conversely, to cause flooding against an enemy or deny that enemy access to the sea. As Ferretti argues, Leonardo's ideas may not have been realised, but his imaginative thinking echoed into the sixteenth century, influencing the extensive river works in Tuscany undertaken by Grand Duke Cosimo I de' Medici and his successors.

In her piece, Monica Azzolini uses the example of St Francisco Borja, one of the many 'disaster saints' that grace the pantheon of Catholic Christendom, to look at how discourses of sanctity served political goals in providing the general populace with a reliable means of consolation in the face of the unknown. Through the story of how a painted image of Francisco Borja started to perspire 'miraculously' in the small Andean town of Tunja in 1627, Azzolini illustrates the relationship between nature and the divine in the deeply Catholic Kingdom of New Granada and suggests that this relationship was framed by local and political interests that supported, promoted and legitimised miraculous events aimed at cementing Borja's saintly reputation with the help of both the Jesuits and the local and global powers of the Spanish Empire.

Finally, Adam Sundberg brings the reader to early eighteenth-century Holland, where dike authorities discovered a hitherto little-known species of shipworm (*Teredo navalis*) burrowed into the wooden revetments that protected coastal dikes. This wood-boring mollusk undermined dike stability, threatening cities and valuable agricultural land. Fearing an existential disaster if dikes failed, authorities enacted a capital-intensive dike reconstruction programme, replacing wooden components with imported stone. This transformation is well known in Dutch water history and considered a pivotal moment in dike modernisation. However, little scholarship has explored the process of adaptation or the broader context of the Dutch experience with the mollusk. As a matter of fact, shipworms were a primary hazard of oceanic travel and mariners had contended with them for centuries. Although dike authorities often emphasised the novelty of the threat, they nevertheless proposed, tested and implemented shipworm 'remedies' derived from this maritime knowledge. Sundberg's chapter thus explores the challenges shipworms presented for both dikes and ships, the limitations of translating maritime adaptations to the coasts and dike authorities' ultimate decisions to accept or reject these strategies. As a result, this integrated perspective presents a richer understanding of how the transfer and adaptation of knowledge could inform economic and political decisions.

In conclusion, it should now be apparent that the discussions surrounding the Lisbon earthquake of 1755 did not appear out of the blue. As the case studies in this volume collectively demonstrate, they were in fact preceded by wide-ranging and intensive reflections on what could trigger disasters, whether there were ways to mitigate their destructive effects, and how they related to contemporary political and socio-economic life. Indeed, the manner in which we now reflect and act on many foundational issues related to nature does not differ as much as we might think from that of our early modern forerunners.

Notes

1 The most detailed interdisciplinary study of the Lisbon earthquake is, probably, *The 1755 Lisbon Earthquake: Revisited*, ed. by Luis A. Mendes-Victor, Carlos Sousa Oliveira, João Azevedo, and António Ribeiro (Dordrecht; New York; London: Springer, 2009).
2 For how the earthquake was perceived by contemporaries, including Voltaire, Rousseau and their peers, see above all *The Lisbon Earthquake of 1755: Representations and Reactions* (Oxford: Voltaire Foundation; Oxford University Press, 2005).
3 Immanuel Kant, 'Natural Science', ed. by Eric Watkins, trans. by Lewis White Beck et al., in *The Cambridge Edition of the Works of Immanuel Kant*, vol. 14 (Cambridge; New York: Cambridge University Press, 2012), pp. 327–73.
4 For the text, its context and subsequent reactions, see Voltaire, 'Poème sur le désastre de Lisbonne', ed. by David Adams and Haydn Mason, in *Complete Works of Voltaire*, vol. 45A (Oxford: Voltaire Foundation: Oxford University Press, 2009). pp. 269–358.
5 See Voltaire, *Candide and Other Stories*, trans. by Roger Pearson (Oxford; New York: Oxford University Press, 2006), in particular chapters 5–6 (pp. 12–15).
6 This image is supported, among others, in Mark Molesky's *This Gulf of Fire: The Destruction of Lisbon, or Apocalypse in the Age of Science and Reason* (New York: Knopf, 2015).
7 Two recent studies have explored how deeply politics, faith and disasters are connected in human imagination: Philip Jenkins, *Climate, Catastrophe, and Faith: How Changes in Climate Drive Religious Upheaval* (Oxford; New York: Oxford University Press, 2021); Niall Ferguson, *Doom: The Politics of Catastrophe* (New York: Penguin, 2021).
8 Seneca, *Natural Questions*, VI, 3, 1. For a detailed study of Seneca's natural philosophical reflections, see Gareth Williams, *The Cosmic Viewpoint: A Study of Seneca's* Natural Questions (New York: Oxford University Press, 2012).
9 On the reception of *Natural Questions* and Seneca's natural philosophy more broadly, see *Seneca e le scienze naturali*, ed. by Marco Beretta, Francesco Citti, and Lucia Pasetti (Florence: Olschki, 2013).
10 On early modern discussions on the origins of earthquakes, see Rienk Vermij, *Thinking on Earthquakes in Early Modern Europe: Firm Beliefs on Shaky Grounds* (London; New York: Routledge, 2021).
11 For some important studies, see Tim Soens, 'Flood Security in the Medieval and Early Modern North Sea Area: A Question of Entitlement?', *Environment and History*, 19, 2 (2013), 209–32; John Morgan, 'The Micro-Politics of Water Management in Early Modern England: Regulation and Representation in Commissions of Sewers', *Environment and History*, 23, 3 (2017), 409–30; *Governing the Environment in the Early Modern World: Theory and Practice*, ed. by Sara Miglietti and John Morgan (Abingdon; New York: Routledge, 2017); Pamela O. Long, *Engineering the Eternal City: Infrastructure, Topography, and*

the Culture of Knowledge in Late Sixteenth-Century Rome (Chicago; London: University of Chicago Press, 2018).

12 *Natural Disasters, Cultural Responses: Case Studies toward a Global Environmental History*, ed. by Christof Mauch and Christian Pfister (Lanham, MD: Lexington Books, 2009); *Historical Disasters in Context: Science, Religion, and Politics*, ed. by Andrea Janku, Gerrit Jasper Schenk, and Franz Mauelshagen (Abingdon; New York: Routledge, 2012); *Historical Disaster Experiences. Towards a Comparative and Transcultural History of Disasters Across Asia and Europe*, ed. by Gerrit Jasper Schenk (Cham: Springer, 2017); Bas van Bavel, Daniel R. Curtis, Jessica Dijkman, Matthew Hannaford, Maïka de Keyzer, Eline van Onacker, and Tim Soens, *Disasters and History: The Vulnerability and Resilience of Past Societies* (Cambridge: Cambridge University Press, 2020).

13 For example, see Matthew Mulcahy, *Hurricanes and Society in the British Greater Caribbean, 1624–1783* (Baltimore: Johns Hopkins University Press, 2008); Sean Cocco, *Watching Vesuvius: A History of Science and Culture in Early Modern Italy* (Chicago; London: University of Chicago Press, 2013); *Disaster Narratives in Early Modern Naples: Politics, Communication and Culture*, ed. by Domenico Cecere et al. (Rome: Viella, 2018); *Une histoire du sensible: la perception des victimes de catastrophe du XIIe au XVIII siècle*, ed. by Thomas Labbé and Gerrit Jasper Schenk (Turnhout: Brepols, 2018); John Henderson, *Florence under Siege: Surviving Plague in an Early Modern City* (New Haven; London: Yale University Press, 2019).

14 Nicholas Scott Baker, *In Fortune's Theater: Financial Risk and the Future in Renaissance Italy* (Cambridge; Cambridge University Press, 2021); *Fate and Fortune in European Thought, ca. 1400–1650*, ed. by Ovanes Akopyan (Leiden; Boston: Brill, 2021).

15 For some recent and most interesting studies on Renaissance Aristotelianism, see above all Craig Martin, *Renaissance Meteorology: Pomponazzi to Descartes* (Baltimore: Johns Hopkins University Press, 2011); Paul Richard Blum, Studies on Early Modern Aristotelianism (Leiden; Boston: Brill, 2012); Craig Martin, *Subverting Aristotle: Religion, History, and Philosophy in Early Modern Science* (Baltimore: Johns Hopkins University Press, 2014); Eva Del Soldato, *Early Modern Aristotle: On the Making and Unmaking of Authority* (Philadelphia: University of Pennsylvania Press, 2020).

16 Ann Blair, 'Mosaic Physics and the Search for a Pious Natural Philosophy in the Late Renaissance', *Isis*, 91, 1 (2000), 32–58; *Physico-theology: Religion and Science in Europe, 1650–1750*, ed. by Ann Blair and Kaspar von Greyerz (Baltimore: Johns Hopkins University Press, 2020); Kaspar von Greyerz, *European Physico-Theology (1650–c.1760) in Context. Celebrating Nature and Creation* (Oxford: Oxford University Press, 2022).

17 William Poole, *The World Makers: Scientists of the Restoration and the Search for the Origins of the Earth* (Oxford: Peter Lang, 2010). For early modern discussions about the Deluge and its role in the history of the Earth, see Lydia Barnett, *After the Flood: Imagining the Global Environment in Early Modern Europe* (Baltimore: Johns Hopkins University Press, 2019). See also Ivano dal Prete's *On the Edge of Eternity: The Antiquity of the Earth in Medieval and Early Modern Europe* (Oxford: Oxford University Press, 2022).

18 Carl Schmitt, *Political Theology: Four Chapters on the Concept of Sovereignty*, trans. by George Schwab (Chicago; London: University of Chicago Press, 2005), p. 5: "Sovereign is he who decides on the exception." For another important

philosophical work on biopolitics and governmental responses to emergencies that was deeply inspired by Schmitt's account, see Giorgio Agamben, *State of Exception*, trans. by Kevin Attell (Chicago; London: University of Chicago Press, 2005). On transformations of the structure of the state and how they affected decision-making, see also Isaac Ariall Reed, *Power in Modernity: Agency Relations and the Creative Destruction of the King's Two Bodies* (Chicago; London: University of Chicago Press, 2020).

19 For Foucault's famous treatments of the matter, see his *Discipline and Punish: The Birth of the Prison*, trans. by Alan Sheridan, 2nd edn (New York: Vintage Books, 1995) and *Security. Territory. Population. Lectures at the Collège de France, 1977–1978*, ed. by Michel Senellart, trans. by Graham Burchell (Basingstoke; New York: Palgrave Macmillan, 2007).

20 The classic account is Marc Bloch's famous *The Royal Touch: Sacred Monarchy and Scrofula in England and France*, trans. by John Edward Anderson (London: Routledge & Kegan Paul, 1973).

Part 1

Examinations

1 Taming the future?

From 'natural' hazards and 'disasters' to a securitisation against 'risks'[1]

Gerrit Jasper Schenk

Introduction

The bishop of Strasbourg expected the worst: not only had he received news of the advance of the Ottoman Turks at the time when religious unrest was increasing in his diocese in connection with the activities of the Protestant Reformers – but also the stars seemed to foretell nothing good. Experienced astrologers had forecast heavy flooding. In a letter to the Strasbourg city council on 11 March 1523, Bishop Wilhelm (III) of Hohnstein (1505–1541) called these threats a *flagella Dei*, a punishment for the sins of religious disunity.[2]

In his letter, Wilhelm asked the Council to permit a huge procession in order to avert the looming danger through this sign of penance.[3] The procession was to involve all collegiate churches, parishes and monasteries in the city and take place on 25 March after they had held the Mass for the Feast of the Annunciation. The Virgin Mary was not only the patron of the cathedral but, as the patron of Strasbourg, had long enjoyed the veneration of the Council and the city.[4]

This list of threats is typical of the early sixteenth century: the Turks pushing their way forward on the Balkans,[5] the conflicts connected with the incipient division in the Church[6] and the forecast flooding were susceptible to interpretation as apocalyptic signs of the coming Last Judgement. These signs were readable both from the 'book of nature', in this case the starry sky, and also from the state of society. According to many contemporaries, they were intertwined, and it may be assumed that the bishop was reacting to a contemporary discourse that climaxed in 1523[7]: Johannes Stöffler, the Tübingen professor of mathematics and astronomy, had calculated the daily positions of the planets from 1499 to 1531 in a table printed in 1499. In succinct sentences, he explained the meaning of the constellations in the different signs of the zodiac. Regarding the constellation of stars in 1524 he wrote:

DOI: 10.4324/9781003029823-3

In the month of February there will be 20 conjunctions [...], of which 16 are to be found in an aqueous sign. For almost the whole earth [...] they will mean an adjustment, a change and a transformation [...][8].

A debate about the advent of a heavy flood in the years before 1524 was the consequence, and over 150 documents are extant by about 60 authors in several languages. Records show that the debate was not confined to scholarly circles. One example, maybe as early as around 1520, concerns Sebastian Brant, the well-known humanist and Strasbourg municipal counsellor. In a perhaps illustrated tract with memorable verses, he contributed to spreading information about the forecast disaster, while initially interpreting it as a warning to do penitence and, as such, preventable.[9]

Since money could be earned with the annual forecasts, and the flood debate promised huge editions in view of the conflicts caused by the stirrings of the Reformation, the publications multiplied immediately before 1524. The title page of a 1523 pamphlet by Johann Virdung of Haßfurt, the court astrologer from the Palatinate, transposes the forecast into a picture (Figure 1.1)[10]: a celestial constellation in the sign of Pisces, defined by astronomers–astrologers with an astrolabe and quadrant in their hand, exercises an apocalyptic influence on the earth. We see a flood, wheat fields and vines ruined by hailstorms, ordinary people rising up against the aristocracy and only a small group of people praying. Virdung held true to the Catholic Church and took a moderate position in the text, warning of the threatening doom and underlining at the same time that penitent behaviour could still mitigate God's wrath.

Yet the Strasbourg City Council advised the bishop against a huge procession because of the unrest caused by the Reformation.[11] It recommended a small one, with the collegiate clergy, parish priests and monks bearing the sacrament and, unaccompanied by the Council, only processing briefly around their churches. Bishop Wilhelm bowed to this advice almost subserviently and, at the same time, asked whether the Council could influence the rebellious guilds so that the procession could take place without a sensation.[12]

The procession seems to have taken place, but so inconspicuously that in the seventeenth century, Strasbourg chronicler Johann Wencker noted, with reference to older sources, 'that it was not held publicly'.[13] The normal case was quite different: the initiative for extraordinary processions came from the Council that asked the collegiate clergy of the city to organise the clerical part of the procession and had the priests read out an appeal to attend from the city pulpits.[14]

The disastrous flood predicted by the astrologers did not materialise in 1524 but there were two major floods in the region, in Strasbourg itself on 6 January[15] and to the north, near Beinheim, in autumn 1524,[16] as we know

SOMNIVM SCIPIONIS   I

℧SOMNIVM SCIPIONIS EX CICERONIS LIBRO
DE REPVBLICA EXCERPTVM.

VM IN APHRICAM VENISSEM:A MANLIO
confule ad quartā legionē tribunus(ut fcitis)militum: ni
hil mihi fuit potius:ꝗ ut Maſſiniſſem conuenirē regem fa
miliæ noſtræ iuſtis de cauſis amiciſſimū:ad quē ut ueni cō
plexus me fenex collachrymauit aliquāto : poſt fuſpexit ad
cœlū: & grates inꝗ tibi fumme Sol ago:uobiſꝗ reliqs cæ
lites:ꝗ, anteꝗ ex hac uita migro:confpicio in meo regno:&
in his tectis Publi.Corneliū Scipionē : cuius ego nomine
ipfe recreor : ita nunꝗ ex animo meo difceſſit illius optimi
atꝗ, inuictiſſimi uiri memoria.Deinde ego illū de regno fuo:ille me de noſtra repub.
percōtatus ē:multiſꝗ uerbis ultro citroꝗ habitis:ille nobis cōfumptus eſt dies . Poſt
autē regio apparatu accepti:fermonē in multā noctem ꝑduximus:cum fenex nihil ni
fi de Aphricā loꝗretur:omniaꝗ; nō folū eius facta:fed etiā dicta meminiſſet:deinde
ut cubitum difceſſimus:me & de uia:& quia ad multam noctē uigilaſſem:arctior ꝗ fo
lebat fomnus complexus ē.Hic ergo mihi(credo equidē ex hoc ꝗd eramus locuti : fit
enim fæpe fere ut cogitationes fermonefꝗ; noſtri pariant aliꝗd in fomno tale:quale de
Homero fcribit Ennius:de quo uidelicet fæpiſſime uigilās folebat cogitare & loqui)
Aphricanus fe oſtendit ea forma:ꝗ mihi ex imagine eius ꝗ ex ipfo notior erat: quē ut
agnoui eꝗ dē cohorrui.Sed ille ades inꝗ aio:& omitte timorē Scipio:& quæ dicā me
moriæ trade.Vides ne illā urbem:quæ parere reipub.coacta:per me renouat priſtina
bella:nec poteſt quiefcere?(Oſtendebat autē Carthaginem:de excelfo & pleno ſtel
larū milluſtri & claro quodā loco)ad quā tu oppugnandā nūc uenis pene miles.Hāc
hoc biénio conful euertes:eritꝗ; tibi id cognomen per te partū:ꝗd habes adhuc hære
ditarium a nobis . Cum autem Carthaginē deleueris: triumphū egeris:cenforꝗ; fue

                                                                    A

Vltro cꝯ
troꝗ;

Figure 1.1 Ambrosius Theodosius Macrobius, *Commentary on Cicero's* Somnium
Scipionis (Venice: Giunta, 1513), fol. 1*r*, Bayerische Staatsbibliothek
München, 2 A.lat.b. 493.

from chronicles and municipal records. But the social and religious unrest hit the Strasbourg diocese hard. Strasbourg became Protestant and in 1525 the Peasants' War here was particularly violent.[17] So the classical crisis ritual of a procession[18] to avert the threat of disaster was only partially effective.

An essential precondition for believing in a *fatum astrologicum*,[19] as illustrated in the Strasbourg case, was the emergence of the concept of *disaster* in circa 1400. In the following (part 1), I would like to explore the discursive field connected with the ambivalence of this concept in the context of the contemporary worldview. The everyday measures in dealing with hazards and disasters suggest that the events' interpretative patterns included belief in both God's providence and astrological fate or *fortuna*'s ambivalence.[20] My contention is that these old ideas starting from the High Middle Ages are supplemented by a view that possible losses would be offset against future benefits. Their origins lay partially in the cosmopolitan milieu of Mediterranean trading cities of the High and Late Middle Ages. In view of the opportunities and risks of marine trading, there emerged a calculating understanding of the possibility of disasters that was conceptualised in contracts and tractates as *resicum/rischio*. At the pragmatic level of everyday trading, this 'probabilistic reasoning', as Giovanni Ceccarelli recently put it,[21] was changing the premodern faith in God's will, the power of destiny and the ambivalence of *fortuna*.

In addition, I would like to show that this process was very slow and contradictory (Part 2). Recent studies have rightly pointed out that merchants from the High Middle Ages always systematically distinguished between the share of actual and feared damage to be attributed to a given factor: a hazard (generally not open to an influence), such as a storm, or consequences of intentional actions, e.g., the decision of a captain to leave a port while well aware of the risk of piracy.[22] By extension, and contrary to the view of sociological research, it has been claimed that Niklas Luhmann's classical distinction between (contingent) danger from the system 'environment' and (calculable) risk within the system was developed already in the premodern era.[23] However, this typological distinction simplifies the famous 'non-simultaneity of the simultaneous', meaning the temporal and spatial interconnection of contradictory approaches to risks and threats in theory and practice, which a historian of ideas or a sociologist only retrospectively attempts to construe as a seemingly logical narrative.[24] Furthermore, these recent studies overlook the fact that the dividing lines between chance (or contingency) and necessity took a different course around the medieval millennium than in the modern era (and were also volatile).[25] What was conceptualised in the modern era as a 'natural hazard' does not correspond to medieval ideas of 'natural' phenomena and their openness to influence. One consequence of the different worldviews of premodern times was a different way of dealing with hazards. There was a broad range of very different ways of averting or overcoming those threats that were only understood to be (contingent) natural hazards in the modern age. A detailed treatment is not possible here because it would take some hundred pages, but a rough outline with two case studies from representative regions (Strasbourg in Alsace and Florence in Tuscany)[26] should be sufficient to illustrate my main arguments.

Finally, I will try – yet in a rather superficial way – to fit this gradual transformation of the attribution of risks into the bigger picture of a premodern 'culture of insecurity'[27] and its partial securitisation (part 3). With Benjamin Scheller, I argue for dating the 'birth of risk' in the late Middle Ages,[28] and also for locating it in the Mediterranean world of transcultural exchange, but outside the world of trade. I suggest that those trends were to be put in relation to a worldview that learned slowly how to differentiate between first and second (and more) causes of 'natural' phenomena as a result of the reception of Aristotle's teachings in the thirteenth century. Besides the practical experiences that have always made many interpretations possible, this differentiation also provided (or first allowed?) a theoretical rationale for a reading of natural phenomena that integrated older ideas of *natura* as God's book[29] and opened up many options for attributing disasters to various possibilities, such as God's will, natural effects, human action or lack of it and, finally, also to chance.

Worldviews and patterns of interpretation: 'Disastro' and 'catastrophe'

My analytic approach uses a wider conceptual history in order to develop the historical semantics of certain terms. With Hans Blumenberg and Franziska Rehlinghaus, I understand the term 'fate' or 'disaster' as an 'absolute metaphor', a play on words, a space that can be filled with different, even contradictory, content depending on the usage and discursive context.[30] The older German, French and English tradition of conceptual and discourse history[31] was often limited to an analysis of the 'Höhenkamm' discourse of central texts from the pen of great thinkers and doers. This approach limits the sphere of intellectual reflections on perennial problems, such as 'fate' and 'free will', to the thinking of a social elite. But we easily overlook the fact that, in everyday life, people acted on the basis of certain ideas and attitudes that were not treated or explicated intellectually, and, above all, were rarely written down by the actors themselves. This makes the often quite banal everyday practice of traders, sailors and millers particularly important as it was in mutual connection with the world of lawyers, theologians and philosophers. It is even probable that practical problems led to the use or even development of certain intellectually loaded concepts. By contrast, it is also true that legal treatises similarly served to shape practical action through the way they conceptualised and systematised problems.[32]

In other words, practical problems and the fitting term are in mutual relationship so that we have to look not only for words and concepts, but also for specific patterns of action as indicators of certain worldviews. Analysing action by illiterate persons would therefore presuppose a widening of the scope of research. The corpus of texts to be studied should consequently be extended to legal pamphlets, contracts, council minutes and even pictures (as in Figure 1.1).

In the present everyday understanding, a disaster or catastrophe is thought to be an extreme situation, the sudden irruption of an extraordinary misfortune with grave consequences like an injury and loss of life of humans and animals, a material damage and a destruction on a large scale.[33] However, this understanding of disaster does not necessarily have an objective linguistic equivalent in other times and cultures.[34] In English, the term 'disaster' has been mostly used to describe these phenomena. Also found in German (*Desaster, Unstern*) and the Romance languages (French *désastre*, Italian *disastro* and Spanish *desastre*), its origins go back to the astro-meteorological prehistory of the term. This corresponds to the idea found in antiquity and the ancient Near East that certain constellations of stars were responsible for a fatal 'turn' in the human world.

As I showed some years ago,[35] the word *disastro* is a thirteenth-century Italian neologism composed of Greek and Latin elements (Latin *dis-*, and the Greek ἄστρον) that arose in an environment of linguistic and cultural exchange, probably in Tuscany.[36] While rendering, perhaps, an Occitanian version of the narrative cycle of the 'Seven wise masters' in his vernacular, an anonymous translator conveyed the meaning of the apparently unknown Old French word *micieffo* to his readers by using the Italian word *disastro*.[37] It was meant to denote the misfortune inflicted on the hero of the story through a band of robbers. The story goes back to a widespread oriental and Indian narrative (the Sindbād cycle), which came to Europe, perhaps, through the Crusaders.[38] The translation was obviously influenced by this originally Greek-Arabic, Persian or even Indian narrative, along with an increased interest in horoscopes that occurred in Europe in the High Middle Ages. Both these factors favoured the view of a fateful coincidence influenced by certain constellations of stars.[39] Furthermore, early evidence likewise suggests that this new word originated among seafarers and merchants in the Mediterranean between the Levant, North Africa, Sicily, Tuscany, the South of France and Catalonia. Their risky experiences at sea, with their sometimes happy, sometimes terrible endings (Italian: *fortuna di mare*), were apparently known in the narrative versions.[40]

The new word rapidly spread in the most varied contexts: in southern Italian vernacular translations of popular retold classical texts (the destruction of Troy, the Aenead) from the beginning of the fourteenth century, in the Pisan *Breve dell'Ordine del Mare* from the middle of the century, the Florentine chronicle of Baldassarre Bonaiuti from the second half of the century and in the Venetian Gospel Concordance of Iacopo Gradenigo from the end of the century.[41] Despite the resulting considerable semantic differences, the astronomical–astrological relation written into the word gives a frame of interpretation, in all contexts, for the adversities termed disasters – they are not under a lucky star and end badly. This connoted a (presumably gradually fading) relationship between macro- and microcosm and the use of the term shows that this relation was primarily between heaven and humankind – from the fate of an individual to an outcome of battles. The very term *disastro* contains a basic relation between the (divine) natural and (human) cultural spheres. It is an open question as to whether there is a connection between the general circumstances of life in the crisis- and plague-stricken fourteenth century and the success of the new term as an interpretative pattern, either as an indication of a fatalist approach to misfortune or, conversely, as an attempt to master contingency in view of the ambivalence of *fortuna* or *providentia Dei*.[42] Be this as it may, this concept of disaster found its way into English, French and Spanish in the sixteenth century at the latest.[43]

While the power of the elements played a major role for ships, it seems that the word 'disastro' first only went back to the concept of an individual misfortune associated with unfavourable stars and was not related to natural disasters. However, the association was easy to make. At the turn of the late Middle Ages, natural explanations for earthquakes, storms and floods began to spread, having recourse to Aristotelian theories. Mediated by Avicenna's commentary on a passage from Plato's *Timaeus* (22c–23b), entitled *De diluviis* ('On Floods') and translated from Arabic into Latin, there was a discussion on whether there could, on natural grounds, be too much of the four elements air, water, fire and earth.[44] The view that a particular constellation of stars results in the 'victory' of one of the elements over the others was subsequently discussed by Albertus Magnus and other scholars as a possible natural cause of the Great Flood and other floods.[45] Through university teaching and popular wind and water flooding forecasts – like the constantly reformulated Toledo letter, in which an apocalyptic 'flood' through wind or water was predicted in a certain sign of the zodiac – this interpretative pattern gradually spread throughout Europe in the fourteenth and fifteenth centuries.[46]

A telling description of this notion of linking unfavourable stellar constellations and natural disasters is found in an illustration of a book on cosmological questions printed in 1513 in Venice, namely, Macrobius' famous commentary on Cicero's *Somnium Scipionis* (Figure 1.2).[47] It visualised what the theoretical concepts had already linked: wise men from

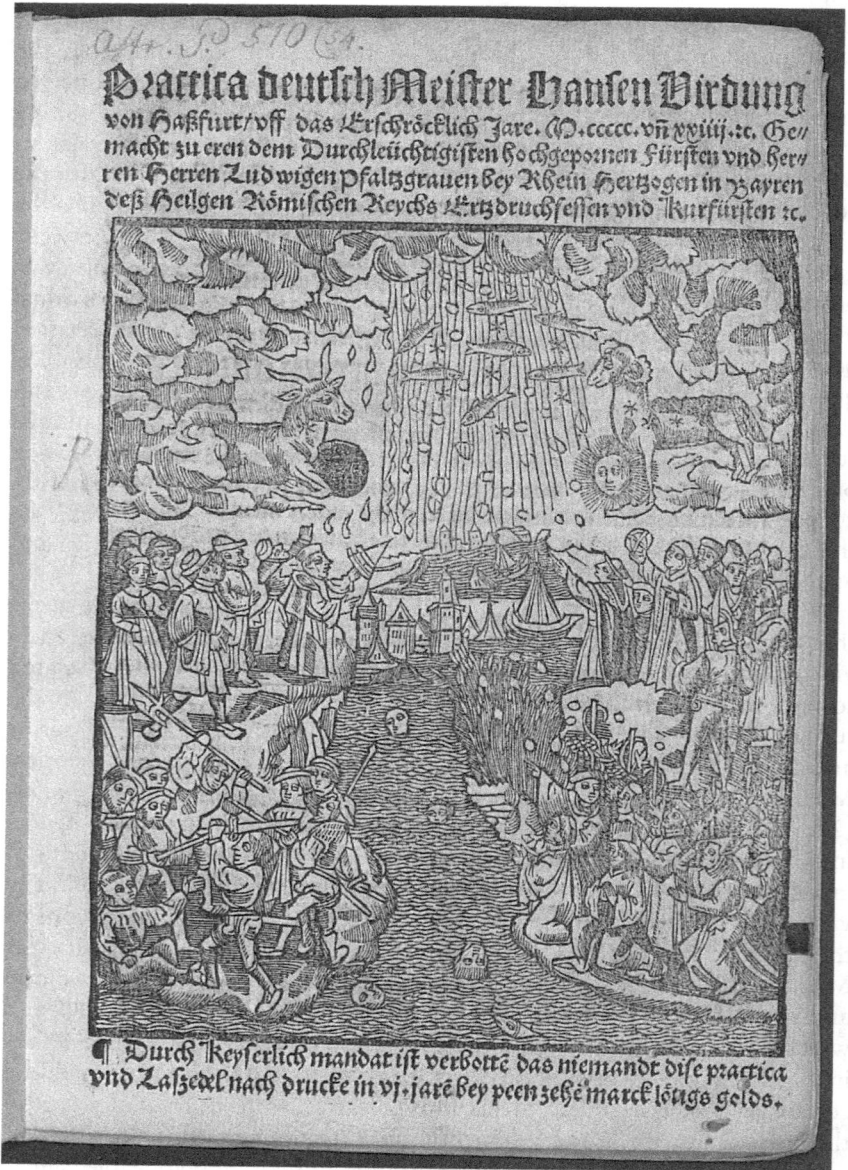

Figure 1.2 Johannes Virdung: *Practica deutsch* (Speyer: Nolt, 1523), frontispiece, Bayerische Staatsbibliothek München, Res/4 Astr.p. 510,54.

the East, indicated by their turbans, calculate the position of the stars between the sun and the moon using technical instruments. Two cherubs in the sky with inflated cheeks, typifying the wind, point to the ancient theory of heating air through incidents in the stellar sphere. The lower half of the picture depicts the effect of the heated air—a collapsing building, shaken by the air expanding in the earth's cavities. The natural disaster 'earthquake' (Latin *terraemotus*) is thus envisaged pictorially as *dis-astro*. At the beginning of the sixteenth century, this abstract concept rapidly left the learned discourse and spread among the general public. Yet neither the Macrobius text of 1513 nor the broadsheet of Virdung from Haßfurt of 1523 used the word 'disastro' to describe what was shown.

In German-speaking areas, the corresponding word *Desaster* did not develop in a similarly influential way to *disastro* in the Italian-speaking world. Instead, the word in use was *Katastrophe*. The history of the term took a different course. While the term *Katastrophe* was likewise semantically charged with astronomico–astrological significance in the Renaissance period, the concept of a sudden and mostly fatal change was differently accentuated due to its origin in dramatic theory and owed itself to the rediscovery and reinterpretation of some ancient facets of the concept of catastrophe.[48]

In antiquity, the Greek word καταστροφή generally stood for turning points, but this broad usage was almost forgotten in the Middle Ages. In the eighth century, diligent monks in the St. Gallen monastery and elsewhere held to the basic meaning of 'turning-point' in their glossaries, probably because of their reading of Jerome. Furthermore, the word remained only as the Latinised Greek medical term for morbid changes in digestion.[49] This reduction in the meaning of catastrophe was not broadened until the emergence of Renaissance humanism. This can be attributed to a conceptual rediscovery of the ancient poetological facets of the term by Desiderius Erasmus of Rotterdam.[50] It was hardly an accident that soon after the abovementioned broadsheets forecasting a disastrous Great Flood started circulating, the Reformer Philipp Melanchthon, a friend of Erasmus, took up the Greek word for a politico-religious upheaval that he expected in 1531 in connection with what contemporaries described as a prodigious appearance of (Halley's) comet.[51] The use of this astrologically inflected term in a Reformation-humanistic context created a terminological and narratological expectation that problems and situations of the present would eventually dissolve at the end of history. This expansion of the meaning of the word, highly charged with medieval astrological and salvific traditions, can also be found in the correspondence of other humanists.[52] The semantics of 'catastrophe' changed, as Olav Briese and Timo Günther aptly put it, very gradually, "from a stage of 'categorical ambivalence' [...] to that which [...] it is today: 'bad case' or 'worst case'".[53] However, in England and France 'catastrophe' remained a word confined more or less to the realm of poetry.[54]

In the debate held in German by Johannes Kepler and Helisäus Röslin, the significance of heavenly phenomena for terrestrial political events, the term 'catastrophe' was fully Germanised and came to describe a turn of events on Earth reflecting motions of the heavens. This use harked back to astronomical terms and stood for a range of extraordinary (according to Röslin, positive, but normally negative) phenomena; it referred to societal processes, however, not natural disasters.

What do these observations mean for an understanding of interpretative models for disasters and related worldviews of fate and *providentia Dei* in the Middle Ages? Faith in the (physical) influence of bad stars on earthly events was by no means incompatible with faith in divine omnipotence.[55] Apparently contradictory interpretative models tended to be ranked or understood in complementary ways, and there was no fundamental questioning of God as the *causa remota* of all events.[56] On the contrary, disasters were wondrous signs from God, *mirabilia mundi* – however terrible – to the orderly world of human beings, who just had to interpret them properly. Understanding these signs as an offer of communication on causes of the disaster and striving to make sense of them did not just open a window to a somewhat more successful strategy for coping with contingencies. In contrast to the spirit of the Apocalypse, it also opened the door to pragmatic dealing with the hazards of the natural as well as the social world.

Theory of practice: Disasters and hazards versus risks

With my case studies on how Florence and Strasbourg handled the risk of flooding, I would like to show how the worldview outlined above fitted in with daily practice.[57] In his famous *Cronica*, Giovanni Villani gives a very detailed report of the Arno flood in 1333, the most devastating flooding of Florence until that of 1966. In it, we find a remarkable sentence. He writes that the astrologers thought that the gravity of the disaster – compared to the less affected Pisa – was primarily the fault of the Florentines themselves and their government:

> When the astrologers were asked why the said flooding had hit Florence more strongly than Pisa, which lies downstream of the Arno, and it should have been greater below than it really was, or in other parts of Tuscany, they replied that the first reason was the lack of farsightedness of the Florentines, as already stated, due to the height of the weirs; and otherwise due to astrology ... that the disfavour and conjunctions [of the stars] was a reason for the greater flooding than in Pisa.[58]

It is possible to interpret this statement of Giovanni Villani as indicating an argument in the contemporary discourse on the causes of the flood. Before the deluge of 1333 and afterwards until the 1380s, the city discussed and mostly prohibited the erection of mills and fishery infrastructure in the urban area and nearby.[59] The reason for these prohibitions of the mills and weirs was the people's fear that they could block the stream and cause the river bed to rise. This could lead to a backwash that would increase the risk to the city in the event of floods.[60]

Historical empiricism shows that the Florentines had accepted the risk of floods from the very inception of their town. Francesco Salvestrini has convincingly shown that there were good reasons for choosing the position of the city at the river, primarily of an economic kind.[61] Yet once it had been built at this location, the city was not so easy to shift anywhere else. In the context of similar consequential decisions, technological historians talk of the 'path dependence' of the development, having to follow a path once you have embarked upon it, or otherwise shouldering very high costs if you decide to leave.[62] Precisely with complicated infrastructural buildings such as bridges, let alone whole cities, this has the consequence that there is less readiness to take risks deriving from natural hazards.[63] The ambivalence of the position by the river was quite clear to the contemporaries of Giovanni Villani and, as Robert d'Anjou, king of Naples, rhetorically asked in his letter of condolence to the *Signoria* in 1333, it was even capable of being explicitly understood as a cost-benefit calculation: 'Why deplore it so greatly if the river – which since the beginning of your city has brought so many joys and such great usefulness – once inflicts harm on you with unusual flooding?'.[64]

The explanations of the flood of 1333 indicate that contemporaries saw the disastrous event as the interplay of divine, natural and human factors. Giovanni Villani's reference to the *mala provedenza de' Fiorentini* as a cause of the force of the floods in Florence shows quite clearly that his contemporaries understood human action as an important factor in what happened, perhaps even the decisive factor. The *provvedenza* of human beings does not contradict the *providentia Dei*, but complements God's intentions and opens quite pragmatic options for action so as to be able to better deal with natural hazards in the future.[65]

In the argumentative tradition of Aristotle, Thomas Aquinas and Albertus Magnus, God was understood as the prime mover of all action (*causa prima*). Yet this did not prevent reflection about causes arising from things of this world themselves (*ex propriis rerum causis*), thus serving as second causes after God.[66] Contemporaries apparently understood the underlying distinction between a physico–philosophical and a theological observation of nature as being in systemic connection. Accordingly, the *providentia Dei* extends not only to the final effect (the flood), but also to the conditions leading to this effect, i.e., to human knowledge, ability and action. Thus, taking measures against future flooding is seen as part of good governance associated with Christian responsibility before God and citizens.[67] This opinion was not intellectually undisputed, as the discussion about the legitimacy of fleeing the plague showed only a few decades later, when Colluccio Salutati condemned an attempt to escape the *providentia Dei* as ridiculous and pointless.[68]

Yet pragmatic action of individuals and society won the day in everyday life. Government action consequently extended to several levels in order to influence the interlocking of the first, second and third causes. Processions of prayer and penance against the divine causes,[69] condemnations of sins against the related human causes,[70] and careful observations of the stars by astrologers against the natural causes were supposed to complement the more pragmatic measures. Furthermore, other experts followed clerics and astrologers to provide answers to the questions of security. In the late Middle Ages, craftsmen and 'engineers' joined the jurists, who had been providing legal answers to many socially determined security questions since the twelfth century.[71] Between the fourteenth and sixteenth centuries, there was a clear trend towards the professionalisation, centralisation and juridification of the handling of 'natural' processes, which I would like to illustrate with a few examples.

One of the very early ideas was to redirect the Arno away from, or even around, the city. Two variants may be discerned here. The older plan was to turn the meandering Arno downstream of Florence into canals in order to achieve faster discharge of water and better navigability. Then, there are later plans for a kind of overflow canal, taking in the Arno flood upstream of Florence and leading around the city. These plans – that were never realised due to the necessary excavation – have often been discussed and are described in detail in the essays by Felicia Else and Emanuela Ferretti present in this particular volume, so I will confine myself here to a few points.

In a letter of 12 August 1487 to Lorenzo de' Medici, the architect Luca Fancelli already proposed canalising the Arno downstream of Florence, but that plan was never realised.[72] Perhaps influenced by Fancelli's proposal, Leonardo da Vinci developed ideas for an extensive regulation of the Arno around 1503/04.[73] Due to its fragmentary preservation on maps, in scattered notes and third-party reports, it is hard to judge whether Leonardo – besides his plans for canalising the Arno below the Porta al Prato via Pistoia into Lago di Bientina – was also thinking of a deviation around the city upstream of Florence.[74] In any case, the Medici water administration, from 1549 known as the *Ufficio dei fiumi*, scrutinised such regulation plans a few years later, after the devastating flood of 11–15 September 1557. Girolamo di Pace from Prato, the over 80-year-old former engineer of the *Ufficio*, presented a long treatise to Duca Cosimo I

(1519–1574) in 1558, featuring a canal with a similar route at some points.[75] He may or may not have known about Leonardo's plans. Despite the change in political systems after the time of Luca Fancelli, the proposals were extremely similar. This continuity of hydraulic plans from the time of the republic until well into the period of the Medici grand duchy cannot be explained only by the constancy of the underlying problem – the threat of flooding on the Arno. With Leonardo Rombai, we can describe the administration of the infrastructure in the city and Contado as 'burocrazia tecnica'.[76] Despite political and institutional change, it seems to have en-sured a continuity of knowledge on the part of the hydraulic specialists extending beyond the end of the republic. Nevertheless, there are distinct differences in the assessment of harmful events. The changed worldview and the new understanding of statecraft consequently led to different conclusions for the handling of Florence's watercourses. But they all deal with the future by drawing conclusions on the basis of past experience and proposing pre-ventive measures.

In the Florence of the Medici dukes, and after the disastrous floods in autumn 1557, the *Ufficio dei fiumi* characterised floods and their conse-quences as *dixordine*.[77] This term indicates the concept of a normal state of the watercourses for the maintenance or restoration of which the administration was responsible. It looked at the disorder of nature through the lens of social policy and wanted – through the professional action of its engineers – to put it in order, human order, brought about by, and for, human beings. This also becomes clear in the undated *Memoria di Averardo da Filicaia al granduca di Toscana* about a method to avoid floods, which may be dated to the second half of the sixteenth century.[78] The author proposed taking preventive measures against the harmful floods of the Arno and Sieve 'because the circumstances that had fallen into disorder can be reordered with the ordering principles of nature itself and a little help from skill'.[79] The sketch accompanying this proposal shows a staggered canalisation of the rivers upstream of Florence to reduce the impact of the rushing waters. What Averardo aimed to achieve was to restore the natural order for the sake of the duke's subjects – and not just in the Boboli Gardens for an exclusive princely ostentation, although, certainly, the comparatively small gardens were easier to landscape with the technical and financial resources of the time than the whole territory of the grand duchy.[80]

Averardo's proposal proves typical for the sixteenth century: on the one hand, it builds on the experiences of the past and continues the geomet-rical approach to nature that was widespread among medieval scholars.[81] On the other, his proposal is so elaborate and abstract that it is far from the everyday detailed work of the engineers in the *Ufficio dei fiumi* and, hence, cannot be understood as practical.[82] The reason for this feature seems to be that Averardo's praiseful narrative was initially motivated by the author's desire to thank the duke for offering him a generous benefaction rather than by the idea to provide a practical guidance.[83]

The much simpler and cheaper solution of deepening the river bed surfaced in Florence long before the disastrous flood of 1966.[84] The chronicler Iacopo Nardi (1476–1563), who was exiled after the return of the Medici, already reported of such actions initiated by Cardinal Giulio de' Medici, later Pope Clement VII (1478–1534). Around 1514, he commissioned the construction of a gate built into the Arno weir at Ognisanti that could be opened when there was a flood and at regular intervals to flush out the sediment from the river bed.[85] This scheme could not be particularly effective. The well-informed chronicler Giovanni Cambi reports on 4 October 1532 that the duke of Florence Lorenzo de' Medici (1510–1537) had the dam in the Arno at Ognisanti lowered in order to deepen the river bed through a stronger flow.[86] The motive for the plan is reportedly the limited functioning of the mills because of the sand deposited there.

To conclude, the range of pragmatic interventions in handling risks was amazing. It indicates that chances and risks were deliberately weighed up. Christian Rohr examined this with the example of the bridge over the River Traun in Wels (Austria) using the 'bridge office invoices'. Towards the end of the Middle Ages, the bridge was constantly damaged by floods or even destroyed, but owing to the carefully kept stocks of timber it was regularly rebuilt. From these documents, Rohr concludes that 'the way people dealt with flooding was highly rational. There were never any complaints and not a single explanation that it was divine punishment'.[87] A similar result could be shown for the much larger and more expensive bridge over the Rhine near Strasbourg, which once again suggests a general pragmatic attitude when grappling with risks through natural hazards.[88]

As far as I can see, no fatalist attitude can be perceived in the sense of a purely passive submission to fate. In the sixteenth century, the extreme position of a fatalist attitude was attributed only to Turkish opponents and Islamic religious enemies.[89] The characterisation of this attitude as *fatum turcicum* against *fatum christianum* was a polemical tool aimed at profiling one's own reasonable chain of arguments with causes and effects according to God's will. By contrast, there were widespread interpretative patterns with which events were understood conceptually in such a way that they were fundamentally accessible to human agency. Arguing on the Old Testament example of pious and suffering Job ultimately pointed to the conception of individual attitude stressed by many preachers, and expressed by Robert of Naples in view of the flooding of Florence in 1333: disasters are interpreted as a test from God that offers the opportunity to exercise exemplary faith.[90] It was a widespread view that governments were basically responsible for looking after their subjects, i.e., to restore and keep order. That was true, as we saw by the example of Florence, precisely in the case of unfavourable stellar constellations. Having this pointed out did not relieve the powerful from responsibility. On the contrary, the assumed connection between macro- and microcosm opened the door for taming risks of the future through predictions and thereby proving the legitimacy of *buon governo* as desired by God. Johannes Fried pointed out years ago that this concept of an intelligible world contributed to the rise and implementation of the natural sciences in the West.[91] However, the fundamental religious bonds to the underlying worldview should not be overlooked. According to the contemporary view, diabolical forces could also play a role alongside God.

From a modern perspective, the link of apparently rational pragmatic steps with those that seem irrational may well appear contradictory at first sight. However, these positions ought to be understood as complementary and not as leading people at the time to a 'cognitive dissonance'.[92] A closer look at the other case, Strasbourg in Alsace, can confirm this notion as well as a contemporary understanding of calculable risk.[93] Far from the modern mathematical definition, the risk was considered as a relation between a perceptible threat and an appropriate reaction.[94] It does not seem to have been a vague, indefinite relation. On the contrary, the reactions of society, from quite practical steps to crisis rituals, relate to real dangers and display a calculated relation to each other. Dangers estimated to be high are met with rapid, stronger responses than those that are considered to be unclear and of less intensity. The penitential practices, such as processions against dangers understood as a sign of God's wrath, for instance potential floods or crop failure, exemplify this feature.

Among the processions recorded in Strasbourg on special occasions from 1275 until the Reformation period (1525), fifteen[95] took place expressly because of bad weather, primarily rain and thereby a threat of flooding, or a real flood.[96] These extraordinarily large intercessory or rogation processions were not only justified by an event interpreted as a divine sign or an imminent threat. It is striking that at the turn of the sixteenth century, the traditional apocalyptic triad of plague, war and inflation was particularly popular as biblically founded.[97] At first sight, this explanation can be suspected of being purely a *topos*, but that would be deceptive. For example, three Strasbourg processions were not only explained by inclement weather, but were put specifically in relation with all three members of the triad, that is, the danger of bad harvests, rising food prices and the risk of epidemics or wars, all perceived as highly realistic scenarios.[98] Two more were held due to extremely hot and dry weather and one to avert a threat of failed harvest.[99]

More processions were held not specifically in response to bad weather, but also (and sometimes even exclusively) to ease evil premonitions and the populace's feeling of vulnerability and danger. At the same time, the city authorities were presented as a provident and caring government. Particularly after the experience of the extreme flooding in the summer of 1480, which had caused great damage, people seem to have felt threatened more frequently.[100] During the usual processions in May 1482, the Council instructed that special prayers be said for God's support against unfavourable weather and price rises.[101] This suggests that there was perhaps a connection between the rumours that were rife about weather magic and fear of hailstorms. After all, during this period, trials were taking place of alleged 'weather witches', for instance, in Lucerne.[102]

The idea that the weather could be influenced in an unnatural way was widespread in this age and region. It was discussed in legal treatises, for example, by Ulrich Molitor (1489),[103] and disseminated in the vernacular by preachers, such as the Strasbourg cathedral preacher Johannes Geiler of Kaysersberg (1509).[104] The great procession on 12 July 1485, ordered jointly by the Cathedral Chapter and the Council mainly as intercession for good weather and, apparently, also against the risk of epidemics, was probably a reaction to a spectacular solar eclipse on 16 March 1485 in Strasbourg. It was understood as a premonition and the population connected it with events that were supposed to follow.[105] The City Council responded to political threats to the city and its government with two more special processions: on 13 May 1517, evidently responding to the Sickingen Feud, and on 1 May 1519, against the stirrings of the Reformation and the rising unrest among peasants.[106] In both cases, the Strasbourg Council felt its regime to be politically vulnerable and reacted with the time-honoured crisis ritual.

In the years directly before the Reformation, the bishop seems to have increasingly understood processions as a means of creating the unity of faith and discipline. Repeatedly he had recourse to special processions.[107] A procession wished by Bishop Albrecht of Pfalz-Mosbach (1478–1506) on 26 June 1503 was very probably less a reaction to the weather than to mysterious signs of the cross, which, according to rumours and reports, had repeatedly appeared on people's clothing and bodies since 1500, and which were interpreted as ominous warnings.[108] They were documented in contemporary writings and interpreted as signs from God, for instance, in the manuscript of the Freiburg lawyer and historical chronicler Jakob Mennel *De signis portentis atque prodigiis* (1503).[109] Processions could thus also respond to rumours, give emotions a direction and channel societal unrest.[110]

Unlike discursive and problem-solving socio-cultural procedures, these symbolically communicative interventions could barely contribute to the lowering of specific vulnerability factors. The case of epidemics illustrates the counterproductive potential of this kind of crisis ritual, which was clear even to contemporaries. When around 1510 the Strasbourg Council requested the Cathedral Chapter to hold a procession because of an epidemic, the medically trained canons refused on the basis of the increased risk of infection in large crowds, assuring the Council that it would be better for the priests to ward off danger through praying harder.[111] Here too, we see a considered response regarding the relation between the risk to be taken and the assumed benefit of the action, even though it is not mathematically defined. No doubt the clerics saw no contradiction between their insight into modes of infection, e.g., through the infectious vapours of sick people (miasms) – an insight resulting from the study of ancient theories and confirmed by natural observations – and their theologically justifiable and politically desired practice of penitence and intercession.

Renaissance securitisation of risks?

I would now like to outline the development of a special semantic field in the period under review, i.e., from the twelfth to the sixteenth centuries, providing some elements for writing a history of 'risk'. Alongside *disastro*, *fatum* and *providentia Dei*, other Latin and Italian concepts can be located at that time, and related to describing (in modern terms) individual and collective risk management. They are mainly terms for future events with an ambivalent character, i.e., either a good or a bad result: Latin *periculum*, *fortuna* and *resicum*, and Italian *azzardo* and *rischio*. They all deserve an examination in the light of an extended history of concepts. I will confine myself to a rough sketch in the context of research to date, namely, by Sylvain Piron and Giovanni Ceccarelli.[112]

In the milieu of Mediterranean maritime trading cities, it was an everyday experience for undertakings at sea to be exposed to numerous unpredictable risks: storms, shipwreck, piracy, mutiny, war and peace. Yet, whoever dared could also win – and the takings from maritime trading were high. The risks at sea were not only termed *periculum*, danger, but also *fortuna di mare*, probably to characterise their ambivalence. The concept of *fortuna* was many-faceted – and tinged over time by the legacy of the ancient goddess of luck, the medieval idea of a capricious *fortuna* on the wheel and by Macchiavelli's *fortuna-occasio* of the Renaissance.[113] The unpredictable gale-force wind that could bring both profit and loss was also called *fortuna*.[114]

Since the chances in sea trading seem to have been very chancy, or aleatory, they became the subject of intellectual debates: were they to be compared with casting dice (Latin *alea*) or a wager, and on what, or against what, were they betting? After the *Naviganti* decree of Pope Gregory IX (1227–1241), theologians and lawyers hotly debated the moral theological question of whether *commenda* contracts to insure risky deals with potentially high profits were to be rejected as usury.[115] Through their nearness to the notorious game of chance, merchants' profits were somewhat suspicious, morally speaking. Consequently, the legitimacy of commercial profit was the topic of thirteenth-century treatises, which justified it as God's just reward because of the uncertainty of the future profit and the risk taken (*periculum sortis*).[116]

Soon a further word with a similar meaning joined *periculum* and *fortuna* – the Latin word *resicum*, which is documented first in 1156 in an entry in the Chartular of the Genoan notary Giovanni Scriba; sometime afterwards, the vernacular word *rischio* is found in contracts.[117] Sylvain Piron claims that this word originally stemmed from Arabic.[118] The concept arose in the Mediterranean contact zone between North Africa, the Hispanic peninsula, the south of France, Italy and Sicily and stems from sailors' dialectal arab word *rzk* for 'chance' and 'favourable coincidence'. By contrast with the more comprehensive concept *fortuna,* its meaning is narrowed to the economic and legal aspect of dangerous but potentially profitable business deals. This semantic narrowing at the same time weakens the aspect of danger and stresses the anticipatory, calculating character of the 'risky' business affairs. It is fitting that roughly the same time saw the emergence of the Old Italian term *zara,* which most probably also stems from Arabic (from *az-zahr/ yasara*).[119] In the Middle Ages, it designated the (originally unhappy or invalid) casting of the dice and, in the figurative sense, a process that puts valuable things at stake and ultimately stands for 'venture' and 'risk' (French *hasart*, English *hazard*).[120] This concept therefore covered the complementary aspect of something incalculably dangerous.

In the fourteenth century, the more specialised concept of *resicum/rischio* became more and more frequent in contracts serving to handle possible loss in maritime trade. The traders developed a kind of 'insurance' against the risks of sea trading (*commenda*).[121] Giovanni Ceccarelli has been able to show that during the fifteenth and sixteenth centuries, premiums of these 'risk insurances' were increasingly calculated according to the risk of loss.[122] The calculation was based on structural experience regarding the ship and the captain, the distance, season of the year and goods, and on more intuitive estimates of the respective situation of pirates, on political and even weather conditions. Ceccarelli calls this 'probabilistic reasoning' – premium calculation attempted to determine risks under market conditions as exactly as possible – and depicts a system of future provision through an economic securing of insecurity.[123] Sometime later, fire and life insurance companies took up this basic idea of maritime insurance.[124]

From the twelfth to the sixteenth century, there were evident and considerable changes in the semantic field of concepts connected to coping with fate, contingency and risk. Perhaps by analogy with Koselleck's conceptual 'Sattelzeit' of the modern age, we can speak of a premodern 'Sattelzeit' in that special field.[125] In the transcultural Mediterranean contact zone, there arose new concepts that, through focusing on certain aspects, led to a semantic differentiation of the field. Previously polyvalent terms like *fortuna* and *periculum* found themselves in competition with the new notions and seem to have lost their semantic ambivalences. All in all, conceptual history points to an increasing securitisation and economisation of everyday life at least in trading towns.[126] 'Taming the future' was therefore a common practice in the period, oscillating between belief in divine (or diabolical) intervention and trust in human agency, and ranging from astro-meteorological practice, prayer and processions to the pragmatic prevention of 'natural' and 'social' hazards – but predominantly without systematically differentiating between hazards originating from 'nature' or 'society'.[127]

An essential precondition for this development was the emergence of a zone of relative security in a context that can be characterised as a 'culture of insecurity' from the crisis-ridden fourteenth century well into the modern age.[128] Large cities were the main islands of security in a sea of dangers. With their city walls and towers, they were stone bastions of safety, and with their fire brigades, granaries and drinking water provision, hospitals and poor-relief funds, they increasingly institutionalised welfare and insurance policies.[129] Here the practices of rulers and administration (as 'good government') firmed themselves into a resistance of danger and construction of security, even if hygienic problems long kept urban mortality at a high level.[130] This environment of relative security enabled a growth in the readiness and, thanks to capital accumulation, also the financial capacity to take risks. This applied particularly to the world of long-distance merchants, who after the 'commercial revolution' of the thirteenth century set out less often on journeys themselves – due to the progressive specialisation of roles in society – and were able to manage their businesses from their *fondachi* (warehouses).[131] A similar joy in risk also applied to other groups in society, in the field of politics and war. Niccolò Machiavelli impressively described the risks his contemporaries were willing to take in his *Principe*.[132]

The divine and human spheres were initially not thought of as separate but as entangled and complementary to each other (similar to the entangled spheres of humans and nature). The main trade book written by the most famous merchant of the early Renaissance, Francesco di Marco Datini from Prato (1335–1410), began with the significant line 'nel nome di dio e di guadagno' ('in the name of God and money-making').[133] Until far into the early modern age, many contemporaries remained convinced that nature and society were related to each other, with God working in creation and communicating with people through natural signs. Even though a mathematical risk calculation only became possible during the Enlightenment with a consequent growth in the 'the taming of chances',[134] from the High Middle Ages people did weigh up interests based on their experience and feeling – certain social groups made sure that this was practised, theoretically justified and legally formalised. Only in the complex late medieval urban society did an experimental change between play and seriousness – both in the commercial marketplace and on the battlefield, and in love as well as in politics – open the way to a practical and theoretical differentiation of danger and risk, of human agency and God's will, of chance and necessity.

Notes

1 This essay is a modified English version of my German article: 'Die Zukunft zähmen? Die Entstehung eines Risikobegriffs in der Sicherheitskultur spätmittelalterlicher Städte angesichts wiederkehrender „Natur"-Gefahren', in *Kulturen des Risikos im Mittelalter und in der Frühen Neuzeit*, ed. by Benjamin Scheller (Berlin; Boston: De Gruyter, 2019), pp. 195–227. I would like to thank the publisher De Gruyter for offering permission to use it.

2 Archives de la Ville et de l'Eurométropole (=AVE) Strasbourg, 1 MR 3, pag. 198 (=fol. 111*r*). His first letter was dated 6 March 1523, but only a later archival note is preserved: Sébastien Brant, 'Annales (Suite)', in *Mittheilungen der Gesellschaft für Erhaltung der geschichtlichen Denkmäler im Elsass/ Bulletin de la Société pour la conservation des monuments historiques d'Alsace*, ed. by Léon Dacheux, 19 (1899), 33–260 (p. 56 and passim). The final order to paricipate, AVE Strasbourg, 1 MR 3, pag. 196 (=fol. 110*r*), is dated 14 March 1523. On this event, see Gerrit Jasper Schenk, 'Spielräume der Macht – Macht der Spielräume? Die performative Herstellung öffentlichen Raumes in Städten zwischen Konflikt und Konsens am Beispiel von Straßburg und Worms im ausgehenden Spätmittelalter', in *Bischofsstadt ohne Bischof? Präsenz, Interaktion und Hoforganisation in bischöflichen Städten des Mittelalters (1300–1600)*, ed. by Andreas Bihrer and Gerhard Fouquet (Ostfildern: Thorbecke Verlag, 2017), pp. 41–73 (pp. 41–4).

3 For a full account of the major processions see Luzian Pfleger, 'Die Stadt- und Rats-Gottesdienste im Strassburger Münster', in *Archiv für Elsässische Kirchengeschichte*, 12 (1937), pp. 1–55; Gabriela Signori, 'Ritual und Ereignis. Die Straßburger Bittgänge zur Zeit der Burgunderkriege (1474–1477)', *Historische Zeitschrift*, 264 (1997), 281–328. See also Sabine von Heusinger, "Cruzgang' und 'umblauf' - Symbolische Kommunikation im Stadtraum am Beispiel von Prozessionen', in *Kommunikation in mittelalterlichen Städten*, ed. by Jörg Oberste (Regensburg: Schnell & Steiner, 2007), pp. 141–55; eadem, 'The Topography of Sacred Space and the Representation of Social Groups: Confraternities in Strasbourg', in *Politics and Refomations: Communities, Polities, Nations, and Empires. Essays in Honor of Thomas A. Brady Jr.*, ed. by Christopher Ocker et al. (Leiden; Boston: Brill, 2007), pp. 67–83; and eadem, 'Zur Durchdringung von Stadtraum mit Herrschaft – Prozessionen in Köln und Straßburg', *Rheinische Vierteljahrsblätter*, 79 (2015), 124–42.

4 Klaus Schreiner, *Maria. Jungfrau, Mutter, Herrscherin* (Munich: Hanser, 1994), pp. 350–54.

5 See Josef Engel, 'Kreuzzug und Türkenkrieg im 16. und frühen 17. Jahrhundert', in *Die Entstehung des neuzeitlichen Europa*, ed. by Theodor Schieder and Josef Engel (Stuttgart: Union, 1971), pp. 274–93 (pp. 274–81).

6 Notably the Knights' Revolt (Sickingen Feud) and Peasants' War of 1525, cf. Marc Lienhard, 'Mentalité populaire, gens d'église et mouvement évangélique à Strasbourg en 1522–23: le pamphlet "Ein brüderlich warnung an meister Mathis ... " de Steffan von Büllheym', in *Horizons européens de la Réforme en Alsace. Mélanges offerts à Jean Rott pour son 65ème anniversaire*, ed. by Marijn de Kroon and Marc Lienhard (Strasbourg: Istra, 1980), pp. 37–62; Reinhard Scholzen, 'Franz von Sickingen als Machtfaktor im Kampf zwischen Mainz, Hessen, Kurtrier und Kurpfalz', *Blätter für pfälzische Kirchengeschichte und religiöse Volkskunde*, 68 (2001), 287–306; Volker Leppin, *Martin Luther* (Darmstadt: Lambert Schneider Verlag, 2010), pp. 125–64, 209–20.

7 The Great Flood forecast of 1524 has been a matter of scholarly interest since the study by Aby Warburg, 'Heidnisch-antike Weissagung in Wort und Bild zu Luthers Zeiten', in *Sitzungsberichte der Heidelberger Akademie der Wissenschaften, Philosophisch-historische Klasse*, 26 (1920), pp. 29–35. See also Paola Zambelli, 'Introduction: Astrologers' Theory of History', in *Astrologi hallucinati. Stars and the End of the World in Luther's Time*, ed. by Paola Zambelli (Berlin; New York: De Gruyter, 1986), pp. 1–28; Ottavia Niccoli, *Profeti e popolo nell'Italia del Rinascimento* (Rome; Bari: Laterza, 1987), pp. 185–215.

8 Johannes Stöffler and Jakob Pflaum, *Almanach nova plurimis annis venturis inservientia* (Ulm: Reger, 1499), fol. 387r. Se also Stefanie Gehrke, 'Kommt eine neue Sintflut? – Astrologen und ihre Prognosen im frühen 16. Jahrhundert', in *Die Sterne lügen nicht. Astrologie und Astronomie im Mittelalter und in der Frühen Neuzeit*, ed. by Christian Heitzmann (Wiesbaden: Harrassowitz, 2008), pp. 80–85 (p. 82).

9 Edited after a manuscript by Strasbourg archivist Theophil Dachtler from the first half of the seventeenth century and based on a lost original, the tract probably had a woodcut illustration: Sebastian Brant, *Kleine Texte*, vol. 1, ed. by Thomas Wilhelmi (Stuttgart; Bad Cannstatt: Frommann-Holzboog, 1998), pp. 621–24. For the date of the publication, see idem, *Kleine Texte*, vol. 2, ed. by Thomas Wilhelmi (Stuttgart; Bad Cannstatt: Frommann-Holzboog, 1998), p. 181 and Joachim Knape and Thomas Wilhelmi, *Sebastian Brant Bibliographie. Werke und Überlieferungen*(Wiesbaden: Harrassowitz, 2015), pp. 321–29 (p. 329). On the role broadsheets of Brant played, see Michael Schilling, 'Die Flugblätter Sebastian Brants in der Geschichte der Bildpublizistik', in *Sebastian Brant (1457–1521)*, ed. by Hans-Gert Roloff, Jean-Marie Valentin, and Volkhard Wels (Berlin: Weidler, 2008), pp. 143–67.

10 See Heike Talkenberger, *Sintflut: Prophetie und Zeitgeschehen in Texten und Holzschnitten astrologischer Flugschriften, 1488–1528* (Tübingen: Niemeyer, 1990), pp. 154–335, 436 and 520; Gabriele Wimböck, 'In den Sternen geschrieben—in die Bilder gebannt: Die Furcht vor der Großen Sintflut im Zeitalter der Reformation', in *AngstBilderSchauLust: Historische Katastrophenerfahrungen in Kunst, Musik und Theater*, ed. by Jürgen Schläder and Regina Wohlfahrt (Dresden: Henschel, 2007), pp. 212–39; Monica Juneja and Gerrit Jasper Schenk, 'Viewing Disasters: Myth, History, Iconography, and Media across Europe and Asia', in *Disaster as Image: Iconographies and Media Strategies*

across Europe and Asia, ed. by Monica Juneja and Gerrit Jasper Schenk (Regensburg: Schnell & Steiner, 2014), pp. 7–40 (p. 29).

11 It was called the lux procession. See see Gerrit Jasper Schenk, 'Lektüren im "Buch der Natur". Wahrnehmung, Beschreibung und Deutung von Naturkatastrophen', in *Geschichte schreiben. Ein Quellen- und Studienhandbuch zur Historiographie (ca. 1350–1750)*, ed. by Susanne Rau and Birgit Studt (Berlin: De Gruyter, 2010), pp. 507–21 (pp. 514–17) and idem, 'Krisenrituale: Vom Nutzen und Nachteil kommunaler Selbstinszenierung am Beispiel Straßburgs im Elsass', in *Kommunale Selbstinszenierung Städtische Konstellationen zwischen Mittelalter und Neuzeit*, (Zürich: Chronos, 2018), pp. 123–54 (pp. 134–37).

12 AVE Strasbourg, 1 MR 3, pag. 198 (=fol. 111r).

13 Brant, 'Annales (Suite)', p. 57.

14 See Pfleger, 'Die Stadt- und Rats-Gottesdienste im Strassburger Münster', 25 and passim.

15 On the January flood of 1524 in the region, see *Heinrich Hugs Villinger Chronik von 1495 bis 1533*, ed. by Christian Roder (Tübingen: Fues, 1883), p. 96; Martin Hille, *Providentia Dei, Reich und Kirche. Weltbild und Stimmungsprofil altgläubiger Christen 1517–1618*, (Göttingen: Vandenhoeck & Ruprecht, 2010), pp. 340–55.

16 Archives departementales du Bas-Rhin (AD), Strasbourg, sér. E 5182 (29) 1–7 on Stollhofen's neighbourly assistance for Beinheim by the Margrave of Baden.

17 Georges Bischoff, *La guerre des paysans. L'Alsace et la révolution du Bundschuh (1493–1525)* (Strasbourg: La Nuée Bleue, 2010), pp. 302–5, 354–58, 471 and passim.

18 See Gerrit Jasper Schenk, 'Ein beliebtes Krisenritual: Prozessionen, in Mensch. Natur. Katastrophe', in *Von Atlantis bis heute. Begleitband zur Sonderausstellung "Mensch. Natur. Katastrophe. Von Atlantis bis heute"*, ed. by Gerrit Jasper Schenk et al. (Regensburg: Schnell & Steiner, 2014), pp. 199–201.

19 Franziska Rehlinghaus, *Die Semantik des Schicksals. Zur Relevanz des Unverfügbaren zwischen Aufklärung und Erstem Weltkrieg* (Göttingen: Vandenhoeck & Ruprecht, 2015), pp. 57–68.

20 On *providentia, fatum* and *fortuna*, see Sibylle Appuhn-Radtke, 'Fortuna', *Reallexikon zur Deutschen Kunstgeschichte*, 10 (2005), 271–401.

21 Giovanni Ceccarelli, 'The Price for Risk-Taking: Marine Insurance and Probability Calculus in the Late Middle Ages', *Journal electronique d'histore des probabilités et de la statistique*, 3, 1 (2007), 1–26 (1) (http://www.jehps. net/juin2007.html, accessed 26 January 2022).

22 See Benjamin Scheller, 'Die Geburt des Risikos. Kontingenz und kaufmännische Praxis im mediterranen Seehandel des Hoch- und Spätmittelalters', *Historische Zeitschrift*, 304 (2017), 305–331, mainly based on Giovanni Ceccarelli's 'Risky Business. Theological and Canonical Thought on Insurance from the Thirteenth to the Seventeenth Century', *Journal of Medieval and Early Modern Studies*, 31 (2001), 607–658 and 'Stime senza probabilità. Assicurazione a rischio nella Firenze rinascimentale', *Quaderni Storici*, 45 (2010), 651–703 as well as Silvain Piron's, 'L'apparition du *resicum* en Méditerranée occidentale, XIIe–XIIIe siècles', in *Pour une histoire culturelle du risque. Genèse, évolution, actualité du concept dans les sociétés occidentales*, ed. by Emmanuelle Collas-Heddeland et al. (Strasbourg: Éditions Histoire et Anthropologie, 2004), pp. 59–76. See also the foundational book by Giovanni Ceccarelli, *Il gioco e il peccato. Economia e rischio nel Tardo Medioevo* (Bologna: Mulino, 2003).

23 Following Niklas Luhmann, *Soziologie des Risikos* (Berlin; New York: De Gruyter, 1991), p. 30 and passim; Scheller, 'Die Geburt des Risikos', 305–7.

24 Cf. Reinhard Koselleck's early work on the problem, *Vergangene Zukunft. Zur Semantik geschichtlicher Zeiten* (Frankfurt: Suhrkamp, 1989), pp. 321–39; in connection with the development of a new security system in the early modern age, see Cornel Zwierlein, 'Sicherheit durch Versicherung: Ein frühneuzeitliches Erfolgsmodell', in *Sicherheit in der Frühen Neuzeit. Norm, Praxis, Repräsentation*, ed. by Christoph Kampmann and Ulrich Niggemann (Cologne; Weimar; Vienna: Böhlau, 2013), pp. 381–99 (p. 395 and passim).

25 On the whole issue of chance in the premodern age, see Wolfgang Behringer, 'Das europäische Konzept des Zufalls, oder Von der Unsicherheit zur Versicherung. Ein Kommentar', in *Sicherheit in der Frühen Neuzeit. Norm, Praxis, Repräsentation*, pp. 459–64; Dominik Perler and Ulrich Rudolph, *Occasionalismus. Theorien der Kausalität im arabisch-islamischen und im europäischen Denken* (Göttingen: Vandenhoeck & Ruprecht, 2000), pp. 127–211. See also footnotes 32 and 68 below.

26 On the fundamental comparability of the two regions and the use of comparisons, see Gerrit Jasper Schenk, 'Managing Natural Hazards: Environment, Society, and Politics in Tuscany and the Upper Rhine Valley in the Renaissance (1270–1570)', in *Historical Disasters in Context: Science, Religion, and Politics*, ed. by Andrea Janku, Gerrit Jasper Schenk, and Franz Mauelshagen (London; New York: Routledge, 2012), pp. 31–53. It is indicative that the first attempts to develop institutional insurances on the model of Italian marine insurances took place in the early seventeenth century in cities like Strasbourg: AVE Strasbourg, série III 93/3. See also Cornel Zwierlein, 'Frühe Formen der Institutionalisierung von 'Versicherung' und die Bedeutung der Versicherungsgeschichte für eine allgemeine Sicherheitsgeschichte', in *Sicherheit in der Frühen Neuzeit. Norm, Praxis, Repräsentation*, pp. 441–58.

27 See Dominik Collet, 'Eine Kultur der Unsicherheit? Empowering Interactions während der Hungerkrise 1770–72', in *Sicherheit in der Frühen Neuzeit. Norm, Praxis, Repräsentation*, pp. 367–80.

28 Following Scheller, 'Die Geburt des Risikos' and against Alain Guerreau, 'L'Europe médiévale. Une civilisation sans la notion de risque', *Risques. Les cahiers de l'assurance*, 31 (1997), 11–18.

29 Schenk, 'Lektüren im "Buch der Natur"', p. 508.

30 According to Hans Blumenberg ('Paradigmen zu einer Metaphorologie', *Archiv für Begriffsgeschichte*, 6 (1960) 7–142 (11)), terms are absolute metaphors, when they prove resistant to the terminological claim and cannot be dissolved into terminology. Cf. also Rehlinghaus, *Die Semantik des Schicksals*, p. 23 and passim with note 46.

31 See on differing national traditions Jörn Leonhard, 'Grundbegriffe und Sattelzeiten – Languages and Discourses: Europäische und anglo-amerikanische Deutungen des Verhältnisses von Sprache und Geschichte', in *Interkultureller Transfer und nationaler Eigensinn: Europäische und anglo-amerikanische Positionen der Kulturwissenschaft*, ed. by Rebekka Habermas and Rebekka von Mallinckrodt (Göttingen: Wallstein Verlag, 2004), pp. 71–86; Hans Joas and Peter Vogt, 'Einleitung', in *Kontingenz und Zufall. Eine Ideen-und Begriffsgeschichte*, ed. by Peter Vogt (Berlin: De Gruyter, 2011), pp. 17–21; Hans Erich Bödeker, 'Reflexionen über Begriffsgeschichte als Methode', in

Begriffsgeschichte, Diskursgeschichte, Metapherngeschichte, ed. by Hans Erich Bödeker (Göttingen: Wallstein Verlag, 2002), pp. 73–121.

32 The development of the insurance industry is a good example. See James Franklin, *The Science of Conjecture. Evidence and Probability before Pascal* (Baltimore; London: Johns Hopkins University Press, 2001), pp. 258–88.

33 For this matter, see Gerrit Jasper Schenk, 'Historical Disaster Experiences. First Steps toward a Comparative and Transcultural History of Disasters across Asia and Europe in the Preindustrial Era', in *Historical Disaster Experiences. Towards a Comparative and Transcultural History of Disasters Across Asia and Europe*, ed. by Gerrit Jasper Schenk (Cham: Springer, 2017), pp. 3–44, in particular pp. 15–22.

34 See Mischa Meier, 'Zur Terminologie der (Natur-)Katastrophe in der griechischen Historiographie: Einige einleitende Anmerkungen', *Historical Social Research*, 32, 3 (2007), 44–56; Anna Akasoy, 'The Man-Made Disaster: Fires in Cities in the Medieval Middle East', *Historical Social Research*, 32, 3 (2007), 75–87, particularly 75–8.

35 Gerrit Jasper Schenk, 'Vormoderne Sattelzeit? 'Disastro', Katastrophe, Strafgericht—Worte, Begriffe und Konzepte für rapiden Wandel im langen Mittelalter', in *Krisengeschichte(n). 'Krise' als Leitbegriff und Erzählmuster in kulturwissenschaftlicher Perspektive*, ed. by Carla Meyer, Katja Patzel-Mattern, and Gerrit Jasper Schenk (Stuttgart: Steiner Verlag, 2013), pp. 177–212; Juneja and Schenk, 'Viewing Disasters', pp. 26–34.

36 See entry on 'disastro' in *Tesoro della Lingua Italiana delle Origini*<http://tlio.ovi.cnr.it/TLIO/> [accessed 26 January 2022].

37 See *Libro dei Sette Savi di Roma: Versione in prosa F*, ed. by Andrea Giannetti (Alessandria: Edizioni dell'Orso, 2012), p. 121.

38 See Udo Gerdes, 'Sieben weise Meister', in *Die deutsche Literatur des Mittelalters: Verfasserlexikon*, vol. 8, ed. by Kurt Ruh et al. (Berlin; New York: De Gruyter, 1992), cols. 1174–89 and Bettina Krönung, 'Fighting with Tales: 1 The Arabic Book of Sindbad the Philosopher', in: *Fictional Storytelling in the Medieval Eastern Mediterranean and Beyond*, ed. by Carolina Cupane and Betina Krönung (Leiden; Boston: Brill, 2016), pp. 365–79.

39 For a good history of astrology in the Middle Ages, see, for instance, Gerd Mentgen, *Astrologie und Öffentlichkeit im Mittelalter* (Stuttgart: Hiersemann, 2005).

40 Burkhardt Wolf, *Fortuna di mare. Literatur und Seefahrt* (Zurich: Diaphanes, 2013), pp. 11–14, 89–109; Piron, 'L'apparition du *resicum* en Méditerranée occidentale', pp. 61–68.

41 For example, Angelo di Capua, *La istoria di Eneas vulgarizata per Angilu di Capua*, ed. by Gianfranco Folena (Palermo: Mori, 1956), p. 222; Guido delle Colonne, *Libro de la destructione de Troya, volgarizzamento napoletano trecentesco di Guido delle Colonne*, ed. by Nicola de Blasi (Rome: Bonacci, 1986), pp. 62, 81, 140, 142, 179; *Statuti inediti della città di Pisa dal XII al XIV secolo*, vol. 3, ed. by Francesco Bonaini (Florence: Vieusseux, 1857), p. 535; 'Cronaca fiorentina, 1537–1555', ed. by Enrico Coppi, *Deputatione di storia patria per la Toscana. Documenti di storia italiana*, 2, 7 (2000), 429; Jacopo Gradenigo, *Gli Quatro Evangelii concordati in uno*, ed. by Francesca Gambino (Bologna: Commissione per i testi di lingua, 1999), pp. 251, 269.

42 Unfortunately, this idea that would call for an examination of the connections to the semantic fields of 'hazard', 'fortune/misfortune', 'risk', 'fate', 'chance', 'providence' cannot be explored further here.

43 On the reception in French and English in the sixteenth century, see *A Lexicon of Latin Derivatives in Italian, Spanish, French, and English. A Synoptic Etymological Thesaurus with Full Indices for Each Language*, ed. by James H. Dee (Hildesheim; Zürich; New York: Olms, 1997), p. 30, note 175. For the French *désastre*, see *Le Trésor de la Langue Française Informatisé* (http://atilf.atilf.fr/, accessed 26 January 2022); for the English *disaster Oxford English Dictionary Online* (http://dictionary.oed.com, accessed 26 January 2022).

44 See Manuel Alonso Alonso's edition of the *Libellus de diluviis* in 'Homenaje a Avicena en su milenario. Las traducciones de Juan González de Burgos y Salomon', *Al-Andalus*, 14 (1949), 306–8, particularly 306. On the textual history, see Jean-Marc Mandosio and Carla Di Martino, 'La 'Météorologie' d'Avicenne (Kitāb al-Šifā' V) et sa diffusion dans le monde latin', in *Wissen über Grenzen. Arabisches Wissen und lateinisches Mittelalter*, ed. by Andreas Speer and Lydia Wegener (Berlin; New York: De Gruyter, 2006), pp. 406–24, particularly p. 420. On the concept of *diluvium* in antiquity, see Jasmin Hettinger, 'Von aqua magna bis diluvium. Eine systematische Annäherung an den Hochwasserbegriff in den antiken lateinischen Schriftquellen', *Orbis Terrarum*, 12 (2014), 109–129.

45 See Schenk, 'Vormoderne Sattelzeit?', p. 189 and passim.

46 On the Toledo letter, see Dorothea Weltecke, 'Die Konjunktion der Planeten im September 1189: Zum Ursprung einer globalen Katastrophenangst', *Saeculum*, 54, 2 (2003), 179–212; Mentgen, *Astrologie und Öffentlichkeit*, pp. 17–158.

47 Macrobius, *En tibi lector candidissime, Macrobius, qui antea mancus, mutilus, ac lacer circu[m]ferebatur, nu[n]c primu[m] integer, nitidus [et] suonitori restitutus, in quo graecae maiestatis dignitas quo ad eius fieri potuit superstes reperit[us]* (Venice: Giunta, 1513), fol. 1r; for the broad reception of Macrobius's commentaries of Cicero's *Somnium Scipionis*, see Bruce Eastwood, 'Manuscripts of Macrobius, *Commentarii in Somnium Scipionis*, before 1500', *Manuscripta*, 38, 2 (1994), 138–55.

48 See Schenk, 'Vormoderne Sattelzeit?', pp. 194–99 with all necessary references.

49 Gerrit Jasper Schenk, 'Dis-Astri. Modelli interpretativi delle calamità naturali dal medioevo al Rinascimento', in *Le calamità ambientali nel tardo medioevo europeo: realtà, percezioni, reazioni. Atti del XII convegno del Centro di Studi sulla civiltà del tardo Medioevo S. Miniato, 31 maggio–2 giugno 2008*, ed. by Michael Matheus et al. (Florence: Firenze University Press, 2010), pp. 23–75, in particular p. 68 and passim; Schenk, 'Vormoderne Sattelzeit?', p. 195 and passim.

50 See Alan Rosen, *Dislocating the End. Climax, Closure and the Invention of Genre* (New York: Peter Lang, 2001), pp. 6–11; Olaf Briese and Timo Günther, 'Katastrophe. Terminologische Vergangenheit, Gegenwart und Zukunft', *Archiv für Begriffsgeschichte*, 51 (2009), 155–95; Vera Fionie Koppenleitner, *Katastrophenbilder. Der Vesuvausbruch 1631 in den Bildkünsten der Frühen Neuzeit* (Berlin; Munich: Deutscher Kunstverlag, 2018), pp. 19–33.

51 See Philip Melanchthon, *Philippi Melanthonis opera quae supersunt omnia*, vol. 2, ed. by Carolus Gottlieb Bretschneider (Halle: Schwetschke, 1835), p. 546; see also Sachiko Kusukawa, *The Transformation of Natural Philosophy. The Case of Philip Melanchthon* (Cambridge: Cambridge University Press, 1995), pp. 124–73.

52 On the debates, see Walther Ludwig, 'Zukunftsvoraussagen in der Antike, der frühen Neuzeit und heute', in *Zukunftsvoraussagen in der Renaissance*, ed. by Klaus Bergdolt and Walter Ludwig (Wiesbaden: Harrassowitz, 2005), pp. 9–64, in particular pp. 18–47.

53 Briese and Günther, 'Katastrophe', 164 (translated from German by the author).

54 *Trésor de la langue française. Dictionnaire de la langue du XIXe et du XXe siècle*, ed. by Paul Imbs, vol. 5 (Paris: Éditions du Centre National de la Recherche Scientifique, 1977), p. 299. In the mid-fifteenth century, François Rabelais uses 'catastrophe' in an astrological context (comet) with a negative connotation, but narratively integrated. see François Walter, *Katastrophen: Eine Kulturgeschichte vom 16. bis ins 21. Jahrhundert* (Stuttgart: Reclam, 2010), p. 16 and passim; Briese and Günther, 'Katastrophe', 163, 165.

55 Gerrit Jasper Schenk, 'Disastro, Catastrophe, and Divine Judgement: Words, Concepts and Images for 'Natural' Threats to Social Order in the Middle Ages and Renaissance', in *Disaster, Death and the Emotions in the Shadow of the Apocalypse, 1400–1700*, ed. by Jennifer Spinks and Charles Zika (Basingstoke: Palgrave Macmillan, 2016), pp. 45–67 (p. 60).

56 Schenk, 'Dis-Astri', p. 73 and passim. For the divine *causa remota*, see Sabine Krüger, 'Krise der Zeit als Ursache der Pest. Der Traktat De mortalitate in Alamannia des Konrad von Megenberg', in *Festschrift für Hermann Heimpel zum 70. Geburtstag am 19. September 1971*, vol. 2 (Göttingen: Vandenhoeck & Ruprecht, 1972), pp. 839–83.

57 Gerrit Jasper Schenk, "… … prima ci fu la cagione de la mala provedenza de' Fiorentini… …' Disaster and 'Life World' – Reactions in the Commune of Florence to the Flood of November 1333', *The Medieval History Journal*, 10 (2007), 355–86; idem, 'Politik der Katastrophe? Wechselwirkungen zwischen gesellschaftlichen Strukturen und dem Umgang mit Naturrisiken am Beispiel von Florenz und Straßburg in der Renaissance', in *Stadt und Stadtverderben. 47. Arbeitstagung in Würzburg, 21.–23. November 2008*, ed. by Ulrich Wagner (Ostfildern: Thorbecke, 2012), pp. 33–76; idem, 'Friend or Foe? Negotiating the Future on the Example of Dealing with the Rivers Arno and Rhine in the Renaissance (ca. 1300–1600)', in *L'acqua nemica. Fiumi, inondazioni e città storiche dall'antichità al contemporaneo. Atti del Convegno di studio a cinquant'anni dall'alluvione di Firenze (1966–2016). Firenze, 29–30 gennaio 2015*, ed. by Concetta Bianca and Francesco Salvestrini (Spoleto: CISAM, 2017), pp. 137–56.

58 Giovanni Villani, *Nuova Cronica*, 3 vols, ed. by Giuseppe Porta (Parma: Guanda, 1991), vol. 3, pp. 15–16 (translation into English by the author).

59 Archivio di Stato, Florence (ASF): *Capitani di Parte guelfa, numeri rossi*, 105, fols. 33r–v, on problems with pent-up water ahead of the mills near Rovezzano (20 June 1382). See also Francesco Salvestrini, *Libera città su fiume regale. Firenze e l'Arno dall'Antichità al Quattrocento* (Florence: Nardini, 2005), pp. 28–30, 74–76; John Muendel, 'Medieval Urban Renewal. The Communal Mills of the City of Florence, 1351–1382', *Journal of Urban History*, 17, 4 (1991), 363–389; Gloria Papaccio, 'Il mulini del Comune di Firenze. Uso e gestione nella città trecentesca', in *La città e il fiume (secoli XIII–XIX)*, ed. by Carlo M. Travaglini (Rome: École Française de Rome, 2008), pp. 61–79.

60 Villani, *Cronica*, vol. 3, p. 11; Osvaldo Cavallar, 'The Wheels of Watermills and the Wheel of Fortune: A 'Consilium' of Donatus Ricchi de Aldighieris',

Rechtsgeschichte. Zeitschrift des Max-Planck-Instituts für europäische Rechtsgeschichte, 13 (2008), 80–116.

61 Salvestrini, *Libera città su fiume regale*, pp. 21–34.

62 See Raymund Werle, 'Pfadabhängigkeit', in *Handbuch Governance. Theoretische Grundlagen und empirische Anwendungsfelder*, ed. by Arthur Benz et al. (Wiesbaden: VS Verlag für Sozialwissenschaften, 2007), pp. 119–31, in particular pp. 120, 126.

63 Jens Ivo Engels and Gerrit Jasper Schenk, 'Macht der Infrastrukturen - Infrastrukturen der Macht. Überlegungen zu einem Forschungsfeld', in *Wasserinfrastrukturen und Macht. Politisch-soziale Dimensionen technischer Systeme von der Antike bis zur Gegenwart*, ed. by Birte Förster and Martin Bauch (Munich: De Gruyter Oldenbourg, 2014), pp. 22–58 (pp. 23–27).

64 Villani, *Cronica*, vol. 3, p. 36 (translation by the author).

65 On *providentia Dei* in its connection with human agency, see Walter Haug, 'Kontingenz als Spiel und das Spiel mit der Kontingenz. Zufall, literarisch, im Mittelalter und in der frühen Neuzeit', in *Kontingenz*, ed. by Gerhart von Graevenitz and Odo Marquard (Munich: Fink, 1997), pp. 151–72; on *prudentia* as part of *providentia futurorum* in Thomas Aquinas and Dante Alighieri, see John Burrow, 'The Third Eye of Prudence', in *Medieval Futures. Attitudes to the Future in the Middle Ages*, ed. by John A. Burrow and Ian P. Wei (Woodbridge: Boydell & Brewer, 2000), pp. 37–48.

66 Thomas Aquinas, *Summa contra Gentiles*, bk. 2, chap. 4, n. 3, p. 26. On the same topic, see Wolfgang Kluxen, *Philosophische Ethik bei Thomas von Aquin* (Hamburg: Meiner Verlag, 1998); Theodor Wolfram Köhler, *Grundlagen des philosophisch-anthropologischen Diskurses im 13. Jahrhundert: die Erkenntnisbemühung um den Menschen im zeitgenössischen Verständnis* (Leiden: Brill, 2000), pp. 137–38, 180–181; Valérie Cordonier, 'La doctrine aristotélicienne de la providence divine selon Thomas d'Aquin', in *Fate, Providence and moral Responsibility in Ancient, Medieval and Early Modern Thought. Studies in Honour of Carlos Steel*, ed. by Pieter d'Hoine and Gerd Van Riel (Leuven: Leuven University Press, 2014), pp. 495–515.

67 See Gerrit Jasper Schenk, ''Human Security' in the Renaissance? Securitas, Infrastructure, Collective Goods and Natural Hazards in Tuscany and the Upper Rhine Valley', *Historical Social Research*, 35, 4 (2010), 209–33. The theological and philosophical discussion was much more sophisticated, however.

68 Coluccio Salutati, *Epistolario*, vol. 2, ed. by Francesco Novati (Rome: Forzani, 1893), pp. 80–83 (dated 1383). See also see Heinrich Dornmeier, 'Die Flucht vor der Pest als religiöses Problem', *in Laienfrömmigkeit im späten Mittelalter*, ed. by Klaus Schreiner (Munich: Oldenbourg, 1992), pp. 331–97.

69 See Richard Trexler, 'Florentine Religious Experience. The Sacred Image', *Studies in the Renaissance*, 19 (1972), 7–41; idem, *Public Life in Renaissance Florence* (Ithaca, NY; London: Cornell University Press, 1991), pp. 354–61.

70 Schenk, 'Dis-Astri', p. 33 with note 33.

71 Marius Sebastian Reusch, Reut Yael Paz, and Stefan Tebruck, 'Juristen als Sicherheitsakteure. Eine Einführung', in *Sicherheitsakteure. Epocenübergreifende Perspektive zu Praxisformen und Versicherheitlichung*, ed. by Carola Westermeier and Horst Carl (Baden-Baden: Nomos, 2018), pp. 95–110.

72 *Luca Fancelli, architetto. Epistolario gonzaghesco*, ed. by Corinna Vasić Vatovec (Florence: Uniedit, 1979), pp. 60–62.

73 Leonardo da Vinci knew Luca Fancelli from his time in Milan: Martin Kemp, *Leonardo da Vinci. The Marvellous Works of Nature and Man* (Oxford: Oxford University Press 2006), pp. 233–34.

74 Along with Emanuela Ferretti's essay on Leonardo in this volume, see Roger D. Masters's detailed but often speculative *Fortune is a River. Leonardo da Vinci and Niccolò Machiavelli's Magnificent Dream to Change the Course of Florentine History* (New York: Free Press, 1999), pp. 93–134.

75 BNC Florence, MS. Landau Finaly 97, fols. 1r–29v; on the manuscript cf. *I manoscritti Landau Finaly della Biblioteca Nazionale Centrale di Firenze. Catalogo*, vol. 1, ed. by Giovanna Lazzi and Maura Rolih Scarlino (Florence: Giunta Regionale Toscana, 1994), pp. 197–98.

76 Leonardo Rombai, 'Prefazione: strade e politica in Toscana tra medioevo ed età moderna', in *Il libro vecchio di strade della Repubblica fiorentina*, ed. by Gabriele Ciampi (Florence: Papafava, 1987), pp. 5–36; idem, 'La "Politica delle acque" in Toscana. Un profilo storico', in *Scienziati idraulici e territorialisti nella Toscana dei Medici e dei Lorena*, ed. by Danilo Barsanti and Leonardo Rombai (Florence: Centro Editoriale Toscano, 1994), pp. 1–41.

77 Schenk, 'Dis-Astri', p. 63 and passim.

78 For an edition, see in Schenk, 'Friend or Foe?', pp. 151–56.

79 Schenk, 'Friend or Foe?', p. 153 (translation by the author).

80 Meinrad von Engelberg, *Die Neuzeit 1450–1800. Ordnung – Erfindung – Repräsentation* (Darmstadt: WBG – Wissenschaftliche Buchgesellschaft, 2013), pp. 109–13; Bruce Edelstein, 'Acqua viva e corrente. Private Display and Public Distribution of Fresh Water at the Neapolitan Villa of Poggioreale as a Hydraulic Model for Sixteenth-Century Medici Gardens', in *Artistic Exchange and Cultural Translation in the Italian Renaissance City*, ed. by Stephen J. Campbell and Stephen J. Milner (Cambridge: Cambridge University Press, 2004), pp. 187–220.

81 Ernst Schubert, *Alltag im Mittelalter. Natürliches Lebensumfeld und menschliches Miteinander* (Darmstadt: Primus Verlag, 2002), p. 129 and passim; Heribert M. Nobis, 'Die Umwandlung der mittelalterlichen Naturvorstellung. Ihre Ursachen und ihre wissenschaftsgeschichtlichen Folgen', *Archiv für Begriffsgeschichte*, 13 (1969), 34–57, in particular 41–45.

82 On the infrastructure administration and its engineers, see the articles in Giovanna Casali and Ester Diana, *Bernardo Buontalenti e la burocrazia tecnica nella Toscana medicea* (Florence: Alinea, 1983); Giovanna Casali, 'La costruzione e riparazione di ponti', in *Costruttori e maestranze edilizie della Toscana medievale. I grandi lavori del contado fiorentino (secolo XVI)*, ed. by Giuseppina Carla Romby (Florence: Le Lettere, 1995), pp. 53–101.

83 See in another context with similar considerations Christian Wieland, 'Grenze zwischen Natur und Machbarkeit. Technik und Diplomatie in der römisch-florentinischen Diskussion um die Valdichiana (17. Jahrhundert)', *Saeculum*, 58 (2007), 13–32. Wieland rightly characterises many hydraulic projects of the seventeenth century as ostentatious performative acts by the ruling elites, not aimed at practical implementation. However, he overlooks the fact that the learned discourse in many – not all – respects differed from the daily practice of local experts and 'engineers'. The exact connection between the political system and the building or maintenance of infrastructure is yet to be studied.

84 See Gerrit Jasper Schenk, 'Die ,Schlammengel' von Florenz 1966. Überschwemmungen des Arno von 1333 bis heute', in *Von Atlantis bis heute*, pp. 186–93 on the 1966 flood.

85 Jacopo Nardi, *Istorie della città di Firenze*, vol. 2, ed. by Lelio Arbib (Florence: Società Editrice delle Storie del Nardi e del Varchi, 1842), p. 74.

86 Giovanni Cambi, *Istorie di Giovanni Cambi cittadino fiorentino*, vol. 2, ed. by Ildefonso di San Luigi (Florence: Cambiagi, 1785), p. 122.

87 Christian Rohr, 'Überschwemmungen an der Traun zwischen Alltag und Katastrophe. Die Welser Traunbrücke im Spiegel der Bruckamtsrechnungen des 15. und 16. Jahrhunderts', *Jahrbuch des Musealvereins Wels*, 33 (2001–2003), 281–327 (327) (translation by the author).

88 Jacques Ungerer, *Le pont du Rhin à Strasbourg, du XIVe siècle à la Révolution* (Strasbourg; Paris: Le Roux, 1952), pp. 9–11; Wilhelm Mechler, 'Die Rheinbrücke Straßburg-Kehl seit 1388', in *Die Stadt am Fluß. 14. Arbeitstagung in Kehl 14.–16.11. 1975*, ed. by Erich Maschke and Jürgen Sydow (Sigmaringen: Thorbecke, 1978), pp. 4–61, in particular pp. 40–48). See reactions to flood damage and preventive measures in the books of the Strasbourg administration: AVE Strasbourg, sér. VII, 1577, fol. 2v (1526), fols. 6v-7r (1529), fols. 7v-8r (1530), fols. 24r-25v (destruction 1555); AVE Strasbourg, sér. VII, 1572, fol. 5v (compensations and provisions as reaction to the flood; June 1530); AVE Strasbourg, AA 2009 nr 39 (repairing the Rhine bridge, November 1555).

89 Rehlinghaus, *Die Semantik des Schicksals*, pp. 111–18 (on *fatum christianum*) and 128–54 (on *fatum turcicum*).

90 Villani, *Cronica*, vol. 3, pp. 26–40; Schenk, "... prima ci fu la cagione de la mala provedenza de' Fiorentini...", 369. See also Philine Helas, '"... und sie bekundeten ihm ihre Teilnahme und trösteten ihn wegen all des Unglücks ...". Die Hiobsgeschichte in der italienischen Malerei des 14. und 15. Jahrhunderts', in *Une histoire du sensible: La perception des victimes de catastrophe du XIIe au XVIIIe siècle. /Eine Geschichte der Sensibilität: Die Wahrnehmung von Katastrophenopfern vom 12. bis zum 18. Jahrhundert*, ed. by Thomas Labbé and Gerrit Jasper Schenk (Turnhout: Brepols, 2018), pp. 69–102.

91 Johannes Fried, *Aufstieg aus dem Untergang. Apokalyptisches Denken und die Entstehung der modernen Naturwissenschaft im Mittelalter* (Munich: Beck, 2001), pp. 106–11, 183–95.

92 Leon Festinger, *A Theory of Cognitive Dissonance* (Stanford: Stanford University Press, 1957).

93 Cf. the recent work by Iso Himmelsbach on the 1480 flood *Erfahrung – Mentalität – Management: Hochwasser und Hochwasserschutz an den nicht-schiffbaren Flüssen im Ober-Elsass und am Oberrhein (1480–2007)* (Ph.D. thesis, Freiburg im Breisgau: University of Freiburg, 2012), pp. 87–104.

94 Mary Douglas, *Risk and Blame: Essays in Cultural Theory* (London; New York: Routledge, 1992), pp. 3–121.

95 See Schenk, 'Die Zukunft zähmen?', p. 217, n. 94. The archival and historical materials document the following events, in chronological order: (flood, after 6 July 1275); *pro serenitate aeris*, late thirteenth century; rain, 2 February 1401; flood, 31 March 1415; rain, 23 June 1438; wet weather; undated, probably 1441; bad weather, 7 February 1479; rain, 26 June 1480; flood, 9 August 1480; bad weather, 12 May 1482; flood, 12 July 1485; bad weather, 23 June 1511; bad weather, 13 May 1517; procession for good harvests, 1 May 1519; fear of flooding, procession on 25 March 1523.

96 On the risks through flooding in the late medieval cultural landscape of the Upper Rhine Vally, see Schenk, 'Politik der Katastrophe?', pp. 49–54.

97 Volker Leppin, *Antichrist und Jüngster Tag. Das Profil apokalyptischer Flugschriftenpublizistik im deutschen Luthertum 1548–1618* (Gütersloh: Gütersloher Verlagshaus, 1999), pp. 96–103.

98 AVE Strasbourg, 1 MR 2, pag. 155 (=fol. 79*v*), procession on 23 June 1438 because of wet weather, linked with price rises; AVE Strasbourg, 1 MR 3, pag. 64 (=fol. 33*r*), procession on 23 June 1511 due to war, bad weather and an epidemic; AVE Strasbourg, 1 MR 3, pag. 196 (=fol. 110), planned procession on 25 March 1523 mentioned above.

99 AVE Strasbourg, Strasbourg, 1 MR 2, pag. 168 (=fol.85*b*), procession on 9 August 1473, initiated by the Council, in response to extreme heatwave and drought of 1473;. AVE Strasbourg, 1 MR 2, pag. 225 (=fol. 115*v*), procession on 23 June 1503, initiated by the bishop.

100 Cf. above note 101 on the 1480 flood.

101 AVE Strasbourg, 1 MR 2, pag. 224 (=fol. 115*r*), a usual procession in the cross or rogation week on 12 May 1482 with additional intercessions against bad weather and price rises.

102 Cf. on the persecution of witches and travelling people, who were accused of weather magic if the economic situation was bad, in Berne and Lucerne in spring 1482, see Valerius Anshelm, *Die Berner Chronik*, vol. 1, ed. by Emil Blösch (Bern: Wyss, 1884), p. 223; *Die Eidgenössischen Abschiede aus dem Zeitraume von 1478 bis 1499*, ed. by Anton Philipp Segesser (Zürich: Bürkli, 1858), p. 120, n. 143: Strasbourg matters were also discussed at this confederal meeting of 20 May 1482 so that a mutual awareness of the events can be assumed at least at the Council level. In Berne, the Council discussed processions to pray for better weather on 17 May 1482, cf. Staatsarchiv Bern, A.II.21, no. 36, pag. 100.

103 Ulricus Molitor, *De laniis et phitonicis mulieribus* (Reutlingen: Otmar, 1489). See Gerrit Jasper Schenk, 'Wunder – Zeichen – Glaube. Unsterne, Prognostiken und Wetterzauber in der Renaissance', in *Von Atlantis bis heute*, pp. 96–101.

104 Rita Voltmer, *Wie der Wächter auf dem Turm. Ein Prediger und seine Stadt. Johannes Geiler von Kaysersberg (1445–1510) und Straßburg* (Trier: Porta Alba, 2006), pp. 845–47 and 945–47; Gerhard Bauer, 'Johannes Geiler von Kaysersberg (1445–1510) und seine Hexenpredigten in der 'Ameise', Straßburg 1516', in *Stregoneria e streghe nell'Europa moderna. Convegno internazionale di studi (Pisa, 24–26 marzo 1994)*, ed. by Giovanna Bosco and Patrizia Castelli (Pisa; Rome: Ministero per i Beni Culturali e Ambientali; Biblioteca Universitaria di Pisa, 1996), pp. 133–67, particularly p. 141. Bauer counts Geiler among the opponents of persecuting witches.

105 AVE Strasbourg, 1 MR 2, pag. 230 (=fol. 118*r*). According to a report by the Rufach chronicler Maternus Berler, *Straszburgische Archiv-Chronik*, ed. by Adam Walther Strobel and Louis Schneegans (Strasbourg: Silberman, 1848), p. 99, a solar eclipse on 6 March 1485 was followed by price rises and the plague. The NASA catalogue (http://eclipse.gsfc.nasa.gov/SEsearch/SEsearchmap.php? Ecl=14850316) reports under the date 16 March 1485 (modern due to the calendar reform) that the solar eclipse must have been visible in southern Alsace.

106 AVE Strasbourg, 1 MR 3, pag. 65 (=fol.33*v*) on the 1517 procession for good weather, good harvest and peace, pag. 158 (=fol. 85*r*) on the 1519 procession for the election of a new emperor and good harvest. For the context, see also Marc Lienhard, 'Mentalité populaire, gens d'église et mouvement évangélique à Strasbourg en 1522–23. Le pamphlet "Ein brüderlich warnung an meister Mathis …" de Steffan von Büllheym', in *Horizons Européens de la Réforme*

en Alsace. Mélanges offerts à Jean Rott pour son 65 anniversaire, ed. by Marijn de Kroon and Marc Lienhard (Strasbourg: Société Savante d'Alsace et des Régions de l'Est, 1980), pp. 37–62; Reinhard Scholzen, 'Franz von Sickingen als Machtfaktor im Kampf zwischen Mainz, Hessen, Kurtrier und Kurpfalz', *Blätter für pfälzische Kirchengeschichte und religiöse Volkskunde*, 68 (2001), pp. 287–306; and Leppin, *Martin Luther*, pp. 125–64, 209–20.

107 Cf. the list in Signori, 'Ritual und Ereignis', 318.

108 AVE Strasbourg, 1 MR 2, pag. 225 (=fol. 115*v*) on the procession for the saving of Christendom.

109 Cf. here, for instance, texts and figures (by the 'Mennel master') in the manuscripts of Jakob Mennel, *De signis, portentis atque prodigiis tam antiquis quam novis cum eorundem typis et figuris* (1503) Österreichische Nationalbibliothek, Vienna, Cod. Vind. Pal. 4417*, fol. 18*v* (sign of cross, June 1503); Württembergische Landesbibliothek, Stuttgart, HB XI 3, fol. 14*r* (sign of cross 1500). On the reports of miraculous signs by Jakob Mennel, Joseph Grünpeck and Sebastian Brant, see Irene Ewinkel, *De monstris. Deutung und Funktion von Wundergeburten auf Flugblättern im Deutschland des 16. Jahrhunderts* (Tübingen: Niemeyer, 1995), pp. 39, 102–6; Andrea Löther, *Prozessionen in spätmittelalterlichen Städten. Politische Partizipation, obrigkeitliche Inszenierung, städtische Einheit* (Cologne; Weimar; Vienna: Böhlau, 1999), p. 239 and passim on the cross miracle processions that also took place in 1503 in Dortmund and Augsburg (although the connection with the 'cross miracles' is not quite certain).

110 On the theodicy problem in this connection, see David K. Chester, 'The Theodicy of Natural Disasters. Christianity, Suffering, and Responsibility', *Scottish Journal of Theology*, 51, 4 (1998), 485–505; idem, 'Theology and Disaster Studies: The Need for Dialogue', *Journal of Volcanology and Geothermal Research*, 146 (2005), 319–28. On processions as means to cope with emotions and unrest, see Gabriela Signori, 'Ereignis und Erinnerung: Das Ritual in der städtischen Memorialkultur des ausgehenden Mittelalters (14. und 15. Jahrhundert)', in *Prozesssionen, Wallfahrten, Aufmärsche. Bewegung zwischen Religion und Politik in Europa und Asien seit dem Mittelalter*, ed. by Jörg Gengnagel, Monika Horstmann, and Gerald Schwedler (Cologne; Weimar; Vienna: Böhlau, 2008), pp. 105–21.

111 Undated entry but presumed ca. 1510, in AVE Strasbourg, 1 MR 12, p. 510 (=fol. 313*r*). On the 1510 procession because of an epidemic, see Berler, *Straszburgische Archiv-Chronik*, p. 217.

112 See above note 21.

113 Simona Cohen, *Transformation of Time and Temporality in Medieval and Renaissance Art* (Leiden; Boston: Brill, 2014), pp. 73–79, 199–243; Thomas Rahn, 'Das Überleben der Güter. Schiffbruch als 'occasio'', *Zeitschrift für Ideengeschichte*, 14, 3 (2020) 31–44. It is well known that Machiavelli used the concept of *fortuna* in his *Principe* systematically.

114 For the (disastrous) gale-force wind of 22–23 August 1456 in Tuscany, see Giovanni di Pagolo Rucellai, *Zibaldone*, ed. by Gabriella Battista (Florence: Edizioni del Galluzzo, 2013), pp. 131–35.

115 Jörg Oberste, *Zwischen Heiligkeit und Häresie. Religiosität und sozialer Aufstieg in der Stadt des hohen Mittelalters 1: Städtische Eliten in der Kirche des hohen Mittelalters* (Cologne; Weimar; Vienna: Böhlau, 2003), pp. 86–88.

116 Ceccarelli, *Il gioco e il peccato*, pp. 404–7.

117 *Il cartolare di Giovanni Scriba. Documenti e studi per la storia del commercio e del diritto communale italiano*, ed. by Mario Chiaudano and Mattia Moresco (Turin: Lattes, 1935), p. 37.

118 See the discussion in Benjamin Zeev Kedar, 'Again: Arabic *rizq*, Medieval Latin *risicum*', *Studi medievali*, 10, 3 (1969), pp. 255–59; Omar B. Bencheikh, 'Risque et l'arabe rizq', *Bulletin de la Société d'Études lexico-graphique et Étymologiques Françaises et Arabes*, 1 (2002), 1–6; Piron, 'L'apparition du resicum en Méditerranée occidentale', pp. 62–68; Scheller, 'Die Geburt des Risikos', 308 and passim. In the Arabic-speaking world, the concept of risk developed in another way, using the (related) concept of 'Gharar', see Abdul-Rahim Al-Saati, 'The Permissible "Gharar" (Risk) in Classical Islamic Jurisprudence', *J. KAU: Islamic Economy*, 16, 2 (1424 AH / 2003 AD.), 3–19.

119 Hariton Tiktin, 'Zur Geschichte von "hasard"', *Archiv für das Studium der neueren Sprachen und Literaturen*, 127 (1911), 162–74; Piron, 'L'apparition du resicum en Méditerranée occidentale', p. 61. A more detailed study of the etymology and history of the concept would be necessary.

120 Tiktin, 'Zur Geschichte von "hasard"', 169–74.

121 See Luca Alexandre Boiteaux, *La Fortune de mer. Le besoin de sécurité et les débuts de l'assurance maritime* (Paris: S.E.V.R.E.N., 1968); Karin Nehlsen-van Stryck, 'Kalkül und Hasard in der spätmittelalterlichen Seeversicherungspraxis', *Rechtshistorisches Journal*, 8 (1989), 195–208; Yadira Gonzales de Lara, 'Institutions for Contract Enforcement and Risk-Sharing: From the Sea Loan to the "Commenda" in Late Medieval Venice', *European Review of Economic History*, 6, 2 (2002), 257–62; Markus A. Denzel, 'Die Seeversicherung als kommerzielle Innovation im Mittelmeerraum und in Nordwesteuropa vom Mittelalter bis zum 18. Jahrhundert', *Richezza del mare, richezza dal mare: secc. XIII-XVIII. Atti della Trentasettesima Settimana di Studi, 11–15 aprile 2005*, ed. by Simonetta Cavaciocchi (Florence: Le Monnier, 2006), pp. 575–609, in particular pp. 578–85; Giovanni Ceccarelli, *Un mercato del rischio. Assicurare e farsi assicurare nella Firenze rinascimentale* (Venice: Marsilio, 2012); Scheller, 'Die Geburt des Risikos', 308–15.

122 Ceccarelli, 'The Price for Risk-Taking', 6–19.

123 Ceccarelli, 'The Price for Risk-Taking', 1. However, calculations were made following informal rules and depended on trust amongst small groups of citizens: idem, '"Tutti gli assicuratori sono uguali, ma alcuni sono più uguali degli altri". Cittadinanza e mercato nella Firenze rinamiscentale', in *Mélanges de l'École Française de Rome*, 125, 2 (2013), 405–20.

124 See Zwierlein, 'Frühe Formen'.

125 See my suggestion in Schenk, 'Vormoderne Sattelzeit?', p. 199 and passim; idem, 'Disastro, Catastrophe, and Divine Judgement', p. 59 and passim.

126 On the concept of securitisation, shaped to a large extent by the so-called Copenhagen School, which saw the field of security as defined primarily by speech acts about what security is, see Trine Villumsen Berling, Ulrik Pram Gad, Karen Lund Petersen, and Ole Wæver, *Translations of Security. A Framework for the Study of Unwanted Futures* (London; New York: Routledge, 2022), pp. 7–11. Whether it was in late medieval cities that people first talked about more security or whether certain actions led to more security is an unanswerable, perhaps irrelevant question. In any case, the increasing thematisation of semantic fields of danger, security and risk is striking.

127 See already Christopher M. Gerrard and David N. Petley, 'A Risk Society? Environmental Hazards, Risk and Resilience in the Later Middle Ages in Europe', *Natural Hazards*, 69 (2013), 1051–79, but limiting their arguments to environmental hazards like earthquakes, volcanic eruptions and extreme weather.

128 See Collet, 'Eine Kultur der Unsicherheit?; Christian Rohr, 'Ein ungleicher Kampf? Sieg und Niederlage gegen Naturgewalten im Spätmittelalter und am Beginn der Neuzeit', in *Inszenierung des Sieges – Sieg der Inszenierung. Interdisziplinäre Perspektiven*, ed. by Michaela Fahlenbock, Lukas Madersbacher, and Inge Schneider (Innsbruck; Vienna; Bozen: StudienVerlag, 2011), pp. 91–99.

129 See Schenk, "Human Security' in the Renaissance?'. The securitisation in towns of course did not reach the extent of modern welfare states, but some origins of Foucault's *dispositif* of security and *gouvernementalité* can be find already here: Christian Jörg, «*Teure, Hunger, Großes Sterben*». *Hungersnöte und Versorgungskrisen in den Städten des Reiches während des 15. Jahrhunderts* (Stuttgart: Hiersemann, 2008) (coping with hunger crises); Gerhard Fouquet, *Bauen für die Stadt. Finanzen, Organisation und Arbeit in kommunalen Baubetrieben des Spätmittelalters. Eine vergleichende Studie vornehmlich zwischen den Städten Basel und Marburg* (Cologne; Weimar; Vienna: Böhlau, 1999), pp. 400–30; Maria Pia Contessa, *L'Ufficio del fuoco nella Firenze del Trecento* (Florence: Le Lettere, 2000); Medard Barth, *Grossbrände und Löschwesen des Elsass vom 13.–20. Jahrhundert mit Blick in den europäischen Raum* (Bühl; Baden: Verlag Konkordia, 1974) (coping with fire disasters).

130 On the process of securitisation, see Jean Delumeau, *Rassurer et protéger. Le sentiment de sécurité dans l'Occident d'autrefois* (Paris: Fayard, 1989), pp. 21–29; Stefanie Rüther, 'Zwischen göttlicher Fügung und herrschaftlicher Verfügung. Katastrophen als Gegenstand spätmittelalterlicher Sicherheitspolitik', in *Sicherheit in der Frühen Neuzeit. Norm, Praxis, Repräsentation*, pp. 335–350; on hygenic problems, see Eberhard Isenmann, *Die deutsche Stadt im Mittelalter, 1150–1550. Stadtgestalt, Recht, Verfassung, Stadtregiment, Kirche, Gesellschaft, Wirtschaft* (Cologne; Weimar; Vienna: Böhlau, 2012), pp. 63–88.

131 Peter Spufford, *Handel, Macht und Reichtum: Kaufleute im Mittelalter* (Darmstadt: Theiss, 2004), pp. 14–44.

132 Machiavelli did not make use of the word 'risk', however.

133 Iris Origo, *Im Namen Gottes und des Geschäfts. Lebensbild eines toskanischen Kaufmanns der Frührenaissance. Francesco di Marco Datini 1335–1410* (Berlin: Wagenbach, 1997), fig. 7.

134 Ian Hacking, *The Emergence of Probability. A Philosophical Study of Early Ideas about Probability, Induction and Statistical Inference* (Cambridge; New York: Cambridge University Press, 1975), pp. 122–33; idem, *The Taming of Chance* (Cambridge; New York: Cambridge University Press, 1990), pp. 1–10; Lorraine Daston, *Classical Probability in the Enlightenment* (Princeton: Princeton University Press, 1988), pp. 112–87.

2 Power, fortune and *scientia naturalis*

A humanist reading of disasters in Giannozzo Manetti's *De terremotu*[1]

Ovanes Akopyan

In December 1456, two strong earthquakes hit southern Italy and killed around 30,000 people. The news about the event reached areas as distant as Muscovy, and the dreadful disasters were reflected in numerous contemporary texts.[2] Some adopted an eschatological perspective and, following a popular trope, regarded the earthquakes as divine punishment, whereas others sought to rationalise the discourse in conformity with contemporary intellectual trends. The most famous description of the 1456 disaster belonged to the second group.[3] In the *De terremotu*, addressed to his patron Alfonso of Aragon, the renowned Italian humanist Giannozzo Manetti aimed to reconstruct an ancient understanding of natural disasters and thus provide a detailed account of how to view earthquakes in the present and, probably, prevent them in the future. This essay demonstrates how, through a painstaking reading of ancient sources, Manetti explored an alleged correlation between the political virtues of the ruler and disasters hitting their territories. By addressing the issue of the ruler's *fortuna* that was at the heart of fifteenth-century Italian humanist discussions, Manetti combined a natural philosophical explanation of earthquakes with political ethics. This approach was characteristic of the line of political thinking that James Hankins has recently presented under the banner of 'virtue politics' of the Italian Renaissance.[4] At the same time, as we shall see, it outlived Manetti and his humanist peers and found its way to, among other things, sixteenth-century dream literature, baroque emblem books and even early twentieth-century Latin poetry. This suggests that Manetti's treatise, while being deeply reflective of a peculiar intellectual context of fifteenth-century Italian Renaissance humanism, reveals one of the standard patterns of thinking on the origin of natural disasters in general and earthquakes in particular.

Giannozzo Manetti obtained exceptional fame among his contemporaries. His numerous intellectual and political activities were commemorated in an extensive and detailed chapter of Vespasiano da Bisticci's *Vite*, in which he is

DOI: 10.4324/9781003029823-4

called 'a man of great authority' and 'laudable virtues'.[5] Manetti was, indeed, one of the foremost early fifteenth-century humanists and political actors. He is now remembered principally as a true *vir trilinguis*, the first major Renaissance scholar able to read Latin, Greek and Hebrew, and the one responsible for the first attempt, after the Vulgate, to translate the Bible into Latin.[6] Along with Lorenzo Valla – whose *Collatio Novi Testamenti* scrupulously investigated the shortcomings of Jerome's version of the New Testament – Manetti laid the groundwork for the rise of biblical humanism in the following century.

Manetti did not limit himself to biblical translations and the study of ancient languages. He also authored numerous treatises, including the *De dignitate et excellentia hominis*, usually – but falsely – considered as a text that intellectually preceded Giovanni Pico della Mirandola's famous *Oratio*, widely known under the expanded title *Oratio de hominis dignitate*.[7] The *De dignitate et excellentia hominis* – composed already under the wing of Alfonso of Aragon in Naples – marked the peak of Manetti's political and courtier career, which, as we shall see below, is inseparable from that of Manetti the humanist.

Alongside his numerous achievements in the field of *studia humanitatis*, Manetti obtained a high political position in the Florentine republic. On various occasions, he was appointed ambassador to the papal court in Rome and the court of Alfonso of Aragon in Naples. However, in the early 1450s, his Florentine career went into turmoil. In order to avoid further tensions with his political opponents that might have ended up in further persecution and even incarceration, he decided to stay first in Rome at the court of Pope Nicholas V and later moved to Naples to become secretary to Alfonso of Aragon.[8] As is well known, while in the service of the Florentine republic, Manetti was critical of Alfonso repeatedly contrasting the Florentine political system to the Naples governed by a tyrannical regime. Such critical remarks did not prevent Manetti from softening his opinion of Alfonso depending on political and personal circumstances. Nor did it stop Alfonso from inviting one of the most renowned humanists of the age to his court; on the contrary, it was pictured as an essential sign of Manetti's 'political' re-conversion. To Alfonso, Manetti dedicated the *De dignitate et excellentia hominis*, in the introduction to which he labelled his new patron the most illustrious and glorious king and consequently portrayed him as an ideal ruler.[9] Hence, it comes as no surprise that in the *De terremotu* commissioned by Alfonso himself and designed to be the official account of the 1456 earthquake, Manetti sought to find a reliable response to a question that presented an issue for fifteenth- and sixteenth-century Renaissance scholars concerned with the origin of natural disasters: how might such calamities occur under the reign of a good ruler? In a period when his fellow humanists reflected on what constitutes a perfect reign and how virtue correlates with politics, Manetti complemented the discussion with a text that scrutinised the political dimension of disasters. Furthermore, this brought him to another problem associated with natural philosophy: how could Aristotle's notion of 'final cause', according to which everything is eventually brought to the Good, be reconciled with disastrous events? The quest for a proper interpretation of these two issues defined Manetti's argument, which, being at first glance strongly Alfonsine, sometimes took a fascinating and somewhat sarcastic form.

The *De terremotu* consists of three books in which, as Manetti stated in the introduction, he aimed to work through different opinions on the origin of earthquakes and describe in detail the effects of the 1456 disaster.[10] In the address to Alfonso, he stated that the terrible earthquakes that occurred 'under your joyful and fortunate reign' and hit Campania, Apulia and other regions of the Neapolitan kingdom caused numerous deaths and destroyed several cities and fortresses.[11] Manetti went on to say that to find out how such paradoxical things could happen, he decided to have a look into ancient and new Latin and Greek sources. He stated that, through the treatise, he would gradually explore the positions of poets, historians, jurists and theologians who think that such events are to be ascribed to the intervention of an omnipotent God into the terrestrial world – something which, Manetti added, contradicts the opinions of astrologers and philosophers, for whom earthquakes are provoked exclusively by natural causes.[12] Manetti did not, however, structure the first book of the treatise in accordance with the conflicting positions; instead, in Books 1 and 2, he presented a mixture of views and complemented this digest with the descriptions of various catastrophes of the past – from biblical times onwards – that were reported in Hebrew, Greek and Latin texts.[13] Book 3, in turn, focuses in particular on the damage the 1456 earthquake produced in the Neapolitan kingdom. Seemingly 'unoriginal' (perhaps, with the exception of the last section, containing first-hand material), the *De terremotu* is, in fact, far from being a mere work of compilation. On the contrary, skilful and careful adjustment of various historical and philosophical accounts allows Manetti to design a coherent and developing narrative and, as a result, make testimonies of the past serve needs of his day.

In accentuating the goal of this work, Manetti compared it to what his great predecessor Seneca had done in the *Naturales quaestiones*. To support the idea of intellectual continuity between the two, Manetti recalled that Book V of Seneca's masterpiece was written in response to an earthquake that hit the same region.[14] Manetti clearly confused Book V (on winds) and Book VI (on earthquakes), with no documented manuscript evidence for this mistake traceable in his personal library holdings.[15] The reference to

Seneca, 'vir Cordubensis', is a clear courtesy to the country of Alfonso of Aragon's roots (i.e., Spain). As Peter Stacey has demonstrated, the figure of Seneca was instrumental in developing the humanist royal ideology in Renaissance Naples and constructing a classicised Hispanic origin of the Aragonese reign in Italy.[16] In the Apennine peninsula, Alfonso was universally regarded as a stranger, a characterisation, rather xenophobic by today's standards, that was reflected in numerous writings about and, particularly, against his regime, including those by Manetti.[17] The image of Seneca, that is, of a 'Spaniard' who nonetheless belonged to the *oikumene* and shared a classical Roman identity, was thus exploited to respond to such accusations and justify Alfonso's political claims and ambitions.

Two more things deserve further examination in the passage at hand. First, Manetti said that 'although the earthquake took place in Naples, we, the Florentines, observed it with our eyes and heard it with our ears'.[18] Although an indication of the magnitude of the earthquake, this statement at the same time served to underline Manetti's Florentine origin, something that he kept stressing throughout the treatise to distinguish himself from the rest of the Neapolitan court. Second, and more importantly, he noticed that the *Naturales quaestiones* was composed during the reign of Nero. Given the aforementioned attempt to stress the intellectual continuity between Seneca's and Manetti's works, this remark might not merely be a chronological fact; it could have also been intended to implicitly hit at Alfonso by comparing him with Nero and, if so, completely inverting the whole of Manetti's eloquence and apparent courtier adoration of the ruler. Thus, taken as a whole, the present passage – the only example of this kind in the *De terremotu* – poses a question about Manetti's attitude towards Alfonso. However, the rest of Manetti's discourse has no other signs of a similar 'behind-the-scenes' political controversy. It sticks to the main aim of accumulating the existing knowledge of the events and, through it, giving a new interpretation of earthquakes with regard to their origin and timing.

The first problem Manetti declared is how to bring into agreement the two sets of arguments, supernatural and natural, on the origin of disasters. In order to do so, Book 1 of the *De terremotu* provides a survey of the sources that touched upon the matter. Manetti meticulously examined ancient accounts that underpinned how Greek and Roman poets and historians recognised the divine as the sole cause for earthquakes. Thus, although in other instances he referred to Seneca as a natural philosopher and the author of the *Naturales quaestiones*, here Manetti placed Seneca alongside Vergil, Ovid and the legendary Orpheus to give additional weight to his opinion.[19] In a similar fashion, he quoted from Lucretius implying that the *De rerum natura* followed the pattern.[20] Given the character of Lucretian philosophy, which challenged the divine design of the universe in its entirety, the presence of his name in the list seems unjustified, to say the least. However, it illustrates that Manetti was well aware of, and responsive to, the reception Lucretius enjoyed after the rediscovery of the *De rerum natura* in the early fifteenth century.[21]

Early personifications of a divine power responsible for the earthly tremors were indebted to the poetic tradition and, therefore, lacked an actual divine *persona* behind them. Following Aulus Gellius, Manetti recalled Neptune, who – along with controlling the sea – was believed to 'move the earth'.[22] Christian theology, on its part, introduced the moral understanding of disasters; however, as highlighted by Manetti's selection, a connection between immoral deeds and divine punishment is not as straightforward as one could imagine. Excerpting the passages concerned with earthquakes from the Old Testament, Manetti presumed that God interferes in human affairs by causing disasters to prevent people from wrongdoing. With the manner he quoted from Scripture, Manetti intended to demonstrate that, unlike previous generations of biblical scholars, he operated with the Hebrew original. However, an analysis of the passages he extracted reveals that Manetti exaggerated his innovativeness. Thus, as regards the Psalms, which he had previously translated anew, causing a debate on the eligibility and scope of biblical humanism, all variants present in the *De terremotu* are Manetti's. At the same time, quotations from other books of the Old Testament follow the Vulgate, in some cases with a few minor syntactic corrections.[23] Regardless of whichever reference technique he applied, through the fragment in question, Manetti continuously linked moral decay to natural calamities. Babylon embodies this link and serves, to use Henri Corbin's famous term, as an emblematic city.[24] Although in large part developing the narrative, with the death of Christ symbolising its apotheosis,[25] the New Testament allows for a slightly more nuanced vision: earthquakes might signal significant changes that ought to manifest divine omnipotence and are not necessarily but just often caused by human sins. This formed the grounds on which theologians relied in interpreting the origin of earthquakes. To give some additional perspective on this point, Manetti mentioned how, in the *Praeparatio Evangelica*, Eusebius of Caesarea expressed a belief that nothing illustrated the collapse of a state better than the reign of tyrants and catastrophes occurring in their territory.[26] Manetti continued by adding that this had become a common way of thinking on what could instigate actual calamities.[27] However, Manetti did not go beyond this assumption and simply acknowledged that it exists. He clearly admitted God's intervention into the terrestrial world but preferred to not elaborate on the topic, leaving the providential character of disasters aside. Instead of raising the delicate topic of how God punishes bad and sinful rulership (which, in this context, would have implied that of Alfonso of Aragon) with disasters, Manetti presented himself as a humanist 'historiographer' whose goal is merely to reproduce a variety of opinions taken from ancient authors and Scripture.

As his record suggests that natural disasters could be considered as a divine attempt to steer the sinful onto the right path, this premise might certainly compromise the political prestige of Manetti's patron. To solve the ambiguous situation, Manetti devoted much more attention to the second cluster of ideas centred on natural philosophy. Compared to the list of quotations he collected from ancient sources and the Bible (and which he refrained from commenting on in depth), this section presents a more elaborate analysis. Manetti harmonised it with a contextualised interpretation of historical disasters, permitting him to find a balance between the temporality of events and divine predestination. This strategy determined Manetti's central argument, which is twofold. First, by providing a large amount of textual evidence on earthquakes, he sought to demonstrate that there is no correlation between calamities and the dignity and virtues of the political figures under whose reign earthquakes occurred or whose lives were affected by them. In the preface to Alfonso, Manetti declared that Alfonso's (and, presumably, any other ruler's) dignity could be not jeopardised by disasters. He went on to support the statement by recalling the historical analogies with Christ and Alexander the Great, whose death and birth, respectively, were manifested with earthquakes.[28] In the main body of the treatise, mainly in Book 2, Manetti mixed together different accounts associated with both good and bad rulers. Thus, Nero resurfaces alongside Augustus – whose reign in accordance with the Christian tradition was the pinnacle (peaceful and prosperous) of pagan antiquity – and the post-Constantinian era, when Christianity had already become the dominant religious movement.[29] With the use of these examples, Manetti highlighted how representative in its ambiguity the historical narrative is and stressed that earthquakes might happen under the reign of any ruler, regardless of their accomplishments or cruelty. Manetti argued consequently that the big picture suggests the hazardous character of disaster.

In addition, as a separate task – but of significant importance for his patron's most pragmatic purposes – he reconstructed all major south Italian disasters, from antiquity to the times of Alfonso.[30] What Manetti did in Book 3 conforms to the way Alfonso of Aragon aimed to represent his power, as of a legitimate and indigenous ruler of southern Italy. Manetti created a long, undivided chain of the region's history, of which Alfonso appears to be an integral part. By applying this artificially established concept of political continuity, the humanist supports Alfonso's claims to be seen as the only successor to the previous governors of the country. Manetti's narrative comprises two important features. First, he held that Calabria and other territories of southern Italy had always belonged to the *oikumene*, the *Grecia Magna*.[31] At the same time, he admitted that the political power and prosperity of the region, which was already in place, increased significantly after the foreign invasion. As Manetti put it, the Norman conquest – which southern Italy witnessed in the eleventh century – became a moment of glory that catapulted the kingdom to its height.[32] With no word said about the two hundred years under the Angevin dynasty that separated the first and the second 'positive' conquests, Manetti purposely omitted the political turmoil into which the Neapolitan state had plunged in the first half of the fifteenth century when, after many years of war, Alfonso defeated René of Anjou and his allies. In Manetti's chronology, Alfonso's rulership crowns what appears to be a tripartite history of the region: antiquity – the Normans – Alfonso. Over the course of all three periods, the country effectively coped with disasters, both natural and political. To exemplify the latter, Manetti looked back at the long-standing opposition of southern Italy to the papacy, a detail that could have resonated well at the Neapolitan court.[33] Thus, at a glance, Book 3 – in which the author recorded the earthquake's consequences, including the number of deaths and destroyed sites – seems fairly technical. In this regard, it resembles a medieval chronicle, albeit composed in elegant humanist Latin. However, a closer look at the text reveals that it goes far beyond the pure statistics and amplifies the political agenda Manetti developed throughout the *De terremotu*.

Considering how he approached the political dimension of disaster, it is not surprising that Manetti did not associate the providential effects or, in Aristotelian terms, the final cause of disasters with the ruler under whom they happen. In his view, the consequences determined by a more elevated and hidden origin may emerge in the long run. Thus, this entails the main question: what causes earthquakes right here and right now? His solution to the issue exposes the two other characteristics of Manetti's argument concerned with the second group of authorities previously mentioned in the introduction to the treatise (i.e., astrologers and philosophers). Manetti structured the respective passages accordingly, partly to comply with his benefactor's interest in the disciplines, mainly in astrology; however, as we shall see below, rather than being a simple hommage to Alfonso and his fascination with predictive practices, the text reflects Manetti's personal opinion on the matter and is more centred on natural philosophy. Manetti operated within a traditional philosophical framework, but the effort to tie the natural origin of earthquakes with the political element of the ruler's *fortuna* indicates that he carefully monitored the contemporary intellectual landscape in order to keep pace with those of his fellow humanists who explored similar issues.

As regards astrology, the first of the two fields, Manetti attempted to sit on two chairs at once. It is well known that at the time, astrology enjoyed the reputation of a legitimate and respected discipline.[34] Along with the university, where astrology was taught as part of the *quadrivium*, it acquired wide support at courts, with rulers wishing to secure their future and astrologers ready to provide them with the consolatory service.[35] The position of a court astrologer was thought to be prestigious, and medieval and later Renaissance rulers often competed with each other over inviting a better candidate under their wing. Thus, the famous thirteenth-century emperor Frederick II of Hohenstaufen, who spent most of his time in southern Italy, hired Michael Scot (a prolific translator and astrologer) and moved him to Palermo. About two hundred years later, in the same region, astrology was again given the green light at the court of the Neapolitan kings of Aragonese roots. Astrological ideas, literature and imagery enchanted Alfonso's and his successors' entourage. This interest resulted in a series of influential texts crowned by Giovanni Gioviano Pontano's literary production.[36] These details explain the inclusion of Manetti's rather brief, a several-pages-long sketch of astrological arguments concerning the cause of earthquakes, meant to adjust to the needs of his patron's court and satisfy Alfonso's sympathy for astrology.

At the same time, however, Manetti's general attitude towards astrology stemmed from his Florentine background. As we have already seen, throughout the treatise, he occasionally underlined his origin. Manetti recalled his celebrated Florentine predecessors, who – like Petrarch, for instance – happened to interact with southern Italian rulers and documented disastrous events of the past.[37] But what is important is that, starting from Dante, who placed the aforementioned Michael Scot and Guido Bonatti (another famed astrologer) in the eighth circle of hell,[38] astrology's position was ambivalent in the Florentine context, with both positive and critical opinions equally present. What formed the core of the discussion was whether astrological predictions could be reconciled with the Christian teaching on free will and whether they limit the divine omnipotence that was believed to be beyond human cognition. In the fifteenth century, the debate intensified, and those who followed Petrarch's ideal of Christian Stoicism expressed their strong reservations concerning the credibility of astrology as a discipline compatible with Christian teachings.[39] Among the doubters, Coluccio Salutati summarised the contradictions between the two in his *De fato et fortuna*. He stated that although God foreknows everything, divine providence is not in conflict with human free choice and does not interfere in provoking terrestrial events, which are accidental. Astrologers, Salutati continued, claim to be capable of controlling and predicting the accidental, though this area of foresight is reserved to God only. Reproducing Augustine's argument in this regard, Salutati firmly denied that any events, either benevolent or malign, depend on the constellation of celestial spheres and could be foreseen by human intellect.[40] Manetti's cursory note on astrology adheres to this trope.

Commonly known as 'mathematicians', astrologers believe that by observing the movement of the spheres and their constellations, they do not compromise the notion of free will. They hold that they are capable of predicting human affairs and social and political events and learning whether one's destiny is inclined towards favourable or adverse fruits of fortune.[41] To justify their ambitions, astrologers came up with the motto of 'the wise will master the stars', the authorship of which Manetti attributed to Ptolemy.[42] However, Manetti continues, while pretending to examine how terrestrial effects correspond to the influences of celestial bodies, astrologers completely fail to provide a trustworthy explanation of the origin of earthquakes. He commented that, instead of digging for an actual cause of the events, astrologers simply suggest that earthquakes have something to do with comets. That is what Ptolemy did in the *Tetrabiblos* when focusing on comets as signs of upcoming disasters and therefore stressing the symbolical, yet not causal, connection between the celestial and the terrestrial. Unlike Ptolemy, who, regardless of his astrological studies, was a respected scholar, other stargazers are deemed much less so. Manetti mentioned Abu Ma'shar and ancient and modern astrologers who ascribed the disastrous effects of earthquakes to the malign Saturn and interpreted them through the planet's position in different signs of the Zodiac.[43] For Manetti, this proved a lack of consistency among the astrologers themselves, since they were not able to define the exact conjunction. Manetti concluded his cursory overview of the state of the art by warning the reader that there is no need to dwell on astrology anymore, as it is, in his words, 'obscure, variable and uncertain'.[44]

After dismissing the validity of astrological interpretation, Manetti turned his attention to philosophy. Alongside a broad outline of philosophical approaches to the topic in question, ranging from Thales of Miletus to Roman scholars, what deserves closer examination is the way Manetti crafted his text. He heavily relied on Seneca's *Naturales quaestiones* and Aristotle's *Meteorologica* in presenting an array of ancient views. It might seem that he borrowed his information from the sources word for word in some cases. However, Manetti's method was not a simple 'cut, copy and paste'. He masterfully guided his reader by pointing to certain minor details in order to navigate them through a variety of opinions. On the one hand, he reproduced ancient accounts; on the other, he occasionally reshuffled, expanded or reduced the quotations to make them appear in a new frame. This serves Manetti's purpose of eventually bringing the reader to what the author acknowledged as the only genuine natural philosophical explanation of disasters (i.e., that given by Aristotle). He shaped the narrative on natural philosophy, which is content-wise the central element of Book 1 of the *De terremotu*, in a manner characteristic of the age of humanism: by extracting and scrupulously working through ancient knowledge to adapt it to contemporary needs and promote his own ideas.

To begin with, Manetti admitted that philosophers expressed contrasting positions on the matter but agreed on the central premise that nature causes earthquakes, and nothing beyond nature.[45] In this regard, Manetti followed a scholar whom he proclaimed to be his role model. In one of the opening passages of Book VI of the *Naturales quaestiones*, Seneca stated that despite what people tend to believe, neither divine anger nor any other form of divine involvement is responsible for any disaster.[46] Although unanimous in what constituted the foundational principle of the events, the philosophers who investigated the physical causes of earthquakes failed to come to an agreement regarding details. Manetti admitted that disagreement is natural to philosophical discourse. Similarly, although philosophers agree on the principles of moral philosophy, what comprises happiness and how it can be achieved cause division. With respect to tremors, various thinkers saw the cause in fire, water or wind, or ascribed the earth shaking to subterranean effects or breath.[47] What is missing from this particular passage by Seneca, which Manetti repeated almost verbatim, is the wind interpretation since antiquity associated with Aristotle. Although Seneca did mention Aristotle in Book VI, the Stagirite was listed in passing as one among the numerous authorities who endorsed a physical breath (i.e., winds) to understand the mechanics of earthly tremors.[48] To the interpretation of subterranean winds introduced in the *Meteorologica*, Seneca opposed a breath 'different from the one preferred by Aristotle' that he complied with the Stoic teaching of pneuma. In contrast to what Seneca did in the *Quaestiones*, solely by adding wind as one of the alternative solutions, Manetti made a clear indication for his learned reader towards which direction he would lean.

Manetti did not structure the natural philosophical section in chronological order. Instead, he went through the five most widespread arguments he previously singled out, discarding them one after another. In all the cases, the material Manetti carefully cited was derived from Aristotle's *Meteorologica* and Seneca's *Quaestiones*, which corresponded to his humanist approach of presenting ancient knowledge in a new package. In line with this reading, Manetti recalled the claims first put forward in the thought of Thales of Miletus, one of the seven sages of Greece, that the subterranean water supports the whole earth, is responsible for its floating, and, by extension, determines all essential effects occurring on the earth's surface, including earthquakes.[49] He likewise directed his reader's attention to the three theories of Democritus, Anaxagoras and Anaximenes, which the Stagirite fiercely denounced in the *Meteorologica*, though Manetti avoided imitating Aristotle's critical tone.[50] Pythagoras and Empedocles personify the interpretation according to which fire causes earthquakes by forcing its way out.[51] What Manetti sought to show is that *en masse* ancient philosophers were concerned with the origin of earthquakes. Most of them – with the exception of Epicurus, whose stance on natural philosophy, according to

Manetti, is hardly understandable – dedicated at least some passages, of greater or lesser length, to the topic at hand. However, out of the entire spectrum, he picked the following two theories as worthy of examination. Both focused on the impact of wind but significantly differed in construing the reasons behind wind's influence.

The first, grounded in a Stoic understanding of pneuma, culminated in Seneca's treatment of *spiritus* or breath.[52] Relying on Seneca's account, and through an instrumental selection of primary evidence, Manetti demonstrates that the majority of philosophers among those who embraced the notion that wind or air is the main driving force for earthquakes were inclined towards a purely physical (that is, originally Stoic) explanation of the effects, devoid of metaphysical connotations. Thus, he placed Seneca's vision in relation to those of historian Callisthenes[53] and philosopher Metrodorus of Chios, whom Seneca criticised in the *Quaestiones* for believing that just as a singing voice causes vibration of the air, the air pressed beneath the earth might reverberate in reaction to the air coming from outside.[54] Manetti also reproduced Seneca's passage concerning another Stoic philosopher, Posidonius, to show the latter's endorsement of pneuma's role in the physical operation of the world and to consequently claim that Posidonius favoured a purely natural wind interpretation.[55] It appears that what Manetti actually strived to achieve by assessing Seneca's position as being one among a wide range of similar approaches was to pave the way for Aristotle. In his eyes, the *Meteorologica* provided the most balanced solution that, while being natural philosophy at its core, leaves enough room for theological complements.

The Stagirite's name does not come up suddenly; indeed, Manetti carefully prepared his reader for the discussion and introduced Aristotle's stance by picturing him as the most authoritative philosopher of antiquity. Manetti thus recalled Plutarchus' unsystematic investigation of philosophical opinions on the origin of earthquakes and, against chronology, placed Aristotle last in order to stress that natural philosophy reached its climax in his thought.[56] Similarly, while calling Avicenna a renowned physician and great philosopher who devoted much attention to the study of natural causes of events, Manetti highlighted that the Islamic scholar simply imitated the Stagirite in many respects. According to Manetti, Avicenna went as far as to entitle his major writings to make them resemble Aristotle's respective works.[57] This, Manetti continued, reminded him of what Cicero said about another faithful disciple and commentator of Aristotle, namely, Theophrastus, and the very (im)possibility of paraphrasing thoughts of foremost Greek philosophers.[58] This remark implicitly suggests that Manetti's primary goal echoed Cicero's statement in the *De finibus bonorum et malorum* and constituted in transmitting, as effectively as possible, an 'authentic' Aristotle and making his teaching accessible for the Latin reader.

After correcting Cicero's expression about Aristotle being second only to Plato and inverting the pantheon of Greek philosophers in favour of the former, Manetti overviewed the Philosopher's corpus to dwell in the end on the *Meteorologica*.[59] Alongside the *Physics* and *On the Heavens*, the *Meteorologica* was a cornerstone of Aristotelian natural philosophy. In this treatise, the Stagirite discussed the entire spectrum of natural and weather phenomena that take place in the sublunar world. Aristotle's in-depth investigation, which ranged from the Milky Way and comets to earthquakes and volcanoes, earned a successful and extensive reception. After the fall of the Roman Empire, the *Meteorologica* received its second life in the Islamic world, with dozens of commentaries on the treatise as a whole or on its portions.[60] In the twelfth century, it was translated into Latin, making Gerard of Cremona's translation one of the first channels through which Aristotelian physics reached the Latin West.[61] A century later, William of Moerbeke – who, as is well known, industriously assisted Thomas Aquinas – prepared a new and improved text from the Greek original, which quickly became the standard point of reference.[62] In the wake of Renaissance humanist reflections and then early modern scientific developments, the *Meteorologica* attracted particular attention.[63] More broadly, its destiny in the Renaissance is rather emblematic, as it reveals how Aristotle's authority had been modified from the fifteenth to the eighteenth century. Earlier attempts to adapt the work's content to the new discoveries and competing natural philosophies were eventually driven out by a gradual decline of Aristotelianism in the course of the seventeenth century and its final replacement in the age of Enlightenment.[64] However, Manetti was far from rejecting Aristotle's authority altogether; instead, his standpoint was built upon the humanist premises of reviving the 'purity' of ancient authorities against their followers' allegedly erroneous and deceptive interpretations.

This feature defined Manetti's treatment of the *Meteorologica*. Not only does the *De terremotu*, in compliance with its title, explain Aristotle's understanding of earthquakes, but it also serves to briefly outline the Aristotelian corpus on physics on the whole – and the place of the *Meteorologica* therein – and recount the Philosopher's principal ideas on the natural causality of terrestrial events.[65] Following Aristotle's stance, Manetti admitted the existence of a connection between the celestial spheres and the earthly world, meaning that through their motion, radiation or similar influences, the planets (and the sun in particular) can determine natural events.[66] Among the manifestations of how the two levels of the universe are interconnected, the most illustrative are the exhalations of air and vapours and their meteorological consequences. Celestial heat produced by the luminaries and the circular character of exhalations account for the transformations of the air depending on its position between the earth's surface and the atmosphere.[67] Two types of evaporation, arising from either earth or water and therefore having different attributes, are responsible for multiple meteorological effects as a result of their interplay with other elements and the area of the sublunary sphere they control.[68] Hence, this general cycle of evaporation covers all four basic elements – earth, water, air and fire – and is of a universal nature. After Aristotle, Manetti maintained that while the celestial bodies are to be regarded as the efficient cause of meteorological occurrences, the material cause is dual. It comprises both exhalation/vapour (which Manetti recognised as the immediate cause) and earth/water (the remote cause) from which the vapours spring.[69]

In the footsteps of the Stagirite, the Italian humanist construed the winds as an outcome of dry and hot exhalations.[70] Building on this assumption, he went on to state that streams of wind get compressed and trapped in the earth; their constant motion beneath the earth in search of an exit disrupts the subterranean stability and ultimately provokes earthly tremors.[71] Considering that the material cause of earthquakes is responsive to seasonal change, Manetti added that earthquakes are more likely to occur in two cases. First, in spring and autumn, an increase in rain and other water-related meteorological effects intensifies the cycle of evaporation. The extra amount of water generates more wind while also filling the potential exits on the surface of the earth, thus preventing the winds from breaking away from the trap. Second, concave areas of the earth remarkably complicate natural evaporations.[72] In this sense, the earthquake appears to be an inevitable and necessary solution to the physical deadlock and acts as vaporisation of the energy clot caused by the winds. It is easy to grasp how, in a similar fashion, the theory under discussion could well be applied to the natural disaster of volcanoes, with the winds believed to be capable of altering the level of the fire underneath.[73]

Manetti asserted that the Aristotelian stance was largely accepted by the subsequent tradition. Thus, 'the Commentator' Ibn Rushd (known in the Latin West as Averroes) developed a three-fold interpretation of the origin of earthquakes that was grounded in the three principles taken from the *Meteorologica*: the concavity of place; the material cause concerned with the heat, dryness and density of exhalations; and the motion of winds.[74] Theophrastus and Menedemus, whom Manetti named as the two brightest and most acclaimed followers of the Stagirite and the peripatetic school in general, unanimously agreed with their mentor.[75] Furthermore, their commentaries amplified Aristotle's argument and, in some instances, clarified its obscurities, which, in turn, secured its large support on the part of Christian authors.[76] At this point, Manetti effectively rounded off his investigation. What started with an exposition of a theological understanding of disasters and continued with a thorough analysis of natural philosophy concluded as a reconciliation of the former with the latter. With the use of two major authorities in theological matters, namely, Eusebius of Caesarea and Thomas Aquinas, the humanist argued that earthquakes – as well as pestilence, thunder or any other calamity – depend exclusively on natural factors and should not be attributed to God's will, as 'God causes nothing bad'.[77] This statement presents, in Manetti's view, a coherent and unified solution and discards the dichotomy between natural philosophical and theological interpretations of disasters. He admitted that it might be in God's will to reveal the upcoming disasters through certain signs (such as the aforementioned comets) or eventually uncover the implications of the divine providence, but, in principle, earthquakes have natural and, therefore, accidental reasons. If so, the ruler's *fortuna* and personality are not responsible for the conditions under which such catastrophes take place.

Certainly, Manetti was not unique in his attempt to explore the possible similarities between the body politic and the body of nature. Whether disasters and good government – or, more broadly, the moral and socio-political state of affairs in a time of their happening – are intertwined has always been a matter of popular beliefs and constant reflections, and the supposition that such events might have a providential character goes on unabated to date regardless of the science.[78] In the early modern period, among the calamities that held a particularly strong political dimension were plagues and floods, that is, those lasting in time and effects.[79] Earthquakes underwent politisation much less frequently. This means that Manetti's account was essentially shaped by contemporary discussions on fate, fortune and political leadership rather than an established political interpretation of earthquakes as such. The Italian Renaissance marked the proliferation of texts that put the ruler's virtue and valour in dialogue with luck, including Salutati's aforementioned *De fato et fortuna*, Poggio Bracciolini's *De infelicitate principum* and *De varietate fortunae* and Giovanni Pontano's *De fortuna*.[80] This tendency was crowned by Machiavelli's famous treatment of *fortuna* in *The Prince* and *The Discourses on Livy*.[81] But none of these works addressed the moral and political aspect of earthly tremors; the only exception appears to be Petrarch's *De remediis utriusque fortunae*, which suggested, in quite standard Christian Stoic terms, that in the face of a devastating earthquake there is nothing left for one than to accept the divine will humbly.[82] In this respect, the *De terremotu* stands out as a rare example of contemporary humanism's specific engagement with earthquakes. Several subsequent texts did strive to politicise this type of disaster independently from one another. Although taken as a whole, they seem to have fallen short of forming an entrenched tradition of 'the politics of earthquakes', these separate works somewhat echo how disasters have popularly been perceived.

To provide some perspective, I shall briefly mention three different testimonies. First, in a chapter on earthquakes in his dream encyclopaedia *Somniorum Synesiorum libri quatuor* published in 1562, Girolamo Cardano affirms that if one has a dream in which earthly tremors occur, it signifies a prince's future arrest and similar political calamities.[83] To the best of my knowledge, neither Neoplatonic scholar Synesius of Cyrene upon whose treatise Cardano formally drew nor Marsilio Ficino who for the first time translated Synesius' *On Dreams* into Latin in the late fifteenth century ever mentioned the political facets of dreams concerned with earthquakes. What, apart from an evident analogy between the two different types of tremors, earthly and political, influenced Cardano's decision to include such a chapter in his lengthy treatise is unclear. It is, however, in line with Cardano's general view that earthquakes, just like other manifestations of nature, are to be regarded as signs of upcoming socio-political events. Cardano alleged the existence of such a connection in several of his writings, but never composed a text devoted specifically to its investigation.[84] The material, scattered across his oeuvre, demonstrates an interpretative pattern, though: substantiated by an adaptation of astrological theories, including the famous theory of 'great conjunctions', according to which great events of the future are revealed through the conjunction of Saturn and Jupiter, Cardano's reflections show his preoccupation with the effects of the unknown and divine upon the terrestrial world.[85]

In the second text, the lavishly illustrated *Meteorologia Philosophico-Politica* (1698), the Austrian Jesuit professor Franz Reinzer sought to interpret how various meteorological effects corresponded metaphorically to politics.[86] Every chapter is also supplemented with a beautiful emblem, usually inserted in the section on politics. Of the twelve *dissertationes*, two, the tenth and the eleventh, are devoted to earthquakes and volcanoes, respectively. There, Reinzer compared earthquakes to laws, capable of shaking the foundation and stability of the state and stressed that what could help endure and re-emerge from the disaster was piety; at the same time, according to him, volcanoes are the reflection of the ruler's temperament, which they should keep under control and not allow to burst. It seems evident that the treatise had a didactic and practical goal: Reinzer's former student and wealthy nobleman Johann Bernhard Caelestin von Rödern, who is alleged to have commissioned the whole publication, presented the *Meteorologia* to Emperor Joseph I, perhaps, with the aim of changing his opinion about the Jesuit order.[87] There was more than just pure pragmatism behind Reinzer's argument, though: largely drawn upon the extensive work of his fellow Jesuits, it was essentially characteristic of the way of thinking that some scholars associate with 'baroque science', namely, in its attempt at unveiling the all-embracing unitary connection of seemingly dispersed parts of the universe.[88]

Finally, the most peculiar instance of the shape the political discourse on earthquakes might take is Giuseppe Giannuzzi's *De Siciliae et Calabriae excidio carmen*. Composed on the occasion of the terrible Messina earthquake of 1908, and strongly inspired by Vergil, it claimed that social misery, political turmoil and human misdeeds were to blame for the disaster. In Giannuzzi's opinion, the earthquake's horrible consequences were revealed to urge people to put themselves onto the right path.[89] This example shows that some intellectual patterns barely change over time, and irrespective of the different time, settings and circumstances of their composition, the works discussed above addressed the same set of issues at the heart of human perception of disasters.

In sum, one might claim, quite fairly, that Giannozzo Manetti's *De terremotu* was not a ground-breaking treatise that proposed a revolutionary interpretation of earthquakes. It is, however, cutting-edge in a different way. In the thorough humanist treatment of the phenomenon, Manetti embodied three foundational ideas that are characteristic of the age in which his work was brought to fruition. First, being in the front line of the contemporary humanist movement, Manetti became one of the first Renaissance scholars who derived advantage from an array of previously unknown ancient texts concerned with natural philosophy. Purporting to reconstruct an ancient understanding of disasters – an attempt that, on a larger scale, reflected the humanist desire to virtually return *ad fontes* and *originem* – Manetti laid down a literary and argumentative feature that outlived him and remained in force at least until the early eighteenth century. Since the fifteenth century, every major account that was engaged with natural disasters presented ancient descriptions as a valuable and relevant starting point for further discussion on the matter. Second, Manetti's meticulous exposition of Aristotle's teaching resonated with the intellectual trends that would define early modern natural philosophy in the following centuries, including a steady replacement of the dominant Aristotelian framework with alternative theories. Finally, Manetti touched upon a critical point of contemporary moral and political philosophy, namely, whether calamities might be considered correlative to the time and political environment in which they occur. It seems that in this regard, the *De terremotu*, clearly characteristic of the period and intellectual environment in which it was written, concurrently articulates a more general and rooted understanding of disasters that largely remains in force until this day.

Notes

1 This project has received funding from the European Research Council (ERC) under the European Union's Horizon 2020 research and innovation programme (grant agreement No. [741374], 2017–2022).
2 For examples, see Lorenzo Bonincontri, 'Annales ab anno 1360 usque ad 1458', in *Rerum Italicarum Scriptores*, vol. 21, ed. by L. A. Muratori (Milan: ex typographia Societate Palatinae, 1732), pp. 7–162; Angeluccio di Bazzano, 'Cronaca delle cose dell'Aquila dall'anno 1436 all'anno 1485', in *Antiquitates*

Italicae Medii Aevi, vol. 6, ed. by L. A. Muratori (Milan: ex typographia Societatis Palatinae, 1742), pp. 883–964; Armando Tallone, 'Un poemetto storico inedito di Antonio Astesano sul terremoto del 1456', *Archivio muratoriano*, 4 (1907), 191–217; Matteo dell'Aquila, *Tractatus de cometa atque terraemotu (Cod. Vat. Barb. Lat. 268)*, ed. by B. Figliuolo (Salerno: Laveglia, 1990); *Catalogue of Earthquakes and Tsunamis in the Mediterranean Area from the 11th to the 15th Century*, ed. by E. Guidoboni and A. Comastri (Rome: INGV, 2005), pp. 625–26. For reflections across the peninsula and their context, see also Jonas Borsch, *Erschütterte Welt. Soziale Bewältigung von Erdbeben im östlichen Mittelmeerraum der Antike* (Tübingen: Mohr Siebeck, 2018), pp. 28–38. The news about the earthquakes in southern Italy got to Muscovy within Theophilus Dederkin's *Letter from Rome, from the Latin World*. This tract interpreted the disaster from a providential point of view stating that it hit Rome and surrounding territories to cure numerous sins in which the Latin world had embroiled. The *Letter* also contains a list of the forty-five cities affected by the earthquake, which might indicate that Dederkin could have used first-hand testimony about the event's repercussions. See Pavel Simoni, *Памятники старинного русского языка и словесности XV – XVIII столетий. Выпуск 3. Задонщины по спискам XV–XVIII столетий. Сборник Отделения русского языка и словесности Российской академии наук* (Saint Petersburg: Rossijskaya gosudarstvennaya akademicheskya tipografiya, 1922), pp. 13–16; Stepan M. Shamin, *Иностранные «памфлеты» и «курьезы» в России XVI – начала XVIII столетия* (Moscow: Ves' mir, 2020), pp. 35–37.

3 Giannozzo Manetti, *De terremotu*, ed. by D. Pagliara (Florence: SISMEL – Edizioni del Galluzzo, 2012).

4 James Hankins, *Virtue Politics: Soulcraft and Statecraft in Renaissance Italy* (Cambridge, MA; London: Harvard University Press, 2019). The *De terremotu* and, in general, the politics of disaster are totally absent from Hankins' work, whilst Manetti's name appears only passingly.

5 Vespasiano da Bisticci, *Vite di uomini illustri del secolo XV* (Florence: Barbera, Bianchi and co., 1859), pp. 444–72 (p. 444). On the Neapolitan period of Manetti's career: ibid., pp. 59–60, 470–71. The *De terremotu* is listed among Manetti's works: ibid., p. 472. On another complimentary biography of Manetti composed in Italian by an anonymous author in Dantesque triplets, see Stefano U. Baldassari and Bruno Figliuolo, *Manettiana. La biografia anonima in terzine e altri documenti inediti su Giannozzo Manetti* (Rome: Roma nel Rinascimento, 2010) (the text – pp. 82–167).

6 Jerry H. Bentley, *Humanists and Holy Whit: New Testament Scholarship in the Renaissance* (Princeton: Princeton University Press, 1983), pp. 57–59; Paul Botley, *Latin Translation in the Renaissance. The Theory and Practice of Leonardo Bruni, Giannozzo Manetti and Desiderius Erasmus* (Cambridge: Cambridge University Press, 2004), pp. 82–114; John Monfasani, 'Criticism of Biblical Humanists in Quattrocento Italy', in *Biblical Humanism and Scholasticism in the Age of Erasmus*, ed. by E. Rummel (Leiden; Boston: Brill, 2008), pp. 15–38 (pp. 30–33); Annet den Haan, 'Giannozzo Manetti's New Testament: New Evidence on Sources, Translation Process and the Use of Valla's "Annotationes"', *Renaissance Studies*, 28, 5 (2014), 731–47; eadem, *Giannozzo Manetti's New Testament: Translation Theory and Practice in Fifteenth-Century Italy* (Leiden; Boston: Brill, 2016).

7 Giannozzo Manetti, *Dignità ed eccellenza dell'uomo*, ed. by G. Marcellino (Milan: Bompiani, 2018). For a detailed analysis of the concept of *dignitas*

hominis in modern European thought and the misunderstandings of Giovanni Pico's *Oratio*, see Brian P. Copenhaver, *Magic and the Dignity of Man: Pico della Mirandola and His* Oration *in Modern Memory* (Cambridge, MA; London: Harvard University Press, 2019).

8 On how Manetti fell from grace with the Florentine rulers and spent the rest of his life in 'voluntary exile', when being at the zenith of his fame, see David Marsh, *Giannozzo Manetti: The Life of a Florentine Humanist* (Cambridge, MA; London: Harvard University Press, 2019), pp. 110–63. Manetti's Roman period, including the ways he engaged with Pope Nicholas V, is reconstructed in Christine Smith and Joseph F. O'Connor, *Building the Kingdom. Giannozzo Manetti on the Material and Spiritual Edifice* (Tempe, AZ; Turnhout: Arizona Center for Medieval and Renaissance Studies in collaboration with Brepols, 2006). On Manetti's biography of the Pope, see Anna Modigliani, 'Il testamento di Niccolò V: la rielaborazione di Manetti nella biografia del Papa', in *Dignitas et excellentia hominis. Atti del convegno internazionale di studi su Giannozzo Manetti*, ed. by S. U. Baldassari (Florence: Lettere, 2008), pp. 231–59. On Manetti's relationship with the Neapolitan court and Alfonso, see above all Stefano U. Baldassari, 'Giannozzo Manetti e Alfonso il Magnanimo', *Interpres*, 29 (2010), 43–95; Stefano U. Baldassari and Brian J. Maxson, 'Giannozzo Manetti, the Emperor, and the Praise of a King in 1452', *Archivio Storico Italiano*, 172, 3 (2014), 513–69.

9 Manetti, *Dignità ed eccellenza dell'uomo*, pp. 44–50. See also Baldassari and Maxson, 'Giannozzo Manetti, the Emperor, and the Praise of a King in 1452', 536–37.

10 Manetti, *De terremotu*, p. 91. Without going into detail, David Marsh provides a brief overview of the treatise's structure, not devoid of minor inaccuracies: Marsh, *Giannozzo Manetti*, pp. 160–63. So does Daniela Pagliara, who interprets, rather unconvincingly, Manetti's argument through the *dignitas hominis* concept: Daniela Pagliara, 'Annotazioni storico-culturali a proposito del *De terremotu*', in *Dignitas et excellentia hominis*, pp. 261–78. Rienk Vermij's *Thinking on Earthquakes in Early Modern Europe: Firm Beliefs on Shaky Grounds* (London; New York: Routledge, 2021), which provides rather sketchy descriptions of most early modern texts on earthquakes, devotes less than ten lines to the *De terremotu*, without delving into Manetti's argument at all and mistakenly suggesting that 'there is hardly any reference to religious interpretations' (p. 50).

11 Manetti, *De terremotu*, p. 89.

12 Manetti, *De terremotu*, p. 89.

13 Manetti, *De terremotu*, p. 135.

14 Manetti, *De terremotu*, p. 90: "Nam si L. Anneum Senecam virum Cordubensem, licet doctissimum philosophum ac Rome ea tempestate sub Neronis imperio commorantem, novus quidam illius temporis terremotus, qui eandem Veteris Campanie regionem vehementer invasisse ac lesisse scribitur, integrum quintum *De questionibus naturalibus* librum ad scribendum de terremotibus".

15 Thus, Manetti kept in his private library a manuscript containing Seneca's *Naturales quaestiones* (now Pal. Lat. 1540 from the Biblioteca Apostolica Vaticana), which has the standard sequence of books. Marsh does not recognise this mistake and follows Manetti, who falsely alluded to Book 5 of Seneca's treatise: Marsh, *Giannozzo Manetti*, pp. 160–61. Harry M. Hine demonstrates, however, that in the medieval manuscript tradition the books

were presented in various orders and numerations and that there is a strong argument to suggest that the original order differed from the one usually accepted today: Harry M. Hine, An Edition with Commentary of Seneca, "Natural Questions," Book Two (New York: Arno Press, 1981), pp. 4–23; idem, 'Translator's Introduction', in Seneca, *Natural Questions*, trans. H. M. Hine, pp. 1–2. See also Bardo M. Gauly, *Senecas* Naturales quaestiones. *Naturphilosophie für die römische Kaiserzeit* (Munich: C. H. Beck, 2004), pp. 53–67. If so, following the original order the book on earthquakes would indeed be Book 5, and not 6. At the same time, the extent to which Manetti could have been familiar with this question remains unclear.

16 Peter Stacey, '*Hispania* and Royal Humanism in Alfonsine Naples', *Mediterranean Historical Review*, 26, 1 (2011), 51–65.

17 Baldassari, 'Giannozzo Manetti e Alfonso il Magnanimo', 55–58.

18 Manetti, *De terremotu*, p. 90: "Quid hi duo, de quibus mentionem fecimus et quos propriis oculis vidimus et quos etiam propriis auribus audivimus, terremotus erga nos Florentinos ac Neapoli sub tuo Regno ... "

19 Manetti, *De terremotu*, pp. 92–95. Cf. Eusebius of Caesarea, *Die Praeparatio Evangelica*, 2 vols, ed. by K. Mras et al. (Berlin: Akademie Verlag, 1982–1983), vol. 2 (*Die Bücher XI bis XV. Register*) (1983), XIII, 12, 4–5; Ovid, *Metamorphoses*, 2 vols, ed. and trans. by F. J. Miller (London: Heinemann; Cambridge, MA: Harvard University Press, 1971), vol. 1 (*Books I–VIII*), I, 283–84; II, 846–51; Virgil, *Aeneid, VII–XII. The Minor Poems*, ed. and trans. by H. Rushton Fairclough (London: Heinemann; Cambridge, MA: Harvard University Press, 1918), X, 100–2; Seneca, 'Hercules Furens', in *Seneca's Tragedies*, 2 vols, ed. and trans. by F. J. Miller (London: Heinemann; Cambridge, MA: Harvard University Press, 1938), vol. 1 (*Hercules Furens. Troades. Medea. Hippolytus. Oedipus*), 81–82, 520–22.

20 Manetti, *De terremotu*, pp. 92–93. Cf. Lucretius, *Titi Lucreti Cari De rerum natura libri sex*, 2 vols, ed. and trans. by H. A. J. Munro (Cambridge: Cambridge University Press, 2010 [1864]), vol. 1, VI, 285–89.

21 Much has been written on the reception of Lucretius and atomism in the Renaissance. For the Italian context, see above all Alison Brown, *The Return of Lucretius to Renaissance Florence* (Cambridge, MA; London: Harvard University Press, 2010); Ada Palmer, *Reading Lucretius in the Renaissance* (Cambridge, MA; London: Harvard University Press, 2014).

22 Manetti, *De terremotu*, p. 95. Cf. Seneca, *Natural Questions*, 2 vols, ed. and trans. by T. H. Corcoran (Cambridge, MA: Harvard University Press, 1972), vol. 2 (*Books 4–7*), VI, 23, 4. See also a fine recent edition by H. M. Hine (Chicago; London: University of Chicago Press, 2010).

23 For some examples, see Manetti, *De terremotu*, pp. 100–1.

24 Manetti, *De terremotu*, p. 102. In his 1976 essay, later translated into English, Henry Corbin coined the term to designate how images that exist in the *mundus imaginalis* of both individual authors and the general public 'precede and impose order on all empirical and sensible perceptions' of specific spaces: Henry Corbin, 'Emblematic Cities: A Response to the Images of Henri Steirlin', trans. by K. Raine, *Temenos*, 10 (1989), 11–24.

25 Manetti, *De terremotu*, p. 105.

26 Manetti, *De terremotu*, p. 106: "Quippe Eusebius in libro *De preparatione evangelica* quodam loco ita scribit: «Quid autem miramur si tyramnorum interdum ministerio effusas hominum iniurias Deus compescit, cum etiam sepius non aliorum opere sed per se ipsum terremotu, peste aliisque huiusmodi,

quibus multas urbes desolates videmus, id factitet." Cf. Eusebius of Caesarea, *Die Praeparatio Evangelica*, vol. 2, VIII, 14, 41. Translated into Latin in 1448 by George of Trebizond, the *Praeparatio Evangelica* was a Renaissance best-seller and survived in 47 manuscripts. Although George 'took great liberties' with the text, sometimes omitting certain passages and paraphrasing other ones (John Monfasani, *George of Trebizond. A Biography and a Study of His Rhetoric and Logic* (Leiden: Brill, 1976), p. 78), every Renaissance scholar after 1448 cited not Eusebius but rather George's relatively loose translation: *Collectanea Trapezuntiana: Texts, Documents, and Bibliographies of George of Trebizond*, ed. by J. Monfasani (Binghamton, NY: Medieval and Renaissance Texts and Studies in conjunction with the Renaissance Society of America, 1984), pp. 721–24; John Monfasani, 'Marsilio Ficino and Eusebius of Caesarea's *Praeparatio Evangelica*', *Rinascimento*, 49 (2009), 3–13 (later reprinted in idem, *Renaissance Humanism, from the Middle Ages to Modern Times* (Farnham; Burlington, VT: Ashgate, 2015).

27 Manetti, *De terremotu*, pp. 106–7.

28 Manetti, *De terremotu*, p. 90: "Ad hec accedebat fortunata ac fausta et felix nativitas tua, que cum dignitate terremotus non immerito idcirco apparuisse videtur, ut futuram Maiestatis tue magnitudinem plane aperteque portenderet, ceu de Iesu Christo ac Alexandro rege ac Platone et de quibusdam aliis admirabilibus viris legisse meminimus." In this sentence, Manetti refers to an earthquake that apparently accompanied Alfonso's birth.

29 Manetti, *De terremotu*, pp. 152–53, 155–56, 161.

30 Manetti, *De terremotu*, pp. 175–78.

31 Manetti, *De terremotu*, p. 178.

32 It is important to note that Manetti allegedly borrowed much of his account on the earlier history of the region from another text commissioned by Alfonso of Aragon, Flavio Biondo's *Italia illustrata*: Marsh, *Giannozzo Manetti*, p. 163; Flavio Biondo, Italia Illustrata: *Text, Translation, and Commentary*, ed. and trans. by C. J. Castner (Binghampton, NY: SUNY Press, 2005–2010), vol. 2 (*Central and Southern Italy*) (2010), pp. 206–12.

33 Manetti, *De terremotu*, p. 177.

34 The most convincing overviews of astrology and magic in the long Middle Ages are Ornella Pompeo Faracovi, *Scritto negli astri: L'astrologia nella cultura dell'Occidente* (Venice: Marsilio, 1996); Jean-Patrice Boudet, *Entre science et nigromance. Astrologie, divination et magie dans l'Occident médiéval (XII–XV siècle)* (Paris: Éditions de la Sorbonne, 2006); Graziella Federici Vescovini, *Medioevo magico: la magia tra religione e scienza nei secoli XIII e XIV* (Turin: UTET, 2008). Among recent publications, see also H. Darrel Rutkin, *Sapientia Astrologica: Astrology, Magic and Natural Knowledge, ca. 1250–1800. Vol 1: Medieval Structures (1250–1500): Conceptual, Institutional, Socio-Political, Theologico-Religious and Cultural* (Dordrecht: Springer, 2019).

35 Among recent publications on astrology at medieval and Renaissance courts, see above all Michael A. Ryan, *A Kingdom of Stargazers: Astrology and Authority in the Late Medieval Crown of Aragon* (Ithaca, NY: Cornell University Press, 2011); Monica Azzolini, *The Duke and the Stars: Astrology and Politics in Renaissance Milan* (Cambridge, MA; London: Harvard University Press, 2013); Steven vanden Broecke, 'Astrology and Politics', in *A Companion to Astrology in the Renaissance*, ed. by B. Dooley (Leiden; Boston: Brill, 2014), pp. 193–232.

36 The most elaborate analysis of the problem is in Matteo Soranzo, *Poetry and Identity in Quattrocento Naples* (Farnham; Burlington, VT: Ashgate, 2014). On Pontano's humanist method in the study of astrology and nature, see also Michele Rinaldi, 'Sic itur ad astra': *Giovanni Pontano e la sua opera astrologica nel quadro della tradizione manoscritta della* Mathesis *di Giulio Firmico Materno* (Naples: Loffredo, 2002). Benedetto Soldati's *La poesia astrologica nel Quattrocento. Ricerche e studi* (Florence: Sansoni, 1906) still remains valid and useful in many respects.

37 Manetti, *De terremotu*, p. 171. Manetti confuses the dates: Petrarch was twice in Naples, in 1341 and 1343, whilst in the *Familiares*, he recalls witnessing two earthquakes both of which occurred in the late 1340s. Manetti also celebrated his most acclaimed Florentine predecessors, namely, Dante, Petrarch and Boccaccio, in a series of *vitae*: Giannozzo Manetti, *Biographical Writings*, ed. by S. U. Baldassari and R. Bagemihl (Cambridge, MA; London: Harvard University Press, 2003), 2–104.

38 Dante Alighieri, *The Divine Comedy. Volume I: Inferno*, ed. and trans. by R. Durling (Oxford; New York: Oxford University Press, 1997), 1:210 (*Inferno*, 20, pp. 115–18). On Dante and the astrologers, see Jean-Patrice Boudet, *Entre science et nigromance*, pp. 13–18.

39 Much has been written on Renaissance Stoicism. For some orientation in the subject, see Jill Kraye, 'Stoicism in the Renaissance from Petrarch to Lipsius', *Grotiana*, 22–23 (2001–2002), 21–45; eadem, 'Stoicism in the Philosophy of the Italian Renaissance', in *The Routledge Handbook of the Stoic Tradition*, ed. by J. Sellars (London: Routledge, 2016), pp. 132–44; John Sellars, 'Renaissance Consolations: Philosophical Remedies for Fate and Fortune', in *Fate and Fortune in European Thought, ca. 1400–1650*, ed. by O. Akopyan (Leiden; Boston: Brill, 2021), pp. 13–36.

40 Coluccio Salutati, *De fato et fortuna*, ed. by C. Bianca (Florence: Olschki, 1985); Charles Trinkaus, 'Coluccio Salutati's Critique of Astrology in the Context of His Natural Philosophy', *Speculum*, 64, 1 (1989), 46–68.

41 Manetti, *De terremotu*, p. 107: "Astrologi namque, qui mathematicorum appellatione comprehenduntur ac vulgo mathematici et appellantur et sunt, ex astrorum cursu lapsuque siderum cunctas res humanas ita regi gubernarique arbitrantur, ut homines secundum diversas, ceu aiunt, constellations varie disponantur diverseque ad singula queque negocia peragenda et ad fortunam prosperam vel adversam inclinentur; non ut omnino penitusque cogantur, ne liberum humane nature arbitrium usquequaque tollatur".

42 Manetti, *De terremotu*, p. 107: "Unde natum est illud quod per cunctas eruditorum virorum scholas quasi loco proverbii celebrator: «Sapiens dominabitur astris», quam quidem sententiam Ptolemeus in *Almagesti* et *Quadripartiti* libris aliis verbis expressit." On the motto and how it got attributed to Ptolemy in the Middle Ages, see Jim Tester, *A History of Western Astrology* (Woodbridge: Boydell Press, 1987), p. 177; Tullio Gregory, Mundana sapientia. *Forme di conoscenza nella cultura medievale* (Rome: Edizioni di Storia e Letteratura, 1992), pp. 316–17.

43 Manetti, *De terremotu*, p. 107.

44 Manetti, *De terremotu*, p. 107: "Sed hec pauca de astrologorum sententia dixisse sufficiat, cum obscura, varia et incerta sit".

45 Manetti, *De terremotu*, p. 108: "Nunc ad philosophos accedamus. Omnes physici, etsi in hoc maxime convenire videantur, quod unusquisque terremotus

naturalis sit ac suapte natura proveniat, de causis tamen eius utpote cuiusdam naturalis rei perscrutari multum, opinionibus suis ab invicem dissenserunt".

46 Seneca, *Natural Questions*, VI, 3, 1. On Seneca's understanding of earthquakes, see Gareth D. Williams, *The Cosmic Viewpoint: A Study of Seneca's* Natural Questions (New York: Oxford University Press, 2012), pp. 213–57.

47 Manetti, *De terremotu*, p. 108. Cf. Seneca, *Natural Questions*, VI, 3, 1: "Causam qua terra concutitur alii in aqua esse, alii in ignibus, alii in ipsa terra, alii in spiritu putaverunt, alii in pluribus, alii in omnibus his".

48 Seneca, *Natural Questions*, VI, 13, 1; 14, 1: "Sunt qui existiment spiritu quidem et nulla alia ratione tremere terram, sed ex alia causa quam Aristoteli placuit".

49 Manetti, *De terremotu*, pp. 108–9. Cf. Seneca, *Natural Questions*, VI, 6, 1.

50 Manetti, *De terremotu*, pp. 109–10. Cf. Aristotle, *Meteorologica*, ed. and trans. by H. D. P. Lee (Cambridge, MA; London: Harvard University Press; Heinemann, 1952), II, 7, 365a.20–365b.20; Seneca, *Natural Questions*, VI, 9, 1; 10, 1. According to Aristotle, Democritus held that since the earth was already full of water, earthquakes occurred when a significant amount of extra water fell besides this. Anaximenes thought that earthquakes were caused by two extreme weather conditions – droughts and showers – during which the earth was tearing apart. Finally, in Anaxagoras' opinion, earthquakes originated when air motion hindered by rain happened to reach the hollows beneath the earth.

51 Manetti, *De terremotu*, pp. 110–11.

52 Manetti, *De terremotu*, p. 113; Seneca, *Natural Questions*, VI, 12, 1–26, 4.

53 Manetti, *De terremotu*, p. 112.

54 Manetti, *De terremotu*, p. 112: "Metrodori Chii, de voce alicuius per dolium cantantis quam aeris speluncis, sub terra pendentibus similem imaginabatur, opinionem penitus omictamus, cum frivola atque omnino futilis a cunctis paulo doctioribus fuisse cognoscatur." Cf. Seneca, *Natural Questions*, VI, 19, 1–2.

55 Manetti, *De terremotu*, p. 112. Cf. Seneca, *Natural Questions*, VI, 17, 3–18, 1.

56 Manetti, *De terremotu*, p. 116.

57 Manetti, *De terremotu*, p. 116.

58 Manetti, *De terremotu*, p. 116. Cf Cicero, *De finibus bonorum et malorum*, ed. and trans. by H. Rackham (London; New York: Heinemann; Putnam's Sons, 1931), I, 2, 6.

59 Manetti, *De terremotu*, pp. 116–17. Cf. Cicero, *Tusculanae disputationes*, ed. by M. Pohlenz (Leipzig: Teubner, 1918), I, 10, 22.

60 Paul Lettinck, *Aristotle's* Meteorology *and Its Reception in the Arab World: With an Edition and Translation of Ibn Suwār's* Treatise on Meteorological Phenomena *and Ibn Bājja's* Commentary on the Meteorology (Leiden; Boston; Cologne: Brill, 1999), pp. VIII–IX, 1–2.

61 Gerard of Cremona's translation was prepared ca. 1260 and made the medieval discourse on meteorology Aristotelian (Haskins, *The Renaissance of the Twelfth Century* (Cambridge, MA; London: Harvard University Press, 1927), pp. 313–14). Henricus Aristippus translated the fourth book of the treatise only (ibid., p. 292). For a general introduction into meteorology in the Middle Ages, see Anna Lawrence-Mathers, *Medieval Meteorology. Forecasting the Weather from Aristotle to the Almanac* (Cambridge: Cambridge University Press, 2020).

62 Both Gerard's and William's translations were published in Pieter L. Schoonheim,

Aristotle's Meteorology *in the Arabico-Latin Tradition: A Critical Edition of the Texts, with Introduction and Indexes* (Leiden: Brill, 2000).

63 Craig Martin, *Renaissance Meteorology: Pomponazzi to Descartes* (Baltimore: Johns Hopkins University Press, 2011).

64 The last decades have witnessed a soaring interest in the role of the Aristotelian tradition in shaping early modern philosophical and scientific ideas. These studies have successfully challenged two outdated positions: that after its revival in the fifteenth century, Platonism was by far the prevailing philosophical school in the Renaissance; and that new science emerged strictly outside Aristotelianism. To name but a few studies that allow for a more nuanced understanding of this subject, see Charles B. Schmitt, *Aristotle and the Renaissance* (Cambridge, MA; London: Harvard University Press, 1983); Paul Richard Blum, *Studies on Early Modern Aristotelianism* (Leiden; Boston: Brill, 2012). On Aristotle's authority in the Renaissance and the ways it was used and challenged in theological and scientific discussions of the time, see Craig Martin, *Subverting Aristotle: Religion, History, and Philosophy in Early Modern Science* (Baltimore: Johns Hopkins University Press, 2014); Eva Del Soldato, *Early Modern Aristotle: On the Making and Unmaking of Authority* (Philadelphia: University of Pennsylvania Press, 2020).

65 Manetti, *De terremotu*, pp. 116–20.

66 Manetti, *De terremotu*, p. 117.

67 Manetti, *De terremotu*, p. 117.

68 Aristotle, *Meteorologica*, I, 4.

69 Manetti, *De terremotu*, p. 117: "Aristoteles nanque harum impressionum omnium causam efficientem corpora celestia esse dicit et sentit. Materialem vero … duplicem propinquam scilicet et remotam esse contendit. Propinqua est exalatio vel evaporatio. Remota aqua et terra, unde vapores illi elevantur".

70 Manetti, *De terremotu*, p. 121: "Itaque materialis ventorum causa est exalatio quedam calida et sicca per solem ab ipsa terra attracta et elevata." Cf. Aristotle, *Meteorologica*, II, 4–6.

71 Manetti, *De terremotu*, p. 122. Cf. Aristotle, *Meteorologica*, II, VIII, 365b.21–366a.5.

72 Manetti, *De terremotu*, p. 122: "Alio etiam modo terremotus generari posse existimatur et creditur, quando ex concussione aquarum, sub concavitatibus terre semper existentium ac continue fluentium, vapores multiplicantur et, supervenientibus iugibus pluviis, terrestres pori ac foramina occluduntur; unde vapores inspissate, cum exire conantur nec ob occlusionem talium foraminum egredi valeant, terram concutere ac per hunc modum movere perhibentur." Cf. Aristotle, *Meteorologica*, II, VIII, 366b.3–15.

73 On how interpretations of the events were largely connected, see Vermij, *Thinking on Earthquakes in Early Modern Europe*. On volcanoes in early modern thought, Sean Cocco, *Watching Vesuvius: A History of Science and Culture in Early Modern Italy* (Chicago; London: University of Chicago Press, 2013); David McCallam, *Volcanoes in Eighteenth-Century Europe: An Essay in Environmental Humanities* (Liverpool: Liverpool University Press; Oxford University Press, 2019).

74 Manetti, *De terremotu*, p. 122: "Quod et Commentator in secundo *Metaurorum* libro plane et aperte asserere et confirmare videtur, ubi inter cetera ita inquit quod ad causationem terremotus tria necessario concurrant: oportet locus scilicet ac materia, et causa impellens. Locus, ut sit multa concavitas; materia, ut sit vapor calidus, et siccus ac crassus; causa impellens, ut sit vehemens quedam

aeris agitatio." The editor of the 2012 publication of the *De terremotu* suggests, without providing sufficient evidence, that under the 'Commentator', Manetti meant Thomas Aquinas. It contradicts the common usage of the word with which medieval and Renaissance scholars labelled Ibn Rushd, who authored two commentaries on the *Meteorologica*. Furthermore, it does not conform to the way Manetti structures his argument by moving, step by step, from members of the Aristotelian tradition to Christian theologians.

75 Manetti, *De terremotu*, p. 123: "Theophrastus et Menedemus, duo clarissimi ceterorum omnium famosissimique Aristotelis discipuli, et tota peripateticorum turba ... ab unico eorum preceptore nullatenus dissenserunt".

76 Manetti, *De terremotu*, p. 123.

77 Manetti, *De terremotu*, p. 123. Cf. Eusebius of Caesarea, *Die Praeparatio Evangelica*, vol. 2, VIII, 14, 53.

78 On how politics, disasters and faith are connected in human imagination, see above all Philip Jenkins, *Climate, Catastrophe, and Faith: How Changes in Climate Drive Religious Upheaval* (Oxford; New York: Oxford University Press, 2021); Niall Ferguson, *Doom: The Politics of Catastrophe* (New York: Penguin, 2021).

79 Along with Michel Foucault's renowned work on how measures against plague were instrumental in the formation of the modern state, see also Samuel K. Cohn, Jr., *Cultures of Plague: Medical Thinking at the End of the Renaissance* (Oxford: Oxford University Press, 2009); Ernest B. Gilman, *Plague Writing in Early Modern England* (Chicago; London: Chicago University Press, 2009); John Henderson, *Florence under Siege: Surviving Plague in an Early Modern City* (New Haven; London: Yale University Press, 2019). On water management and governmental responses to floods, see among other things John Morgan, 'Understanding Flooding in Early Modern England', *Journal of Historical Geography*, 50 (2015), 37–50; idem, 'The Micro-Politics of Water Management in Early Modern England: Regulation and Representation in Commissions of Sewers', *Environment and History*, 23, 3 (2017), 409–30; Pamela O. Long, *Engineering the Eternal City: Infrastructure, Topography, and the Culture of Knowledge of Late Sixteenth-Century Rome* (Chicago; London: University of Chicago Press, 2018).

80 Poggio Bracciolini, *De infelicitate principum*, ed. by D. Canfora (Rome: Edizioni di Storia e Letteratura, 1998); idem, *De varietate fortunae*, ed. by O. Merisalo (Helsinki: Suomalainen Tiedeakatemia, 1993); Giovanni Gioviano Pontano, *De fortuna*, ed. by F. Tateo (Naples: La Scuola di Pitagora, 2012).

81 Guido Giglioni, 'Fate and Fortune in Machiavelli's Anatomy of the Body Politic', in *Fate and Fortune in European Thought, ca. 1400–1650*, pp. 95–117 (the essay also contains the most pertinent literature on the subject).

82 Petrarch, *De remediis utriusque fortunae* (Paris: Boucher, 1547), II, 91, ff. 341*v*–343*r*. There is no modern critical edition of this work. Vermij (*Thinking on Earthquakes in Early Modern Europe*, pp. 52–53) mistakenly references it as Chapter 92 (not 91) and does not indicate that it is from the second book of the *De remediis*.

83 Girolamo Cardano, *Somniorum Synesiorum libri quatuor. Les quatre livres des songes de Synesios*, 2 vols, ed. and trans. by J.-Y. Boriaud (Florence: Olschki, 2008), I, pp. 144–46.

84 This idea is present throughout Cardano's writings, from the early *De arcanis aeternitatis* to some more mature works. See Girolamo Cardano, 'De arcanis aeternitatis', in *Opera*, vol. 10 (Lyon: Huguetan & Ravaud, 1663), IX, p. 15;

idem, *De rerum varietate libri XVII* (Basel: Petri, 1557), XIV, 72, p. 545; idem, 'In Claudii Ptolemaei libri quatuor de astrorum iudiciis', in *Opera*, vol. 5, pp. 173–74.

85 On Cardano's astrology, including the socio-political applications of the theory of great conjunctions, see above Anthony Grafton, *Cardano's Cosmos. The Worlds and Works of a Renaissance Astrologer* (Cambridge, MA; London: Harvard University Press, 2001 [1999]), in particular pp. 127–55; Germana Ernst, '"Veritatis amor dulcissimus": Aspects of Cardano's Astrology', in *Secrets of Nature. Astrology and Alchemy in Early Modern Europe*, ed. by William R. Newman and Anthony Grafton (Cambridge, MA; London: MIT Press, 2001), pp. 39–68; Anthony Grafton and Nancy Siraisi, 'Between the Election and My Hopes: Girolamo Cardano and Medical Astrology', in *Secrets of Nature*, pp. 69–131; Ornella Faracovi, 'The Return to Ptolemy', in *A Companion to Astrology in the Renaissance*, pp. 87–98; Jonathan Regier, 'Shadows of the Thrown Spear: Girolamo Cardano on Anxiety, Dreams, and the Divine in Nature', *Early Science and Medicine*, 28, 1 (2023), 95–119. On how Cardano's interest in astrology got him into trouble, see Jonathan Regier, 'Reading Cardano with the Roman Inquisition: Astrology, Celestial Physics, and the Force of Heresy', *Isis*, 110, 4 (2019), 661–79.

86 Franz Reinzer, *Meteorologia philosophico-politica in duodecim dissertationes* (Augsburg: Wolf, 1698), pp. 216–31, 234–35, 239–44. Reinzer's work has been largely overlooked in the scholarship on early modern natural philosophy and science. For some general introduction, see the two rather descriptive studies of the *Meteorologia* Christoph Meinel, 'Natur als moralische Anstalt – Die *Meteorologia philosophico-politica* des Franz Reinzer, S.J., ein naturwissenschaftliches Emblembuch aus dem Jahre 1698', *Nuncius*, 2, 1 (1987), 37–94; Christian Peters, 'Atmospheric Pressure: Natural Philosophy, Political Didactics and the Exigencies of Praise in Franz Reinzer's *Meteorologia Philosophico-Politica* (1698)', in *Emblems and the Natural World*, ed. by Karl A.E. Enenkel and Paul J. Smith (Leiden; Boston: Brill, 2017), pp. 351–80.

87 The *Meteorologia* was dedicated to Emperor Joseph I, who was hostile to the Jesuit order, by von Rödern on the occasion of a public disputation about the whole of philosophy that marked the end of his philosophical studies at Linz (where Reinzer, the *praeses* at van Rödern's *defensio*, taught). As usual in the early modern period, the *praeses*, who presumably was von Rödern's teacher, was the real author of the *Meteorologia*. It is also worth noting that *The Meteorologia* was not the text that von Rödern defended at the public disputation. Appended to the end of the volume are the theses he presented for the debate, and they have nothing to do with the content of the work itself. On early modern dissertations as a genre of scientific literature, see above all Ku-ming (Kevin) Chang, 'From Oral Disputation to Written Text: The Transformation of the Dissertation in Early Modern Europe', *History of Universities*, 19, 2 (2004), 129–87.

88 Ofer Gal and Raz Chen-Morris, *Baroque Science* (Chicago; London: University of Chicago Press, 2012).

89 Giuseppe Giannuzzi, *De Siciliae et Calabriae excidio carmen*, ed. and trans. by G. Laudizi (Lecce: Pensa, 2012). On the poem and its author, see Dirk Sacré, '*De Siciliae et Calabriae excidio carmen*: Giuseppe Giannuzzi's Neo-Latin Poem on the Italian Earthquake of 1908', in *Vergil und das antike Epos. Festschrift für Hans Jürgen Tschiedel*, ed. by Stefan Freund et al. (Stuttgart: Steiner, 2008), pp. 525–44.

3 Thinking with the flood

Animal endangerment and the moral economy of disaster

Lydia Barnett

Concern about the danger humans pose to other animal species has rightly become a major theme of environmental activism, science, literature and law, as the climate crisis worsens and accelerating biodiversity loss stokes scientists' fears that we are living through a 'Sixth Mass Extinction' in the history of life on earth.[1] In her book *Imagining Extinction* (2016), the eco-critic Ursula K. Heise analyses narratives and images of animal endangerment in the Anthropocene, identifying elegiac, tragic, apocalyptic and even comic elements in the stories we tell one another about the negative impact of anthropogenic environmental change on nonhuman animals. Heise challenges us to ask 'What stories do we tell, and which ones do we not tell' about animal endangerment, inviting us to interrogate the complex cultural beliefs that motivate concern – or lack thereof – for animals at risk.[2]

In the early modern period in Europe, during an earlier period of climate crisis known as the Little Ice Age, the biblical story of Noah's Flood was one of the most important stories people told themselves about the ways humans endangered animals in times of environmental catastrophe. Rather than being a purely happy story about humans saving animals from the clutches of a planetary disaster, Noah's Flood was, in the eyes of artists, scientists and religious writers in early modern Christian Europe, equally a story about mass animal death as the tragic consequence of human actions. During a period of accelerating environmental harm associated with natural resource extraction and deforestation in Europe's overseas colonies, as well as agricultural enclosure and the origins of a fossil fuel economy in Europe, the story of Noah's Flood served as a key cultural framework for articulating anxieties about the strained relationship between humans, God and nature in times of ecological crisis.

European scientists largely refused to countenance the idea of species extinction before the turn of the nineteenth century, but representations of Noah's Flood from the sixteenth, seventeenth and eighteenth centuries reveal that early modern Europeans did take very seriously the notion of

DOI: 10.4324/9781003029823-5

animal endangerment. Noah's Flood, regarded as a catastrophe sent by God but sparked by human sin, was an especially potent means of imagining the unintended but deadly consequences of human behaviour on animal life. Christian traditions of Scriptural interpretation held that the Flood justly killed all humans except Noah and his family as punishment for their sins. However, the Flood also killed countless animals who had not sinned and had therefore done nothing to provoke the disaster. The problem of innocent animals killed in a large-scale disaster of humanity's own making appears as a repeated motif in Flood science, art and scholarship from the long seventeenth century, which I explore in the first half of this essay. The second half brings to light a related motif in medical and religious writing regarding the Flood as the turning point in history when humans began killing and eating animals in large numbers, in a global shift from vegetarianism to carnivorism.

This essay argues that the biblical story of Noah's Flood was a rich site for imagining the negative impact of human behaviour on nonhuman animals in early modernity. While previous scholars have shown that the Flood story played a key role in articulating emerging notions of species conservation and biodiversity, I argue that it was at the same time a significant and underappreciated source of ideas about animal endangerment.[3] Moreover, tracking early modern stories, theories and images of mass animal death during and after the Flood suggests that presentist concerns for animal welfare did not play a driving role in emerging notions of animal endangerment at human hands. Rather, the pointless killing of sinless animals in a global deluge seems to have struck a chord with early modern scholars and artists who believed strongly in a moral economy of disaster, in which the scale of punishment was precisely tailored to the scale of the sin that provoked it. Overlapping debates about whether animals began facing widespread predation by humans immediately after the Flood were more explicitly tied to presentist concerns about food, health and diet, but here again, the question of carnivorism's origins does not appear to have been primarily motivated by concern for the welfare of eaten animals, but rather for the physical and spiritual health of eating humans.

In many ways, the moral economy of disaster that animated Flood science, art and scholarship reflects the religious anxieties and intellectual commitments of post-Reformation Europe. This essay brings together texts and images from across media, genres and places in early modern Europe, including Italy, England, Scotland, Ireland, France, the Low Countries, Switzerland and the German lands. Casting a wide net reveals that ideas and images relating to biblical disaster and animal endangerment circulated across national, confessional and linguistic borders in the Republic of Letters. I pay particular attention to authors from Italy and England in order to highlight the exchange of ideas between northern and southern Europe and between Protestant and Catholic Europe, not in order to suggest that the Republic of Letters made such distinctions meaningless but rather to show how the preoccupation with anthropogenic animal endangerment, and the moral economy of disaster that undergirded it, spoke to religious, scientific, medical and aesthetic concerns that cross-crossed the Continent in the long seventeenth century.

The accusatory cow: Art, philosophy and the roots of animal endangerment

Noah's Flood was one of the most popular topics of research and debate in early modern European science and scholarship.[4] While the carrying capacity of the ark was widely debated across the Republic of Letters – a vein of scholarship wonderfully emblematised by Athanasius Kircher's detailed diagram of Noah's Ark as a kind of floating natural history cabinet full of taxonomized animals – so too were topics that have received less attention from modern historians but equally consumed early modern writers, such as the postdiluvial pathways of human and animal migration and the size, shape and scale of the Flood itself.[5] All of these issues were linked by a common concern with animals impacted by a natural disaster that was designed to punish humankind.

Within the context of Flood scholarship, which reached the height of its popularity in the century between 1650 and 1750, a lively discussion ensued about the spiritual ethics of animal endangerment. The Swiss naturalist and collector Johann Jakob Scheuchzer, who played an important role in popularising the idea that fossils were relics of living things who died in the Flood, referred to fossils as 'innocent victims who perished in the great *bouleversement* of the Deluge'.[6] Animals were innocent victims of a natural disaster that was 'design'd to punish Men alone, because Men alone are capable of Sin', declared Scheuchzer's contemporary, the Swiss biblical scholar and journalist Jean Le Clerc.[7] The Flood was provoked by human sinfulness; nonhuman animals were incapable of sin; therefore the countless animals who drowned in the Flood were the tragic casualties of humanity's transgressions. While God was obviously responsible for sending the Flood, it was humanity who bore moral responsibility for causing the disaster, and thus indirectly, for the loss of animal life. In the words of Scheuchzer's longtime correspondent Louis Bourguet, another avid fossil collector and Flood enthusiast, 'the part of the Human Race that lived before the Catastrophe was the cause of that event'.[8] Notable works of biblical criticism, universal history, natural philosophy and natural theology from the long seventeenth century expressed a sense of deep unease about the way that human misdeeds inadvertently led to the mass death of innocent animals, who perished in Noah's Flood through no fault of their own.

The roots of early modern thinking about animals imperilled by planetary catastrophe are multiple and complex, stretching back centuries and across the domains of philosophy, religion, literature and art. Devotional literature of the later Middle Ages that encouraged Christians to have compassion for animals, such as the hagiography around St. Francis of Assisi, anticipated the more specific worry that human sin put animals at risk.[9] A sixteenth-century dialogue between the Italian humanists Francesco Pucci and Fausto Sozzini on the subject of immortality and original sin exemplifies this trend well. Responding to Pucci's contention that God created all living things immortal until Adam introduced mortality to the world through his Fall, Sozzini objected that God would not have punished animals by rescinding their immortality, since they were not the ones who had sinned.[10] Their debate, which took place in letters exchanged in 1577–1578, raised the possibility that the negative consequences of sinful behaviour might endanger the lives not only of humans but of nonhuman creatures as well – in this case, imperilling the very conditions of immortal life. Pucci and Sozzini's epistolary debate, situated at the confluence of Renaissance humanist philosophy and early Reformation-era theology (Sozzini's name and legacy would later lend themselves to the anti-Trinitarian doctrine of Socinianism), shows how discussions of biblical history and interpretation inspired reflections on animal endangerment in early modernity.

Not all expressions of concern for the harms inflicted on nonhuman animals were made in reference to biblical history or to theological questions about the nature of sin. In the following century, another exchange of letters between two learned philosophers, the French natural philosopher René Descartes and the Cambridge Platonist Henry More, discussed animal endangerment in the context of present-day dietary practices. Descartes' theory that animals lacked speech, reason and thought was, he claimed, preferable to the alternatives in part because it exonerated people from feeling guilty about killing animals for meat.[11] While Descartes himself is frequently held up in modern scholarship as the origin and poster child of demeaning Western attitudes towards animals, his remark to More in 1649 gestures towards the existence of early modern people who experienced guilt or at least unease about killing and eating animals – a topic to which I return in the final section.

Meanwhile, literature offered another site for early modern Europeans to work through their feelings about the dangers they posed to animal life. A minor tradition in early modern Dutch poetry, for example, featured anthropomorphised animals lamenting their imperilment by human behaviour. In one anonymous manuscript poem from the seventeenth century, whales protest the deaths of their brethren at the hands of Dutch whalers and vow to seek vengeance; in another poem of 1607 by P. C. Hooft, cows trapped in a village fire lament their own impending deaths and accuse their human caretakers of not doing more to save them.[12] These poetic whales and cows identify human behaviour as a threat to animal life, their rhetorical voices underscoring the injustice of animals dying from neglect or from intentional slaughter.

But the specific concern with human sin as an unintentional cause of animal death on a planetary scale seems to have emerged most forcefully in relation to the story of Noah's Flood. Moreover, it seems to have appeared in visual art before becoming a major preoccupation of natural philosophy. Around the turn of the sixteenth century, paintings, woodcuts, frescoes and engravings of Noah's Ark began to proliferate in greater numbers and, significantly, they began to include animals more prominently.[13] Along with images of happy animals boarding the ark under Noah's pastoral care (Figure 3.1) – the picture that most likely comes to mind to someone in the twenty-first century when prompted to imagine 'animals' and 'ark' – there also appeared far darker images of animals flailing in the floodwaters alongside human sinners as the ark left them all behind.

Figure 3.1 Roelant Savery, *Noah's Ark*, c. 1625, oil on canvas. National Museum of Warsaw.

Figure 3.2 Jan van Scorel, *The Flood*, c. 1530, oil on panel, 109×178cm, (P001515). Madrid, Museo Nacional del Prado. Copyright of the image: © Photographic Archive Museo Nacional del Prado.

The Dutch painter Jan van Scorel's oil painting *The Deluge* (c. 1530) (Figure 3.2) is an early example of Flood art featuring a wide range of animals as well as humans in the frantic crowd of the soon-to-be-drowned. In the foreground, people desperately cling to horses and camels while a doomed menagerie of foxes, monkeys, owls, cats, dogs, goats, cows and a single, terrified elephant looks on and the ark sails inexorably away into the distance. While horses made frequent appearances in contemporary art depicting the victims of the Flood, the sheer variety of animals on display in van Scorel's *The Flood* – European and foreign, wild and domestic – suggests that humanity's sinful behaviour put *all* animals at risk, on a global scale.

Artistic representations of animals who did not get to board the ark positioned humans as a danger to animals, not as their saviours. The tenor of these artworks is tragic, even accusatory. The Flemish painter Jan Brueghel the Elder's *The Deluge and the Ark of Noah* (1601) (Figure 3.3) embellishes van Scorel's theme of animal endangerment by making a single, individuated animal the focal point of his composition.[14] Like van Scorel's earlier treatment, Brueghel's oil painting on copper represents the moment after the fortunate few have boarded the ark and those who have been left behind make futile attempts to escape their fate. Amidst the panicked melee of humans and animals, a single cow stands apart, seemingly stock still, in

Figure 3.3 Jan Brueghel the Elder, *The Deluge and the Ark of Noah*, 1601, oil on copper. Kunsthaus Zürich, The Betty and David Koetser Foundation, 1986.

the bottom centre of the composition. The animal's coat is a luminous white, suggesting purity and innocence, while its large dark eyes look out beyond the frame directly at the viewer. The cow's stillness and steady gaze suggest a profound awareness of what is happening, of its own mortality, of the imminent deaths of every living creature around it, in a disaster of humans' – but crucially, not animals' – own making. The largest figure in the painting, dead centre and bright white, Brueghel's cow is in many ways the protagonist of this Flood scene, attracting the viewer's eye and sympathy even as its level gaze seemingly implicates the viewer and indeed all of sinful humanity in the catastrophe to come.[15]

Brueghel painted *The Deluge and the Ark of Noah* not long after his return from Rome, where he had ample occasion to view Michelangelo's *Deluge* (1508–1512) on the ceiling of the Sistine Chapel.[16] Brueghel's treatment of the subject follows Michelangelo's and van Scorel's in placing the ark in the distance, sailing away from the viewer, while placing the Flood's imminent victims in the foreground.[17] Artistic depictions of the onset of Noah's Flood at just the moment when disaster strikes and 'the damned recognise that the ark will forever be inaccessible to them', in the words of art historian Molly Faries, became popular around the turn of the sixteenth century.[18] Coinciding with widespread fears of an apocalyptic global flood, sparked by the prophecies of Girolamo Savonarola in the 1490s and the predictions of several Italian and German astrologers that a second Noah's Flood would arrive in 1524, these artworks may also reflect more diffuse cultural anxieties about the status of the one true Church – which Noah's Ark had long symbolised in the Christian tradition – in the wake of the Lutheran Reformation.[19] Sixteenth-century artworks depicting the drowned and the damned certainly register an intense fascination with the wages of sin and a conviction that divine punishment for sin could find material expression in the natural world.

Brueghel and van Scorel departed significantly from Michelangelo in the prominent inclusion of nonhuman animals among the panicked crowds of the left behind. Doomed animals changed the meaning of 'drowned and the damned'-themed artworks from rather straightforward depictions of human sinners getting their just desserts to more complicated meditations on the unintended effects of human sin on nonhuman animals. Biblical painting provided rich terrain for articulating themes of animal innocence, human sin and catastrophe. While concerns about animal endangerment were expressed in a variety of contexts in premodernity, it was particularly in reference to the biblical disaster of Noah's Flood – and to a lesser extent, the earlier biblical 'disaster' of the Fall – that the notion of human behaviour accidentally imperilling animal life on a mass scale was most clearly and emphatically articulated.

'Innocent dumb creatures': Science, scholarship and the moral economy of disaster

Natural philosophy, universal history and biblical criticism opened a space for further reflection on animal endangerment in the later seventeenth and early eighteenth centuries. Scholars across Europe took up the vexing problem of innocent animals who, like Brueghel's cow, were casualties of a natural disaster targeting humankind. An uneasy sense of responsibility for animal death shines through especially forcefully in the argument advanced by scholars in the Netherlands, Germany, Switzerland and England that Noah's Flood did not cover the entire planet and thus spared millions of animal lives. Scholarly debates about the scale of the Flood reveal not only the increasing uptake of animal endangerment as a subject for science but also a deeper moral logic of disaster, in which the scale of a disaster needed always to be congruent to the scale of the sin that precipitated it.

Initially, though, the argument in favour of a limited Flood had nothing to do with animals. One of the first early modern writers to argue for restricting the scale of the Flood was the itinerant Dutch scholar Isaac

Vossius, who held positions as city librarian of Amsterdam, tutor and court librarian to Queen Christina of Sweden and, after Christina's conversion to Catholicism, canon at St. George's Chapel in Windsor, England. In his widely debated work of chronology *On the True Age of the Earth* (1659), written during a multi-year sojourn in the Hague, Vossius provocatively challenged the idea of a Universal Deluge on the grounds that human habitation of the earth was very limited in the time of Noah. 'The human race held only a modest part of Asia' by that point in human history, and had likely spread no further than the lands around Mesopotamia. Therefore, Vossius argued, there was no need for the Flood to cover the whole planet, 'and in fact it would be absurd to say that the places where no men lived should also feel the effects of the punishment on mankind'. God, who 'does not make needless miracles', would obviously have crafted the scale of the disaster to match the scale of the sin that provoked it, and so it was most reasonable to conclude that only 'the inhabited world was overwhelmed' in Noah's Flood.[20] Deeply sceptical that there was sufficient water to cover the entire face of the earth, Vossius contended that a Universal Deluge would have been both physically difficult and morally unnecessary in a time when humanity was not yet universally distributed across the globe.

Vossius was immediately attacked in print by a chorus of scholars, mostly German and English, who were keen on defending the full global extent of Noah's Flood as well as a more literal reading of Sacred Scripture.[21] The negative reaction was due in part to the similarities between Vossius' arguments and those of the widely reviled French Reformed scholar Isaac La Peyrère, whose polygenist *Men Before Adam* (1655) was published anonymously in Amsterdam just a few years before.[22] *Men Before Adam* made the case that 'the Deluge came only upon the Land of the Jews, and not upon the whole world', leaving China, the Americas and other inhabited parts of the planet completely unscathed. Insisting on a small-scale Flood was part of La Peyrère's more general argument that the Pentateuch was a record of Jewish history only and that the different races of humankind each had their own separate creation and histories.[23] La Peyrère's polygenist and polycentric account of early human history was almost universally condemned by Jews, Protestants and Catholics alike.[24] The tide of scholarly opinion in the later seventeenth century was turning decisively in favour of the theory of the Universal Deluge as an anchor for Christian narratives of universal history, as emblematised in the Anglican natural philosopher Thomas Burnet's controversial but hugely influential *Sacred Theory of the Earth* (published in two volumes between 1681 and 1688).

Nevertheless, Vossius' proposal of a small-scale Flood was taken up and elaborated on by other Protestant scholars across northern Europe, who sought to rescue it from its association with the heresies of La Peyrère by insisting that a limited deluge would nevertheless have been sufficient to kill all people living on earth at the time, save Noah and his family.[25] Georg Caspar Kirchmaier (or Kirchmayer), a professor of philosophy at Wittenberg, expanded the idea into a full-length dissertation, *On the Universality of the Deluge*, published in Latin in Geneva in 1667. While he acknowledged that the Book of Genesis seemed to state plainly that Noah's Flood covered 'the entire face of the earth', Kirchmaier argued that this should be read as 'hyperbolic synecdoche' for a flood which in fact covered 'only the land of Asia', as was typical for the 'Hebrew idiom and usage' of the Pentateuch.[26] Moreover, a spatially limited flood could have killed all humans while sparing many more innocent animals. 'The punishment of mankind was the true cause of the Flood', Kirchmaier reminded his readers, so why should animals be 'subject to the same penalty, when they have no understanding of sin?' Kirchmaier mobilised the trope of animal innocence in order to render the emerging theory of a limited deluge more palatable.[27]

As Vossius' proposal of a less-than-global Flood gained traction, the collateral damage to animals implied by the theory of a Universal Deluge became a more prominent part of the critique. The Anglican minister (later bishop of Worcester) Edward Stillingfleet offered a lengthy defense of a scaled-down Flood in his popular work of natural theology, *Sacred Origins* (1662), arguing that it would have spared the indigenous animals of the Americas from the deadly effects of Old World sin. Published in London just three years after Vossius' *Age of the Earth*, Stillingfleet's *Sacred Origins* echoed Vossius' argument that a punishing flood sweeping over unpopulated territory was pointless and irrational, to which he added the further consideration that it would have caused unnecessary loss of animal life. 'What reason would there be that in the opposite part of the globe, viz. America, which we suppose to be unpeopled then, all the living creatures should there be destroyed because men had sinned in this?' he asked. Drawing on by-then well-established tropes of the Americas as a modern Eden, Stillingfleet contrasted the sinful Old World with a sinless New World, inhabited only by animals. Given that animals had established a planetary presence by the time of Noah while humans had not, Stillingfleet reasoned, 'The only probability then left for asserting the universality of the Flood, as to the Globe of the earth, is from the destruction of all living creatures together with man'. But that was clearly nonsense since animals and other living creatures were not the targets of divine vengeance.[28]

A staunch promoter of rational religion and the integrity of the Church of England, Stillingfleet seems to have felt that the best way to safeguard belief in a just, wise and benevolent God was to insist on a minimally destructive Flood that, in sparing sinless animals and the spotless New World, offered a strict though pleasing congruence between the scale of sin and the scale of the ensuing disaster. Stillingfleet insisted on this congruence by posing the rhetorical question: 'And what reason can there be to extend the Flood beyond the occasion of it, which was the corruption of mankinde?'[29] The same moral reasoning undergirded Jean Le Clerc's discussion of animals drowned in Noah's Flood in his 1696 commentary on the Pentateuch. A widely read biblical scholar and influential journalist in the Republic of Letters born in Geneva but resident in Amsterdam for most of his adult life, Le Clerc insisted that the spiritual incapacity of animals to sin should have guaranteed protection from the Flood to the greatest extent possible. 'If any Beasts are involved', he wrote, 'they perished for no other reason, but because they happened to be in the same place where Men lived'. Animal death was unavoidable, Le Clerc conceded, given the nature of the disaster, but it would only have been justified if the animals drowned along with the Flood's true targets. Like Vossius and Stillingfleet, Le Clerc insisted on scalar congruence between sin and disaster, arguing that it was 'absurd to say, That the Effects of the Punishment which was inflicted upon Mankind, should be exerted where no Men were to be found'.[30]

The idea that the scale of any given disaster should be the same as the scale of the community of sinners who provoked it – a moral economy of disaster, if you will – runs through disaster discourse in early modern Europe, not only in discussions of Noah's Flood. 'National Sins deserve National Judgments' preached the Anglican minister Richard Willis to the Societies for Reformation of Manners in the wake of a 'dreadful Storm' in 1703, 'and unless God have some other wise Ends of his providence to serve by sparing such a Nation, will very likely bring them down upon it'.[31] The nation that sinned together was punished together. The scale of the ensuing disaster was in fact a useful index of the scale and magnitude of the preceding sin. In the case of Noah's Flood, champions of the Universal Deluge felt that it was necessary to insist on the planetary scale of the disaster because the sin that provoked it was common to all of humanity; universal sin necessitated universal judgement, and thus a universal disaster. Burnet insisted that the biblical Flood, like the Apocalypse to come, was necessarily global. According to St. Peter, Burnet claimed, it would be just as ludicrous to say 'that the Conflagration shall be only National, and but two or three countries burnt in the last Fire, as to say that the Deluge was so'.[32] Partisans of a less-than-global Flood, meanwhile, argued that the collateral damage to nonhuman entities in a Universal Deluge would have been so extensive as to violate the principle of congruent sin and punishment. Instead, they argued, a very-large-but-not-planetary Flood would satisfy the clear moral imperative of punishing all of humanity for their sins without causing needless devastation to nonhuman nature. In private correspondence with the Swiss naturalist

Bourguet, the Veronese scholar Ottavio Alecchi compared the idea of a Universal Deluge to raising the entire Mediterranean Sea just to destroy one little village on the shores of Lago di Garda in the foothills of the Alps: an exercise in pointless excess.[33] The punishment should fit the crime, they felt; the scale of a disaster should be in proportion to the scale of the community of people who provoked it. A deadly deluge tailored to the size of human occupation of the earth in biblical times beautifully exemplified the moral economy of disaster.

Even staunch advocates of a Universal Deluge acknowledged, with a certain measure of unease, that a global flood would have caused needless animal death on a planetary scale. In his influential history of the ancient world *The Primitive Origination of Mankind* (1677), the English jurist Sir Matthew Hale conceded that 'the Infinite Power of God might have destroyed those Evil Men by a Pestilence as well as by a Flood, without detriment to the harmless Brutes or Birds'. God was omnipotent and could have easily crafted a global catastrophe in which only humans were killed and the 'harmless brutes' were spared. Why God had in fact chosen to send an indiscriminate flood instead of a more precisely targeted pandemic was hard to understand, Hale acknowledged, but he felt sure that this decision reflected the 'Goodness and Bounty' of God, 'though the particulars thereof be hid from us'.[34] In other words, Hale could think of no good reason why a biblical disaster sent to punish sinful humanity should also punish sinless animals, effectively conceding the critique made by Vossius, Kirchmaier, Stillingfleet, Le Clerc and others. Even as Hale defended the universality of the Flood, he acknowledged that the mass death of innocent animals in Noah's Flood was hard to square with the moral economy of disaster and with notions of divine justice.

Continuing to press on the theological problem of an overly murderous global flood, the anonymous English author of a short philosophical tract, *Two Essays Sent in a Letter from Oxford* (1695), pointed out that a Universal Deluge would also have caused pointless harm to the physical planet, yet another part of God's Creation that was incapable of sin. Published under the initials 'L.P.' (a possible reference to La Peyrère), the first of the two essays opined that the theory of a Universal Deluge

is not agreeable to the usual methods of Providence, nor to the Wisdom of the Divine Nature; for what design could there be in destroying all the innocent dumb Creatures, and the Beauty of the Creation, in the Uninhabited Parts ... for the sake only of a few Wanton and Luxurious Asiaticks, who might have been drown'd by a Topical Flood, or by a particular Deluge, without involving all the Bowels of the whole Mass, and the remote Creatures upon the face of the Earth, in their Ruin.[35]

Elaborating on Vossius' argument about the superfluity of a global flood when only part of Asia was populated, and echoing Hale's observation that a small-scale Flood was more consonant with the doctrines of divine omniscience and benevolence, L.P. put forward the additional consideration that a 'particular deluge' would have caused much less damage to the earth itself, as well as far less loss of animal life. L.P. was writing in reaction to recent English works of earth history detailing the destruction of nonliving nature in the global eco-catastrophe of Noah's Flood. Burnet argued that the Flood turned the earth into a 'great Ruine', transforming it from a perfect sphere with a temperate springlike climate to a 'rude and ragged' planet beset by winter storms and hurricanes and pockmarked with cavernous oceans and jagged mountains.[36] The naturalist John Woodward's *Essay Toward a Natural History of the Earth* (1695) claimed that the Flood 'was not sent only as an Executioner to Mankind: but that its prime Errand was to reform and *new-mold the Earth*', specifically by reducing the fertility of the earth's soil and thereby forcing men into agricultural labour.[37] For Burnet and Woodward, the destruction of the planet's surface, interior and climate was part of the suite of punishments meted out by God to humankind at the time of the Flood.

Many of Burnet and Woodward's contemporaries in Britain and on the Continent found the idea of an earth-destroying Flood deeply troubling, in large part because (as they argued) the earth itself was not a spiritual agent and therefore not a target of divine vengeance. The Scottish physician, mathematician and satirist John Arbuthnot attacked Woodward's theory by saying, 'This is turning Nature outside inward'.[38] From Italy, the naturalist and physician Antonio Vallisneri rebuked Woodward in very similar language in his 1721 book on fossils and the Flood, saying: 'Our Lord God wished to castigate Mankind, not to turn the entire earth inside out'.[39] The earth, like its animals, was not capable of sin and should not have been targeted for destruction. The moral economy of disaster was repeatedly invoked by people on both sides of the debate, used by some to justify theories of the Flood as an eco-catastrophe that destroyed soil fertility and caused the climate to change, while being invoked by others to argue that the Flood's impact was exclusively limited to human beings and perhaps the domesticated animals standing nearby.

While concern about animal endangerment runs as a common thread through Flood art, science and scholarship of the long seventeenth century, it does not appear that a presentist concern for animal welfare was the primary – or even the secondary or tertiary – driving force behind it. None of the authors cited above advocated for animal rights in any kind of modern (or even early modern) way, nor did they argue in favour of a vegetarian diet or condemn animal slaughter, as did some of the authors discussed in the next section. Rather, the notion of animals killed by (or spared from) a biblical disaster seems to have been of interest primarily as a way of thinking about the spiritual burdens shouldered by humans in their capacity as accidental agents of environmental destruction.

'The antediluvian diet': Vegetarianism, carnivorism and eco-catastrophe

The biblical story of Noah's Flood inspired a second major way of thinking about animal endangerment: as the historical origin of widespread carnivorism. The Italian vicar, jurist and biblical scholar Giovanni Maria Chiericato, Vallisneri's contemporary in the learned city of Padua, summarised this alleged historical shift from vegetarianism to carnivorism succinctly: 'Before the Flood meat was not permitted, only herbs and fruit, which, after the Flood, were replaced by meat'.[40] The idea that people began eating animals in significant numbers only in the wake of Noah's Flood dated from the time of the Church Fathers.[41] It became a popular subject of scholarly debate in the early modern period, when this scrap of biblical history was mobilised into full-blown theories regarding how and why a massive deluge could have led to the widespread slaughter and consumption of nonhuman animals.[42] Inspired by new scientific theories of the Flood as an eco-catastrophe that dramatically changed the global environment, natural philosophers, medical writers and biblical scholars alike debated the idea that the turn to carnivorism was precipitated by an environmental crisis affecting the food supply in the immediate aftermath of Noah's Flood.

That vegetarianism reigned supreme prior to Noah's Flood was a common, if controversial, idea among scholars and writers in Christian Europe.[43] From the sixteenth century onwards, English and Italian medical advice manuals regularly repeated the idea that a plant-based diet was partly if not wholly responsible for the longevity of the biblical patriarchs and the giant stature of antediluvian men.[44] In *Acetaria: A Discourse of Sallets* (1699), the English naturalist John Evelyn repeated the commonly held notion that Noah was 'the first man to whom the Concession of eating Flesh was granted' by God in the immediate aftermath of the Flood.[45] Perhaps inspired by the medical lectures he attended in Padua as well as the more plant-forward foodways he observed in Italy while on the Grand Tour, Evelyn postulated that prior to the Flood, 'the Antediluvian Diet' was happily and healthfully limited to 'Fruit and wholesom Sallets'.[46] Pondering the mysterious causes of antediluvian longevity, the Paduan physician Vallisneri noted that drinking wine and eating meat did not become common until after the Flood.[47] Dairy products, however, might have been allowed, according to Vallisneri's colleague Francesco Bianchini, who cited multiple authorities of pagan and Christian antiquity in his *Universal History* (1697) attesting that antediluvian men may have consumed milk and other dairy products alongside their main diet of grains, fruits, seeds and vegetables.[48] The alleged virtues of 'the Antediluvian Diet' inspired at least some early modern Europeans to try it themselves. The English mining entrepreneur Thomas Bushell, a disciple of Sir Francis Bacon, adopted a vegetarian diet 'like to that of our long-liv'd Fathers before the flood', which he conceived of, in true Baconian fashion, as 'a perfect experiment upon my selfe for the obtaining a long and healthy life'.[49]

Several authors who supported the idea of a post-disaster turn to carnivorism nevertheless felt that it was unhealthy, unnatural and contrary to God's original intentions for mankind. Evelyn's contemporary John Ray, who became the most celebrated naturalist of his generation after leaving holy orders in 1662, praised antediluvian vegetarianism and lamented the postdiluvial rise of carnivorism in his acclaimed *History of Plants* (1686–1704):

> Certainly Man by Nature was never made to be a Carnivorous Creature, nor is he arm'd at all for Prey and Rapin, with gag'd and pointed Teeth and crooked Claws, sharpned to rend and tear: But with gentle Hands to gather Fruit and Vegetables, and with Teeth to chew and eat them: Nor do we so much as read the Use of Flesh for Food, was at all permitted him, till after the Universal Deluge.[50]

God did not originally intend for humans to eat (other) animals, according to Ray, and they are not naturally suited for it like other beasts of prey. In a 1732 work of natural theology, the Church of Ireland minister Patrick Delany pushed this line of reasoning further to suggest that animals as well as humans were vegetarians before the Flood, noting that 'it is beyond all controversy, that the stomachs of all carnivorous animals, are fitted for the digestion of fruits and vegetables'.[51] In a 1724 tract promoting a plant-centred diet, the eighteenth-century English medical writer George Cheyne suggested that God made meat-eating licit after the Flood precisely in order to shorten human lifespans.[52] God's postdiluvian permission to eat animal flesh was no favour to humanity, Cheyne suggested, but rather of a piece with the divine punishment represented by the Flood. Cheyne's claim was a striking parallel to Burnet and Woodward's earlier arguments about climate change and soil sterility as the natural means by which God's punishment outlasted the Flood itself. The disaster of Noah's Flood pushed God's creatures into an unnatural carnivorism and remained a key part of humanity's ongoing suffering for their sins.

The causal link between the biblical flood and the onset of widespread carnivorism was widely debated by early modern writers. Many explanations relied on changes in nature wrought during Noah's Flood that in turn caused human dietary patterns to change. The best explanation anyone could fathom was that changes to the earth's soil or climate caused plants to become less nourishing after the Flood, making animal flesh a necessary supplement to the human diet. Erasmus' 1526 dialogue *Concerning the Eating of Fish* features a butcher saying to a fishmonger that 'the new earth produced better and more nutritive things than it does now, since it has grown old and almost past bearing'.[53] Invoking the popular Renaissance notion of the world's decline, Erasmus' fictional butcher speculated further that the Flood caused the world's climate to worsen, prompting postdiluvian people to incorporate meat into their diets. 'The Deluge brought on a cold climate', claimed the butcher, 'and even today we see that those who live in cold climes are greater eaters than those of the hotter climates. The Flood destroyed or at least spoiled the products of the earth'.[54] The idea that the Flood ushered in a new, harsher climatic epoch would continue to grow over the next two centuries, and among its virtues was that it offered a plausible explanation for the postdiluvian shift to meat-eating.[55]

Other scholars felt that a planetary decline in soil fertility and botanical nutrition was a more likely culprit. Quoting Woodward's influential theories of the Flood as a vector of permanent and global soil decline, the Scottish cleric Patrick Cockburn alleged that God granted the eating of animals to Noah and his descendants because the Flood dramatically reduced the fertility of the planet's soil, making it impossible to sustain a growing human population on a plant-based diet. In *An enquiry into the truth and certainty of the Mosaic deluge* (1750), Cockburn also speculated that there was more arable land before the Flood, which in addition to reducing the fertility of the soil around the world had also turned much good farmland into deserts, marshes and mountains where crops could not grow.[56] The Flood's status as a global eco-catastrophe could be easily mobilised into explanations as to why the era of widespread vegetarianism so quickly gave way to universal carnivorism.

Chiericato expounded on the connection between flooding, soil fertility and agricultural output that could have made it necessary for people to begin consuming animals for food in a post-disaster landscape. In *The First Age of the Earth, or Discourses on Sacred Genesis* (1708), he reasoned that the 'long lives of the Ancients' was enabled in large part by the nourishing and substantial food that grew from the soil of the antediluvian earth. But the Flood's salty water rendered the soil 'half sterile', and so 'God in his benevolent wisdom expressly conceded the use of animal flesh for food on account of the deficient sustenance of plants'.[57] Chiericato used the term *resa salmastra* to describe the brackish, muddy soil ruined by the 'salty waters' of the Flood, perhaps reflecting contemporary concerns about saltwater flooding the land and ruining crops.[58] The links between flooding, soil fertility and food supply were all too obvious to someone who lived his entire life in the floodplains of the Veneto. Certainly, in his view, this toxic combination spelled doom for the happy plant-based diet of the earth's earliest inhabitants. 'The vigour and substance of the earth was gravely weakened by the waters of the Flood', Chiericato speculated, and because of this 'God in his Divine Mercy conceded that the human race could eat the flesh of animals'.[59] The salty Flood ruined the primaeval Earth but laid the groundwork – quite literally – for a new age of human predation on animals to begin.

Explanations for the shift from antediluvian vegetarianism to postdiluvian carnivorism depended on the increasingly popular idea in science and scholarship that Noah's Flood was an environmental catastrophe with long-durational consequences. Some proposed that Noah's Flood profoundly altered the planet's climate or soil chemistry which, in turn, caused a major shift in human food consumption patterns. Other scholars disputed this narrative of botanical, climactic and soil decline in the aftermath of the Flood. Evelyn rejected the idea that plants had been 'divested of those Nutritious and transcendent Vertues they were at first endow'd withal'.[60] Plants were just as nourishing now as ever, and even if they were not, the advance of human 'Industry and Skill' in agriculture and horticulture ever since then would more than make up for any natural nutritional loss.[61] Others who were ready to entertain the theory of nature's postdiluvial decline nevertheless felt that nature rebounded not long after. Vallisneri claimed that after a period of acute eco-crisis, nature 'returned to its first state, and that now the air, earth, fruit, herbs and grains are just as they were before the Deluge'.[62] The liveliness of the debate on antediluvian vegetarianism and postdiluvian carnivorism is a testament to the force of the idea that Noah's Flood was an eco-catastrophe whose impact on the natural world might also have permanently transformed human–animal relations.

The contrast between Evelyn's love of salad and Erasmus' embrace of carnivorous pleasures reveals the stakes of these scholarly debates about biblical foodways for everyday meat-eating in the early modern period. Erasmus' dialogue between the butcher and the fishmonger was an implicit critique of the Catholic Church's imposition of frequent fasting days throughout the liturgical calendar, during which meat was prohibited and fish was generally substituted. The Erasmian butcher's claims about the necessity of carnivorism in a postdiluvian world dovetailed with medical knowledge about the healthy properties of animal flesh, especially in the treatment of certain illnesses. Surveying the arguments that had been made against the notion that meat-eating was unknown before the Flood, Chiericato noted that contemporary physicians frequently prescribed meat to promote health and vigour, which surely would have made the Methuselahs of the antediluvian age even more long-lived.[63] Meanwhile, Evelyn's insistence that the life-giving properties of soil and plants had not declined significantly since the time of Noah's Flood was part of his more general argument that the 'Antediluvian diet' continued to be both physically healthier and spiritually preferable to the meaty diet of modern times. The vegetable advocacy of *Acetaria* placed it firmly in a long tradition of Italian and European writing on food, health and diet, from Alviso Cornaro's *On the Sober Life* (1558) to Antonio Cocchi's *On the Pythagorean Life* (1743), which advocated a plant-based diet as the one most likely to promote wellness and longevity.[64] Medical and religious ideas about dietary health in the early modern present coincided in these debates about human, animal and plant life before and after the biblical Flood.

The version of animal endangerment explored in these texts on meat-eating and the Flood was more obviously related to presentist concerns about human–animal relations than were the scholarly debates about animal innocence and the scale of the Flood discussed in the previous section. Once again, however, these scholarly discussions about the diluvial origins of carnivorism seemed more strongly motivated by anxieties over the spiritual status of human beings than by a concern for animal welfare. Chiericato cited Ovid on the soul-destroying effects of carnivorism, claiming that 'killing animals in order to eat their meat signifies such cruelty or such barbarity of the soul that it is not agreeable to rational man, who should rather extend mercy towards the beasts'.[65] Indeed, more than anything else these debates about the biblical origins of meat-eating spoke directly to early modern preoccupations with food and diet occasioned by the Reformations.[66]

Eating meat was spiritually significant for early modern Christian Europeans. In Catholic Europe, consuming meat (especially sheep) was symbolically linked with consuming the Eucharist, so much so that sheep consumption in Counter-Reformation Rome sextupled between 1580 and 1630 with the explicit encouragement of popes, papal physicians and medical writers with close ties to the Holy See.[67] In the early days of the Zwinglian Reformation in Zürich, eating meat on fasting days became a way of asserting independence from Rome.[68] Quotidian meat-eating in the Reformed tradition, rather than something to be celebrated, came to be viewed as a reminder of human fallibility and corruption, as contrasted with the spiritual innocence and vegetarian practices of Adam and Eve.[69] Even prior to the Reformations, religious writers in the Latin West vigorously debated the merits of eating animals in the pursuit of a holy life, with religious authorities variously encouraging meat-eating and recommending abstinence from meat to religious orders of monks, nuns and friars.[70] Whichever position they took on the matter, Christian Europeans agreed that eating animals was fraught with moral and spiritual meaning.

It is not surprising, then, that differing views about the spiritual significance of eating animals found expression in the scholarly debates about the origin of meat-eating in the disaster of Noah's Flood.

Conclusion

In early modern Europe, the biblical story of Noah's Flood offered a richly generative framework for thinking about humanity's impact on non-human nature on a planetary scale. Representations of innocent animals imperilled in and by Noah's Flood in art, theology, history, science and medicine may have played an unacknowledged but significant role in forging the conceptual links between ecological catastrophe and animal endangerment, with human behaviour imagined as the underlying cause of both. Early modern representations of mass animal death in ancient history do not neatly anticipate modern theories of species extinction or biodiversity loss, and the story of Noah's Flood no longer anchors scientific discussions of animal endangerment in man-made environmental crises. Nevertheless, they form an important chapter in the long history of catastrophic thinking.

Notes

1 David Sepkoski, *Catastrophic Thinking: Extinction and the Value of Diversity from Darwin to the Anthropocene* (Chicago; London: University of Chicago Press, 2020), pp. 4–5.
2 Ursula K. Heise, *Imagining Extinction: The Cultural Meanings of Endangered Species* (London; Chicago: University of Chicago Press, 2016), p. 4.
3 Janet Browne, 'Noah's Flood, the Ark, and the Shaping of Early Modern Natural History', in *When Science and Christianity Meet*, ed. by David C. Lindberg and Ronald N. Numbers (Chicago: University of Chicago Press, 2003), pp. 111–38. There are several other 'thinking with' books to which I am indebted in my own thinking with and about animals and the Flood: Stuart Clark, *Thinking with Demons: The Idea of Witchcraft in Early Modern Europe* (Oxford: Clarendon Press, 1997); *Thinking with Animals: New Perspectives on Anthropomorphism*, ed. by Lorraine Daston and Greg Mittman (New York: Columbia University Press, 2005); and Sepkoski, *Catastrophic Thinking*.
4 Don Cameron Allen's *The Legend of Noah: Renaissance Rationalism in Art, Science, and Letters* (Urbana, IL: University of Illinois Press, 1949) remains a classic point of reference; more recently, see Claudine Poulouin, *Le temps des origines: l'Eden, le Déluge et 'les temps reculés' de Pascal à L'Encyclopédie* (Paris: Champion, 1998).
5 Exceptions include Giuliano Gliozzi, *Adamo e il nuovo mondo: La nascita dell'antropologia come ideologia coloniale: dalle genealogie bibliche alle teorie razziali (1500–1700)* (Florence: La Nuova Italia, 1977).
6 Johann Jakob Scheuchzer, *Physique Sacrée, Vol. 1* (Amsterdam: Pierre Schenk and Pierre Mortier, 1732), p. 73.

7 Jean le Clerc, *Twelve Dissertations Out of Monsieur Le Clerk's Genesis* (London: R. Baldwin, 1696), p. 156. I did not have the opportunity to consult the Latin original of Le Clerc's *Pentateuchus, sive Mosis prophetae libri quinque* (1696); all quotes are taken from the contemporaneous English translation by the poet Thomas Brown. On the complicated publication history of Le Clerc's biblical commentaries, see Dmitri Levitin, *Ancient Wisdom in the Age of the New Science* (Cambridge: Cambridge University Press, 2015), pp. 200–2.

8 Louis Bourguet, 'Cours sur la Providence' (1733)', Bibliothèque publique et universitaire de Neuchâtel, Fonds Louis Bourguet, MS 1243, f. 65.

9 On animals in Franciscan hagiography, see David Salter, *Holy and Noble Beasts: Encounters with Animals in Medieval Literature* (Cambridge: D. S. Brewer, 2001), pp. 25–52. The topic of animal suffering was also taken up in late-medieval devotional literature and drama, as discussed in Lisa J. Kiser, 'Margery Kempe and the Animalization of Christ: Animal Cruelty in Late Medieval England', *Studies in Philology*, 106 (2009), 299–315.

10 The role of animals in Pucci and Sozzini's debate is analysed in Cecilia Muratori, 'La caduta dell'uomo e la sofferenza degli animali nella "Disputatio" tra Francesco Pucci e Fausto Sozzini', *Bruniana & Campanelliana*, 17, 1 (2011), 139–49 (144).

11 René Descartes to Henry More, February 5, 1649, in *Descartes: Correspondance avec Arnauld et Morus*, ed. by Geneviève Lewis (Paris: Vrin, 1953), pp. 126–27.

12 Johan Koppenol, 'Noah's Ark Disembarked in Holland: Animals in Dutch Poetry, 1550–1700', in *Early Modern Zoology: The Construction of Animals in Science, Literature and the Visual Arts*, ed. by by Karl A.E. Enenkel and Paul J. Smith (Leiden; Boston: Brill, 2007), pp. 522–25. The literary historian Karl Steel finds an analogous tradition of animal lamentations in English poetry, dating back to late Middle English and continuing through the seventeenth century. Steel, 'Huntings of the Hare: The Medieval and Early Modern Poetry of Imperiled Animals', in *The Palgrave Handbook of Animals and Literature*, ed. by Susan McHugh et al. (Cham: Palgrave Macmillan, 2021), pp. 141–52.

13 While Noah's Flood and Noah's Ark were popular subjects in medieval art, nonhuman animals only began populating Flood scenes in significant numbers in the early modern period. Marrigje Rikken, 'Exotic Animal Painting by Jan Brueghel the Elder and Roelant Savery', in *Zoology in Early Modern Culture: Intersections of Science, Theology, Philology, and Political and Religious Education,* ed. by Karl A.E. Enenkel and Paul J. Smith (Leiden; Boston: Brill, 2014), p. 411. For an overview of early modern Flood art, see *Visions du Déluge de la Renaissance au XIXe siècle* (Paris: Réunion des Musées Nationaux, 2006).

14 Rikken credits an earlier 1596 Brueghel painting of Noah's Ark as one of the first treatments of the subject to reorient the focus from the human figures to the animals: Rikken, 'Exotic Animal Painting by Jan Brueghel the Elder and Roelant Savery', p. 411.

15 The painting in fact depicts a trio of cows, including a dark greyish-brown cow in the far-left corner standing on the same plane as the white cow, looking obliquely at the viewer, while a third cow in the middle distance tries desperately to haul itself out of the rising waters and onto dry land. Why cows? One answer has to do with the significance of the cow in sixteenth-century Dutch art and culture specifically: 'To talk about Holland is to talk about cows. Dutch prosperity often was – and still is – symbolized by rich livestock and an abundance of milk and cheese'. Koppenol, 'Noah's Ark Disembarked in

Thinking with the flood 95

Holland: Animals in Dutch Poetry, 1550–1700', p. 467. The more general importance of human-cow relationships in everyday life is discussed in Erica Fudge, *Quick Cattle and Dying Wishes: People and Their Animals in Early Modern England* (Ithaca, NY: Cornell University Press, 2018).

16 Arianne Faber Kolb, *Jan Brueghel the Elder: The Entry of the Animals into Noah's Ark* (Los Angeles: Getty Publications, 2005), 5; 82.

17 Van Scorel was likewise strongly influenced by Italian art and by Michelangelo's *Deluge* specifically: Molly Faries, 'A Woodcut of the Flood Attributed to Jan van Scorel', *Oud Holland*, 97 (1983), 5–7.

18 Faries, 'A Woodcut of the Flood Attributed to Jan van Scorel', 6.

19 On the astrological predictions of a second global flood, see Paola Zambelli, "Fine del mondo o inizio della propaganda? Astrologia, filosofia della storia e propaganda politico-religiosa nel dibattito sulla congiunzione del 1524," in *Scienze, credenze occulte, livelli di cultura: Convegno internazionale di studi* (Florence: Olschki, 1982), pp. 291–368. On Savonarola's earlier predictions of the same, see Donald Weinstein, *Savonarola: The Rise and Fall of a Renaissance Prophet* (New Haven: Yale University Press, 2011), pp. 103–4. On the symbolic equation of Ark and Church, which predated the Reformations but took on new significance thereafter, see Norman Cohn, *Noah's Flood: The Genesis Story in Western Thought* (New Haven: Yale University Press, 1996), pp. 26–31. Brueghel's patron Cardinal Borromeo was a leading figure in the Catholic Reformation, in which 'the ark's symbolism was especially relevant ... and the idea was promoted that only the Church could provide salvation, quite as Noah alone had saved mankind'. Kolb, *Jan Brueghel the Elder: The Entry of the Animals into Noah's Ark*, pp. 5–6.

20 Isaac Vossius, *Dissertatio de vera aetate mundi* (The Hague: Adriani Vlacq, 1659), pp. 53–4. All translations are mine unless otherwise noted.

21 Allen, *The Legend of Noah*, p. 87

22 Contemporaries compared Vossius' position on the Flood to La Peyrère's in spite of the former's efforts to distance himself from the latter. Sir Matthew Hale, for example, observed that 'though he [Vossius] reprehend the *Praeadamitae* ... yet he seems to mince the Universality of the Flood'. Hale, *The Primitive Origination of Mankind* (London: William Godbid, 1677), p. 186. Vossius was not, however, as unorthodox a thinker as La Peyrère and was moreover a far more careful and learned textual critic. Anthony Grafton, 'Isaac Vossius, Chronologer', in *Isaac Vossius (1618–1689) between Science and Scholarship*, ed. by Eric Jorink and Dirk van Miert (Leiden; Boston: Brill, 2012), pp. 67–71.

23 Isaac La Peyrère, *Men Before Adam* (London: [n. pub.], 1656), p. 243. The English translations of La Peyrère's two-part work, which appeared in London several months after their appearance in Amsterdam, were likewise published anonymously, though La Peyrère's authorship was never in doubt.

24 Richard H. Popkin, *Isaac La Peyrère (1596–1676): His Life, Work and Influence* (Leiden; New York: Brill, 1987), pp. 80–1.

25 The confessional dimensions of the debate on the scale of Noah's Flood were complex and variable across the early modern period, which I explore at greater length in *After the Flood: Imagining the Global Environment in Early Modern Europe* (Baltimore, MD: Johns Hopkins University Press, 2019), especially chapters 1 and 2. Generally speaking, Protestant authors were more willing to champion the idea of a small-scale Flood in print than were their

Catholic counterparts, even as they encountered fierce opposition from their Protestant colleagues who increasingly insisted on the universality of the deluge. See also Michel Bligny, 'Il mito del diluvio universale nella coscienza europea del Seicento', *Rivista storica italiana*, 85 (1973), 53–5.

26 Georg Caspar Kirchmaier, *De diluvii universalitate dissertatio prolusoria* (Geneva: Petrum Columesius, 1667), p. 65.

27 Kirchmaier, *De diluvii universalitate dissertatio prolusoria*, p. 46.

28 Edward Stillingfleet, *Origines Sacrae, or a Rational Account of the Grounds of Christian Faith* (London: R[obert] W[hite], 1662), p. 540.

29 Stillingfleet, *Origines Sacrae*, p. 539.

30 This is a nearly verbatim English translation of Vossius' original Latin formulation (quoted in note 20 above); meanwhile Le Clerc's English translator compared him favourably to Stillingfleet as an interpreter of Scripture. Le Clerc, *Twelve Dissertations*, 'Preface to the Reader,' [p. 4]; pp. 156–57.

31 Quoted in Josiah Woodward, *An Account of the Progress of the Reformation of Manners*, 14th edn (London: Joseph Downing, 1706), pp. 50–51.

32 Thomas Burnet, *The Theory of the Earth*, Vol. 1, 3rd edn (London: R[oger] N[orton], 1697), p. 17.

33 Ottavio Alecchi to Louis Bourguet, 1712, quoted in Ivano dal Prete, *Scienza e società nel Settecento Veneto* (Milan: Franco Angeli, 2008), p. 223. While the published defenses of a scaled-down Flood came overwhelmingly from Protestant scholars, several Catholic scholars, including Alecchi and the mathematician Jacopo Riccati, made the case in private correspondence. Jacopo Riccati to Antonio Vallisneri, June or July 1719, in *Carteggio (1719–1729)*, ed. by Maria Laura Soppelsa (Florence: Olschki, 1985), pp. 66–70.

34 Hale, *The Primitive Origination of Mankind*, p. 188.

35 L.P., *Two Essays Sent in a Letter from Oxford* (London: R. Baldwin, 1695), pp. 13–14. The overt anti-Semitic and anti-Asian sentiment of L.P.'s discussion of the animal problem highlights the ways that Flood science and scholarship engaged the emerging racial imaginary of early modern Europe. See Colin Kidd, *The Forging of Races: Race and Scripture in the Protestant Atlantic World, 1600–2000* (Cambridge: Cambridge University Press, 2006), chap. 3; David N. Livingstone, *Adam's Ancestors: Race, Religion, and the Politics of Human Origins* (Baltimore, MD: Johns Hopkins University Press, 2008); and Barnett, *After the Flood*, chap. 2.

36 Burnet, *The Theory of the Earth*, pp. 74–5.

37 John Woodward, *An Essay Toward a Natural History of the Earth* (London: Ric[hard] Wilkin, 1695), p. 93.

38 John Arbuthnot, *An Examination of Dr. Woodward's Account of the Deluge* (London: C. Bateman, 1697), p. 17.

39 Antonio Vallisneri, *De' corpi marini, che su' monti si trovano*, 2nd edn (Venice: Domenica Lovisa, 1728), p. 49.

40 Giovanni Chiericato, *La prima età del mondo overo ragionamenti sopra la sacra Genesi*, 2nd edn (Venice: Andrea Poletti, 1708), p. 165.

41 Michele Cutino, 'Les animaux dans le récit du déluge (Gen 6–9) et ses interprétations patristiques', in *La restauration de la creation: Quelle place pour les animaux?*, ed. by Michele Cutino, Isabel Iribarren, and Françoise Vinel (Leiden; Boston: Brill, 2018), pp. 63–75.

42 Scholarly debate about the ethics of meat-eating and the treatment of animals flourished in humanist circles in the Renaissance and continued robustly throughout the early modern period. See Cecilia Muratori, *Renaissance*

Vegetarianism: The Philosophical Afterlives of Porphyry's On Abstinence (Cambridge: Legenda; Modern Humanities Research Association, 2020); *Ethical Perspectives on Animals in the Renaissance and Early Modern Period,* ed. by Cecilia Muratori and Burkhard Dohm (Florence: SISMEL – Edizioni del Galluzzo, 2013); and Tristram Stuart, *The Bloodless Revolution: A Cultural History of Vegetarianism from 1600 to Modern Times* (New York; London: Norton, 2006); and *Animals and Humans: Sensibility and Representation, 1650–1820,* ed. by Katherine M. Kinsey (Oxford: Voltaire Foundation; Oxford University Press, 2017). However, the overlap between debates about animal ethics and debates about the biblical Flood is not well understood; the following section attempts a brief sketch of some of those connections.

43 Erica Fudge, 'Saying Nothing Concerning the Same: On Dominion, Purity, and Meat in Early Modern England', in *Renaissance Beasts*, ed. By Erica Fudge (Urbana, IL: University of Illinois Press, 2004), p. 71.
44 The growing vogue for vegetables in dietary advice manuals is discussed in David Gentilcore, *Food and Health in Early Modern Europe: Diet, Medicine, and Society, 1450–1800* (London: Bloomsbury, 2016), pp. 124–31.
45 John Evelyn, *Acetaria: A Discourse of Sallets* (London: B[enjamin] Tooke, 1699), p. 157.
46 Evelyn, *Acetaria*, pp. 161, 124.
47 Vallisneri, *De' corpi marini, che su' monti si trovano*, p. 107.
48 Francesco Bianchini, *La istoria universale provata con monumenti*, 2nd edn (Rome: Antonio de' Rossi, 1747), pp. 156–57.
49 Thomas Bushell, *Mr. Bushel's Minerall Overtures* ([London?]: [n. pub.], 1659), p. 5.
50 John Ray, *Historia Plantarum* (London: Typis Mariae Clark, Prostant apud Henricum Faithorne, 1686), p. 46. I have chosen to quote from the colorful, contemporary English translation by John Evelyn in *Acetaria*, pp. 171–72. Ray returned to discuss the topic of Noah's Flood at much greater length in *Miscellaneous Discourses Concerning the Dissolution and Changes of the World* (London: Samuel Smith, 1692).
51 Patrick Delany, *Revelation examined with Candour*, vol. 1 (Dublin: S. Powell, 1732), p. 176.
52 Anita Guerrini, 'A Diet for a Sensitive Soul: Vegetarianism in Eighteenth-Century Britain', *Eighteenth-Century Life*, 23, 2 (1999), 36.
53 Desiderius Erasmus, 'Concerning the Eating of Fish', in *The Essential Erasmus*, ed. and trans. by John P. Dolan (New York: New American Library, 1964), p. 281. I am grateful to Bradford Bouley for this reference.
54 Erasmus, 'Concerning the Eating of Fish', p. 282.
55 Lydia Barnett, 'The Theology of Climate Change: Sin as Agency in the Enlightenment's Anthropocene', *Environmental History*, 20, 2 (2015), 217–37.
56 Patrick Cockburn, *An Enquiry into the Truth and Certainty of the Mosaic Deluge* (London: Printed for C. Hitch and M. Bryson, 1750), pp. 170–71, 176.
57 Chiericato, *La prima età del mondo overo ragionamenti sopra la sacra Genesi*, p. 122.
58 Elisabeth Crouzet-Pavan, *Sopra le acque salse*, 2 vols. (Rome: École Française de Rome, 1992); Karl Appuhn, 'Friend or Flood? The Dilemmas of Water Management in Early Modern Venice', in *The Nature of Cities*, ed. by Andrew C. Isenberg (Rochester, NY: University of Rochester Press, 2006), pp. 79–102.
59 Chiericato, *La prima età del mondo overo ragionamenti sopra la sacra Genesi*, p. 166.

60 Evelyn, *Acetaria*, pp. 163–64.
61 Evelyn, *Acetaria*, p. 124.
62 Vallisneri, *De' corpi marini, che su' monti si trovano*, p. 109.
63 Chiericato, *La prima età del mondo overo ragionamenti sopra la sacra Genesi*, p. 167.
64 Gentilcore, *Food and Health in Early Modern Europe*, pp. 124–30.
65 Chiericato, *La prima età del mondo overo ragionamenti sopra la sacra Genesi*, p. 166.
66 'Food was at the center of both the break of Protestants with Rome and the Roman response to the Reformation'. Bradford Bouley, 'Digesting Faith: Eating God, Man, and Meat in Seventeenth-Century Rome', *Osiris*, 35 (2020), 45.
67 Bouley, 'Digesting Faith', 43–4.
68 Christopher Kissane, *Food, Religion and Communities in Early Modern Europe* (New York: Bloomsbury, 2018), ch. 4–5.
69 Fudge, 'Saying Nothing Concerning the Same', pp. 73–75.
70 Diane M. Bazell, 'Strife among the Table-Fellows: Conflicting Attitudes of Early and Medieval Christians toward the Eating of Meat', *Journal of the American Academy of Religion*, 65, 1 (1997), 73–99.

4 Flood, fire, and tears

Imagining climate apocalypse in Scheuchzer's *De portione* (1707/08)[1]

Sara Miglietti

In the winter of 1707/1708, a brief manuscript reached London's Royal Society from Switzerland. Its author, Johann Jakob Scheuchzer (1672–1733), was a town physician in Zurich and a rising star of the international Republic of Letters, having recently gained access to some of the most important scholarly societies of the time – from the Prussian Academy of Sciences to the French Académie des Sciences, Bologna's Accademia degli Inquieti, and (since 1703) the Royal Society itself. Entitled *De ignis seu caloris certa portione Helvetiae adsignata* ('On the specific amount of fire or heat assigned to Switzerland'), the manuscript belonged to a series of papers (seven in total) that Scheuchzer sent to London between 1703 and 1708, possibly with hopes of publication in the Society's *Philosophical Transactions*.[2] Though all of these papers dealt with aspects of Switzerland's natural history – Scheuchzer's main research focus around those years – the *De portione* differed from the other six in both content and form. Instead of offering an informative first-hand account of specific features of the Swiss environment (such as waters or vegetation), it revolved around a simple but striking thought experiment: what would happen if the climate of Switzerland became warmer and the Alpine glaciers began to melt? Would Switzerland become milder, richer and more productive, or would climate warming be the gateway to an environmental and human disaster of terrifying proportions?

For Scheuchzer, the answer was clear. Though many perceived it as a nuisance, Switzerland's cool climate was actually a God-given gift that enabled the little country, nested in the mountainous heart of Europe, to perform a fundamental function in what we would call the continent's 'ecosystem'.[3] Its glaciers served as Europe's main hydrological reservoir, bringing life and fertility to all nations: without them, the delicate equilibrium on which plants, animals and humans relied for their existence would quickly unravel, and Europe itself would turn into a parched and torrid desert. Before reaching that stage, however, the continent would first go through a phase of extensive flooding, as the Alpine meltwaters rushed violently down the mountains, causing rivers to break their banks

DOI: 10.4324/9781003029823-6

and coastal waters to rise and overflow. Climate warming – described by Scheuchzer as an increase in the 'amount of fire' (*ignis quantitas*) assigned to a given region – would bring disaster not just through 'fire' (i.e., excessively high temperatures) but also through flood.[4]

First rediscovered in 2014, Scheuchzer's *De portione* is an intriguing piece of early modern natural-philosophical prose that still awaits in-depth investigation. In our annotated edition of 2015, William Barton and I took a first step in this direction by highlighting the place of this text in two important northern-European traditions, that of 'physico-theology' on the one hand and that of Neo-Latin mountain writing on the other.[5] More recently, I suggested that some of the central issues raised in Scheuchzer's manuscript – notably, the providential order of creation and the unforeseen dangers that can result from humanity's tendency to want to 'improve' nature for its own purposes – find important parallels in later works by Scheuchzer, such as the *Jobi physica sacra* ('Job's Holy Physics', 1721) and the more extensive *Physica sacra* of 1731–1735, also known as *Kupfer-Bibel* ('Copper-Bible') for its over 750 copper plates.[6] In this essay, I aim to build on some of this earlier work to examine a further aspect of Scheuchzer's manuscript that has thus far remained unaddressed, namely, its creative reuse of a seminal 'disaster narrative' in the Western literary tradition: Seneca's account (in what is now the third book of *Natural Questions* [henceforth *NQ*]) of the great flood that would one day destroy the earth and pave the way for its subsequent regeneration.[7]

Destruction by water (*kataklysmos, diluvium*) was one of two ways in which Stoic philosophers like Seneca maintained that the world was cyclically unmade – the other one being *ekpyrosis* or *conflagratio* (destruction by fire).[8] As Gareth Williams and others have shown, Seneca's description of the cataclysm in *NQ* 3 is not only a fine piece of natural-philosophical writing – an elegant tapestry of allusions to (among others) Virgil's *Aeneid*, Lucretius's *De rerum natura*, and Ovid's *Metamorphoses* (particularly Book 1, in which Ovid narrates the mythical deluge sent by Jupiter to punish a rebellious humankind and rejuvenate the world through Deucalion and Pyrrha).[9] It is also a powerful reflection on the relationship between humankind and the cosmos, painting a stark contrast between the fleetingness of all things human ('A single day will destroy the human race; all that the long indulgence of fortune has cultivated, all that it has lifted to eminence above the rest')[10] and an everlasting natural order that cyclically proceeds to destroy itself only to rise again, purer and stronger, from its own ashes ('when parts of the world must pass away and be abolished utterly so that all may be generated from the beginning again, new and innocent …').[11]

Overall, Seneca's text raises questions about humankind's place in nature that resonated deeply with Scheuchzer. The Swiss scientist referred to Seneca's *Natural Questions* several times in his oeuvre,[12] but nowhere is this intertextuality stronger or more significant than in the *De portione*. It is not simply that this text, like *NQ* 3, occupies a liminal space between science, fiction and prophecy, projecting the reader's mind upward to an ideal 'cosmic viewpoint' and forward to events that can only be described through a tentative act of scientific imagination.[13] As I hope to show in this essay, Scheuchzer found in Seneca not just a source of literary inspiration for his thought experiment on climate change, but also a philosophical springboard for thinking proactively about the relationship between divine providence, human mores and environmental disaster. A close comparison of Scheuchzer's manuscript with Seneca's flood narrative in *NQ* 3 will reveal both Scheuchzer's literary and intellectual debts to the Roman philosopher and strong elements of originality: looking at how Scheuchzer reframed Seneca's text in light of his particular concerns (as a Christian scientist, a Swiss patriot and a member of the supranational Republic of Letters) will enable us to appreciate how, in early Enlightenment Europe, the reception and transformation of classical models still functioned as a fundamental form of original cultural production. Conversely, an analysis of Scheuchzer's *De portione* will allow us to add a new dimension to the history of Seneca's long-lasting European *fortuna*.

Scheuchzer's manuscript must indeed be situated in a longer arc of Western engagements with *Natural Questions*, which has been recently and excellently reconstructed by Fabio Nanni and Daniele Pellacani.[14] While their survey inevitably omits Scheuchzer's manuscript (which was yet to be rediscovered at the time), it helps us recall that Scheuchzer was neither the first nor the last to be drawn to the disaster narratives in *NQ* 3: from the early thirteenth century onwards, these Senecan texts found special favour with Christian theologians and natural philosophers who sought to make sense of them in the light of Holy Scripture on the one hand and of changing philosophical paradigms on the other.[15] Specifically, Seneca's description of the cataclysm was often read in conjunction with the biblical account of Noah's Flood in Genesis 7, whereas the Stoic theory of *ekpyrosis* could be approached through Christian eschatological texts such as 2 Peter 3 (which similarly foretold a universal conflagration in the end times).[16] Particularly significant in this respect is the case of the Flemish scholar Justus Lipsius (1547–1606), well known for his role in the early modern reception of Stoicism.[17] In the early 1600s, Lipsius published not only an influential edition of Seneca's *Natural Questions* (first printed in 1605 and still authoritative in Scheuchzer's time),[18] but also a systematic summary of Stoic natural philosophy in which he sought to reconcile Stoic doctrines of *interitus mundi* with Christian orthodoxy (*Physiologia Stoicorum*, 1604).[19] Scheuchzer's *De portione* belongs in this wider tradition, but with two important differences: Scheuchzer's interpretation of *NQ* 3 looks not backward to the biblical Flood (an unrepeatable event of sacred history),[20] but forward to an ill-defined but possible future[21]; and his primary goal is not to Christianize Seneca's cataclysm (as others had done before him) but to craft a cautionary tale about humanity's relationship with nature.

Of course, this does not mean that the *De portione* does not entertain an equally important relationship with the Bible as with Seneca's *NQ* (and other ancient pagan texts). Nor does it mean that Scheuchzer's Christian background is irrelevant for understanding this manuscript. I shall return to the first point in the essay's conclusion to consider how Scheuchzer's imaginary disaster in the *De portione* compares to his (explicitly Bible-centred) description of Noah's Flood in his later *Physica sacra*. As for the second point, a fundamental claim of this essay is that the *De portione* cannot in fact be understood in isolation from Scheuchzer's engagement with theological themes such as the impact of sin on human nature and behaviour (including in and towards the natural world); the perfection of God's original creation and the havoc wrought to the earth by Noah's Flood; and the complex relationship between God's providence, the regularities of the natural order and human freedom. While all of these themes remain largely implicit in the *De portione*, they form the essential backdrop against which the manuscript ought to be read, not least because Scheuchzer's original addressees – his colleagues at London's Royal Society – shared with him an intense interest in such 'physico-theological' issues (although we shall see that their outlook differed from Scheuchzer's in several important respects). Before considering how Scheuchzer's *De portione* appropriates and transforms Seneca's flood narrative in the service of contemporary debates on humankind and God's creation, let us first examine in some detail the two texts and their main similarities and differences.

'Nothing is difficult for nature'

A bird's eye view of *NQ* 3.27–30 (which contains Seneca's flood narrative) immediately allows one to appreciate its strong influence on Scheuchzer's *De portione*. Two aspects appear especially relevant: the overall framing and pace of the account, and the adoption of a 'cosmic viewpoint' that challenges traditional anthropocentric views of nature.[22] I shall begin with a brief summary of Seneca's text before analysing its relationship to Scheuchzer's manuscript.

Seneca introduces his flood narrative at the very end of Book 3, presenting it as a sort of digression within a broader discussion on 'terrestrial waters' (the main subject of the book). At 3.26, Seneca briefly considered the process whereby 'all standing and enclosed water naturally purges itself', observing for instance that 'the sea … drags from the depths dead bodies, litter, and similar debris of shipwrecks, and purges itself of them, not only in storms and waves but also when the sea is tranquil and calm'.[23] At this point, he turns to examining another form of self-cleansing that nature cyclically performs on a much larger scale: 'But this subject warns me to wonder how a great part of the earth will be covered by water when the fated day of the deluge comes …'.[24] Seneca's subsequent description of the cataclysm, which occupies several pages, revolves around five central ideas: first, the inevitability of the catastrophe; second, its recurrent nature; third, its rapid, violent and unstoppable character; fourth, its universal scope and impact; and, lastly, its providential purpose.

Seneca states repeatedly that the great flood is part of a cosmic rhythm of growth, decay and regeneration that has characterised the universe since its inception and that is fatalistically inscribed in the laws of its existence ('Whether the world is an animated being, or a body governed by nature, like trees and plants, there is incorporated in it from its beginning to its end everything it must do or undergo … among these changes was flood, which occurs by a universal law just as winter and summer do').[25] From this perspective, the cataclysm is a natural phenomenon, fully in step with the laws of the cosmos (even though its destructive progress at times disrupts these very laws)[26] and morally neutral insofar as it occurs by sheer necessity. That said, throughout the passage Seneca also occasionally uses language that suggests a complex relationship between the flood and human corruption, paving the way to a different (moral and human-centred) interpretation of the entire process. First of all, his assertion that 'it seems best to god for old things to be ended and *better* things to begin' (*cum deo visum ordiri meliora, vetera finiri*) implies that the 'old' things must be terminated not merely out of decrepitude but because they have lost their original goodness.[27] This point is confirmed shortly afterwards, when Seneca argues that the flood will occur 'whenever the end comes for human affairs, when parts of the world must pass away and be abolished utterly, so that all may be generated from the beginning again, new and innocent, and no tutor of vice survives [*nec supersit in deteriora praeceptor*]'[28]; and again towards the end of the book, when we learn about the aftermath of the great disaster:

> When the destruction of the human race is completed and the wild animals, into whose savagery men will have passed, are equally extinct, again the earth will absorb the waters … Every living creature will be created anew and the earth will be given men ignorant of sin, and born under better auspices. But their innocence, too, will not last, except as long as they are new. Vice quickly creeps in. Virtue is difficult to find …[29]

The flood is portrayed in these passages as a destructive, yet ultimately beneficent force that wipes away vice and corruption and temporarily restores a golden age of purity and innocence on earth. What exactly is the connection between the cataclysm and human mores remains unclear: two main interpretations are possible – the flood is either specifically sent as a punishment against human depravity, as in the Greek myth narrated by Ovid or in the biblical account of the universal deluge; or it is a purely natural process of periodic cosmic regeneration, not unlike the scene of marine self-cleansing described in *NQ* 3.26. In the former case, the corruption of human mores is to be seen as the *occasion* or *moral cause* of the flood; in the latter, it is merely a symptom of cosmic decay – an inevitable phenomenon, predetermined by a universal natural law and for which humankind bears no special responsibility. Both interpretations have found supporters over the centuries, and the issue cannot be discussed at length here.[30] But regardless of whether one favours the fatalistic explanation (human corruption as a symptom) or the more strongly moralistic one (human corruption as cause or occasion), the text unquestionably weaves the two domains of nature and morality together, showing how they go hand in hand and cannot fail to affect one another for better or worse.[31]

This key point is also made in other ways: for instance, Seneca frequently employs parallel to show how specific processes in the realm of nature are mirrored in those of human life and industry: 'A long time is needed so that a child, once conceived, may come to be born. The tender infant is reared only with great toil ... But how with no effort it is all undone! It takes an age to establish cities, an hour to destroy them. A forest grows for a long time, becomes ashes in a moment'.[32] Yet there is one crucial way in which nature and humankind differ from each other, and that is exactly what the cataclysm helps to reveal: nature is always in control, even and indeed 'especially when she rushes to destroy herself'[33]; whereas humans, caught in the whirlwind of a disintegrating cosmos, are ruthlessly 'crushed' and 'submerged'.[34] The harder they try to assert their agency – 'propping up' their collapsing houses, 'shaking down' food from the few trees that still stand – the more 'pointless' are their efforts:[35] nothing can protect a helpless humanity from nature's overwhelming power ('Nothing is difficult for nature ... for destruction she comes suddenly with all her violence ... Whatever nature changes from the present state, that will be enough to destroy all mortals').[36]

'Imperceptible increases'

Seneca's entire description of the cataclysm conveys the sense of nature's relentlessness, by showing how small beginnings quickly snowball into an unstoppable chain of consequences. It all starts with dark skies and heavy rain: 'At first excessive rain falls. There is no sunshine, the sky is gloomy with clouds, and there is continuous mist, and from the moisture a thick fog which no winds will ever dry out'.[37] This is followed by a series of disasters in close succession (note Seneca's skilful use of adverbs to pace the passage and emphasise nature's inexorable progress):

> *Next*, there is a blight on the crops, a withering of fields of standing grain as it grows without fruit. *Then* all things sown by hand rot and swamp plants spring up in all the fields. *Next* the stronger plants also feel the blight ... Houses sag and drip ... Nothing is stable. After the clouds have massed more and more, and the accumulated snows of centuries have melted, a torrent which has rolled down from the highest mountains carries off forests that are unable to cling fast and tears boulders free from their loosened structures and rolls them along, washes away villas and carries down sheep and owners intermixed. The smaller houses are plucked up by the torrent which carries them off as it passes. *Finally*, the torrent is diverted violently against larger dwellings and drags along cities ad peoples ... *Eventually*, as the torrent passes along it is increased by the absorption into itself of several other torrents and spreads out in scattered devastation on the plain. *Finally*, it pours out in all directions, loaded with the vast stuff of nations.[38]

By this point, what originally began as a fairly localised phenomenon has already escalated to a crisis of pan-European proportions, as Europe's three main rivers – 'the Rhone, the Rhine, and the Danube' – leave their beds and 'roll forth with unchecked force':

> The Danube no longer touches the base nor even the middle of the mountains but attacks the summits themselves, carrying with it the mountainsides it has flooded, the crags it has hurled apart, and the promontories of whole regions which have their foundations weakened and are detached from the continent. The Danube, finding no exit – for it has itself closed off all passages for itself – returns in a circle and envelops in one whirlpool the great circuit of lands and cities.[39]

As rain continues and moisture accumulates in the air, 'the sky becomes heavier ... and for a long time reaps disaster from disaster'. A thick darkness falls upon the world: 'a dreadful and terrible night' interrupted by 'flashes of awful illumination'. The sea, 'increased by the influx of the rivers', pushes inland and 'advances its own shores' until it meets the contrary current of the torrents rushing into it: swirling backwards, it overflows the surrounding land, till 'everything in sight is filled with water'.[40] The few survivors – cut off from each other and 'miserable beyond the realisation of misfortune' – find refuge on top of the highest mountains: everything else is 'swallowed up' by the rising waters; 'the entire earth is swimming'.[41] Within the space of a few paragraphs, Seneca's gripping *crescendo* takes us from a minor episode of strangely persistent rainfall to the 'great shipwreck' of the entire world.[42]

It suffices to place this text side by side with Scheuchzer's *De portione* (readable in full in the Appendix below) to appreciate the strong resemblances between them. Seneca's idea that small changes in the natural order can lead 'by imperceptible increases' (*incrementis fallentibus*) to the 'destruction of all mortals' (*exitium mortalium*)[43] is also the driving concept behind Scheuchzer's manuscript, which similarly describes a catastrophe in 'successive waves'.[44] Like Seneca, Scheuchzer posits an initial cause (in his case, an increase in temperature in the Alpine region) and then builds a powerful climax by showing how each subsequent step in the chain of effects inexorably proceeds from this initial anomaly. As the heat increases, the Alpine glaciers ('those immense expanses of ice and snow on the highest Alpine summits', §6) will 'melt more abundantly' than usual: then the brooks that collect their meltwaters – usually a source of fertility for the valleys below – will 'overflow' and 'rush down the walls of the mountains so violently that the meadows will be completely covered by a huge amount of sandy mud and by large rocks' (§11). Overwhelmed with excess waters, the four great rivers fed by the Alps (the Rhone, the Rhine, the Danube and – in addition to this Senecan triad – the Po) will 'break the banks that now contain them, with considerable damage for the Germans, French, and Italians, through whose lands they flow' (§12); and having reached the sea, they will cause it to rise and overflow, provoking coastal flooding and the mass migration of 'whole thriving nations … away from their homes' (§16). Meanwhile, the humid exhalations from the inundated ground will gather in thick dark clouds, giving rise to terrifying weather events – from intense lightning and thunder to incessant rain and devastating hailstorms (§20–21). Then, once the Alpine glaciers will have completely dissolved, a new and even more destructive phase of the disaster will begin. As 'springs and brooks' dry up once and for all, Switzerland's verdant meadows, 'beneficial for both animals and mankind', will turn to 'parched and sterile fields' (§24). Animals will die of thirst (§17). People will experience a heat 'as intense as between the Tropics', magnified by the reflection of the sun's rays on bare mountain slopes (§25). Eventually, they will have no choice but to leave: Switzerland will have become just as uninhabitable as 'the desert interior of Arabia or the sandy heart of Africa' (§17).

The cosmic viewpoint

Scheuchzer's description of the imaginary disaster neatly divides up into two phases – destruction by water and destruction by 'fiery heat' (*calidum igneum*) – mirroring to some extent Stoic doctrines of *interitus mundi* as well as two famous disaster scenes in Ovid's *Metamorphoses* (Deucalion's flood in Book 1; Phaeton accidentally setting the earth on fire in 2.201–271) and biblical accounts of judgement by water (Genesis 6–7; 2 Peter 2:5 and 3:6) and fire (e.g., Genesis 19; 2 Peter 2:5 and 3:10–12; Revelation 20:9–10). Among all these possible models, however, Scheuchzer's manuscript seems especially indebted to Seneca's *NQ* 3. Aside from a similar structure and pace, the two narratives share an emphasis on specific scenes of destruction – from the rushing streams and rolling boulders (*NQ* 3.27.7; *De portione*, §11) to the rotting plants and inundated grazing land (*NQ* 3.27.4–5; *De portione*, §11, §23); from the extreme weather events caused by excessive moisture (*NQ* 3.27.10; *De portione*, §20–21) to the destruction of human dwellings (*NQ* 3.27.6-7; *De portione*, §11–12) and the displacement of entire populations forced out by flooding and rising sea levels (*NQ* 3.27.11; *De portione*, §16–17). Both accounts emphasise the contrast between the crushing force of nature and the powerlessness of humans who desperately try to save their lives and livelihoods (*NQ* 3.27.11; *De portione*, §16–17, §30); both mention their astonishment and terror at the magnitude of the destruction (*NQ* 3.27.10, 3.27.12; *De portione*, §20–21). At the same time, neither narrative is narrowly focused on the fate of humankind alone. Like Seneca, Scheuchzer is keen to stress the common destiny that unites all living beings in both good and bad fortune: the 'beneficial' ecosystem that formerly ensured the happiness of humans, plants and animals is wiped away in his thought experiment by a disaster that equally affects all creatures (*De portione*, §17, §22, §24).

This decentring of the reader's perspective – from a narrowly anthropocentric outlook to a more holistic one – is yet another important *trait-d'union* between Scheuchzer's manuscript and Seneca's flood narrative. If Seneca's goal in *NQ* 3.27–30 was (to put it with Gareth Williams) to 'redirect' the 'gaze' of his readers from a 'localised mindset' of self-importance towards a 'cosmic viewpoint' of 'enlightened detachment', one can say that Scheuchzer's *De portione* pursues Senecan goals by Senecan means, insofar as it exploits the 'shock effect' of a disaster narrative to 'move us toward a revised perspective on our place in the universal whole'.[45] The objective for both philosophers is to operate a deep conversion in their readers' self-perception, making them simultaneously more aware of the limits inherent in the human condition and more capable of overcoming them through an act of spiritual and intellectual imagination.[46] For Seneca, as for other Stoics, it is only when humans take full stock of their humble standing in nature that they gain access to a loftier form of consciousness, enabling them to transcend the limits of their individuality and to 'plunge' themselves 'into the totality of the cosmos'.[47] Similarly for Scheuchzer, the adoption of a 'view from above'[48] is the key to unlocking an understanding of the world as an ecologically interconnected whole, providentially designed by an all-good, all-powerful God. As we shall see in the next section, the *De portione* is at the same time a performative display of this 'view from above' and a demonstration of what happens when humans fail to adopt it.

'Foolish wishes'

Scheuchzer's entire thought experiment is framed as a response to the 'foolish wishes' (*vota ineptissima*, §5) allegedly expressed by some unspecified individuals for a warmer climate in the Alpine region:

> A perverse reasoning leads to a desire for greater heat in many parts of Switzerland, and especially in mountainous areas, so that crops could mature more quickly, perpetual snows could be melted, summers could be prolonged, winters could be shortened, and so that we would enjoy so many other comforts that the bitter cold now denies us. (§4)

When one considers that Scheuchzer's paper was written in the middle of the so-called 'Little Ice Age' – immediately before one of the coldest winters on record (1708/1709) and in a period in which advancing glaciers were causing major disruption to Alpine economies – one can appreciate how the desire for warmer temperatures would indeed have been widespread among local residents.[49] In putting pen to paper, Scheuchzer was arguably setting himself an unpopular task: to demonstrate that the Swiss climate was just as warm as it should be and that those who thought otherwise did so out of ignorance and arrogance.

To prove these points, Scheuchzer chose to accept the premise of a hypothetical increase in temperatures and to think with his readers through its likely consequences: 'To please those who make such foolish wishes, let us assume that Switzerland could receive more heat than we now actually perceive, and let us see what would happen as a consequence, provided all other conditions on Earth remained the same' (§5). The first effect, Scheuchzer agrees, will be the rapid melting of glaciers and perennial snows. Yet this will hardly prove the blessing that his opponents imagine: 'Then what? The number of Alpine meadows will grow, turning these bleak-looking places into a blooming, garden. Hold your step, unfair judge ...' (§7–9). As the next few pages proceed to contrast this Edenic image with scenes of death, desolation and disaster, one central idea becomes embedded in the reader's mind: in dealing with nature, one has to be careful what one wishes for.

Though the desire for a warmer climate may appear reasonable at first sight (as it certainly did to early eighteenth-century Alpine residents struggling with harsh conditions), it is in fact a 'perverse reasoning' (*perversa ratio*) that flies in the face of both faith and science. Faith teaches that God has 'created all things according to fair weight, number, and measure [*iusto ... pondere, numero, et mensura*]' (§1), meaning that 'a fixed and perfectly fair share' of solar heat has been assigned to the earth as a whole and to each of 'its individual parts' (§2). Science strengthens this belief by teaching how Switzerland's cool climate and mountainous environment fulfil a vital function in the continent's ecosystem ('this tiny corner of Europe from which so many great Europeans countries receive the chief cause of their fertility', §3) and how this delicate ecological balance would be disrupted by a variation in temperatures. Wishing for greater heat is therefore impious madness, as faith and science are united in upholding the providential character of the Swiss climate: 'Let us praise with full mouths the infinite wisdom of God and his most powerful goodness, which blesses our regions ... with sufficient heat, well-proportioned for our land and the whole of Europe' (§31). As the manuscript draws to a close, this physico-theological hymn summarises one of its most important takeaways: in order to deal fairly with God's creation, one has to undergo a drastic change of perspective – from local to supranational, from fragmented to ecological, from human to divine.[50] By seeking to transform the way in which readers view nature, the *De portione* ultimately aims to effect a *moral* change in how they relate to God's creation in thought, word and deed.

The key problem from Scheuchzer's perspective is that humans tend to claim for themselves the right to judge nature and to rectify its alleged faults. Their standpoint in doing so is invariably selfish and parochial: thus the 'climate improvers' in his manuscript wish for a warmer Switzerland because they dream of a stronger local economy ('crops could mature more quickly', 'the number of Alpine meadows will grow') and of a more comfortable lifestyle ('we would enjoy so many other comforts that the bitter cold now denies us'). Such desires, though perfectly natural, are also a manifestation of humanity's self-centredness and short-sightedness – the expression of a human nature that has been radically corrupted by sin. This point, which Scheuchzer will develop in his *Physica sacra* along broadly Calvinist lines, is essential for understanding the anthropological outlook that underpins his manuscript.[51] In their current state of cognitive and moral decay, humans are spontaneously inclined to make hasty, ill-informed judgements that entail a series of unforeseen and often undesirable consequences. The apocalyptic scenario constructed in Scheuchzer's thought experiment is an example of what happens when one evaluates reality from this natural place of selfishness, partiality and self-reliance, instead of embracing God's will in trust and obedience. This does not mean that the *De portione* calls on readers to renounce their intellect for blind faith. Rather, it is an open invitation to use one's intellect differently – to move, that is, from a human-centred to a cosmic mindset. In suggesting that such a mental shift would lead both to a better grasp of the inner workings of nature and to a more worshipful attitude towards God's greatness, Scheuchzer was simultaneously building on a contemporary tradition of thinking science in conjunction with religion and subverting this very tradition in subtle ways.

'For our land and the whole of Europe'

To appreciate the intervention that Scheuchzer was trying to make in contemporary debates, one has to remember who the manuscript was originally addressed to: Scheuchzer's colleagues at London's Royal Society.[52] At the time when Scheuchzer was writing, the Royal Society was undoubtedly one of the most important hubs of what was then known as 'physico-theology' – a tradition with distant precedents in sixteenth-century forms of 'pious philosophy' and 'Mosaic physics', but which especially thrived at the turn of the eighteenth century thanks to Royal Society Fellows such as John Ray (1627–1705), John Woodward (1665–1728) and William Derham (1657–1735).[53] Scheuchzer, who had been elected Foreign Fellow of the Society in 1703, was on particularly friendly terms with Woodward, entertaining with him a lively correspondence that included the exchange of fossils and other natural specimens. As previously mentioned, many of the papers that Scheuchzer sent to the Royal Society in the years 1703–1708 testify to a similar impulse to share new empirical knowledge in the form of natural-historical observations or weather diaries. The *De portione*, however, is of a wholly different nature. Strictly speaking, it offered neither new information nor a novel interpretation of existing data. Rather, it presented itself as a physico-theological manifesto that challenged dominant ways of practising physico-theology in England at the time.

As Ann Blair and Kaspar von Greyerz have recently shown, physico-theology traditionally involved a utilitarian, teleological and anthropocentric approach to nature. It was predicated on the premise of a divinely ordained cosmos in which everything was designed 'to ensure the well-being of humankind and the subservience of nature to that end'.[54] Most English physico-theologians not only accepted, but indeed radicalised these assumptions, with some (like John Ray) going so far as to theorise humanity's right and duty to 'improve' and 'ameliorate' God's creation.[55] As I have argued in more detail elsewhere, Scheuchzer's *De portione* can be viewed as an oblique yet powerful critique of this particular use of physico-theology as a handmaid to contemporary discourses of 'improvement'.[56] The goal was ambitious and – if one has to judge from the paper's lukewarm reception in London – ultimately unsuccessful, even though William Derham's *Physico-Theology*, first published a few years later (1713), contains an attack against the ideology of improvement that may owe something to Scheuchzer's *De portione*.[57] Be it as it may, a consideration of the specific readership to which the manuscript was originally addressed is fundamental for understanding some of the choices that Scheuchzer makes in it – including specific points on which his description of the cataclysm differs from Seneca's flood narrative in *NQ* 3. Three aspects in particular are worthy of attention here: the cause of the disaster, its scale and consequences, and the role of human agency in it.

An obvious first discrepancy between the two narratives consists in the physical cause that sets in motion the whole disaster. For Seneca, it is 'excessive rainfall', although in a later passage he clarifies that precipitation alone could never accomplish such large-scale destruction and that several other factors must be involved (the most important of which is the elemental transmutation of earth into water).[58] For Scheuchzer, the initial physical cause is climate warming and the consequent melting of glaciers (a phenomenon that Seneca does mention in his account, but rather as an effect than as a cause of the disaster).[59] Scheuchzer's emphasis on glaciers was not only fully in line with his interests and expertise at the time of the *De portione* (Swiss natural history being his main field of inquiry in the 1700s;[60] it also afforded him an opportunity to promote his research agenda among an international public, by making the hypothetical melting of the Alpine glaciers a matter of continental rather than purely local concern ('considerable damage for the Germans, French, and Italians', §12). Scheuchzer's attempt to elevate Swiss natural history to a position of centrality in the international scientific debate is mirrored in his manuscript's constant insistence on the *physical* centrality of Switzerland in Europe's geography and ecosystem ('this tiny corner of Europe from which so many great European countries receive the chief cause of their fertility', §3; 'our Switzerland is not only the fertile repository of Europe's waters for distribution but also its storehouse of clouds and winds', §19; 'heat well-proportioned for our land and the whole of Europe', §31). In this sense, Scheuchzer's perspective is different from that of Seneca: his is not the nationless identity of the cosmopolitan Stoic philosopher, equally at home anywhere in the world, but rather the patriotic (though not parochial) identity of the early modern natural historian, deeply attached to his homeland yet also a proud citizen of the supranational Republic of Letters.[61]

This point may also explain a second discrepancy between the two accounts, namely, their different scale. Whereas the planetary dimension of Seneca's cataclysm is fundamental not only on a metaphysical level (the disaster brings about the destruction and regeneration of the whole world) but also on a moral and spiritual one (in Williams' words, 'what matters is Seneca's imaginative construction of a cosmic mind-set ... a form of consciousness that ranges unfettered *over all ages and territories*'),[62] Scheuchzer's disaster is a far-reaching catastrophe that fails however to reach truly global proportions. Throughout the *De portione*, Scheuchzer's focus is decidedly on Europe, and indeed his only mentions of non-European countries ('the desert interior of Arabia', 'the sandy heart of Africa', §17) are disparaging comparisons meant to inspire dread and disgust. Scheuchzer's Eurocentrism, though undoubtedly connected to broader aspects of his worldview,[63] makes special sense in the context of a paper whose primary objective is to speak certain physico-theological truths to a primarily European public.

One of these central truths, as previously mentioned, is the vital importance of rethinking humanity's relationship with nature from a cosmic, as opposed to a narrowly anthropocentric, perspective. While Scheuchzer and Seneca were in perfect agreement on this point, the stakes were much higher for the Swiss scientist, who lived in an age in which humanity's 'footprint' was growing exponentially and the idea of a manmade environmental disaster was increasingly easy to entertain. As Lydia Barnett has recently shown, one of the effects of the physico-theological discussions of the late seventeenth century had been to popularise a notion of the biblical Flood as a global catastrophe for which humanity bore a direct responsibility. Though this responsibility was usually understood in moral and spiritual terms rather than as a form of actual physical agency, there was undoubtedly a new emphasis in this period on the 'anthropogenic' nature of the Flood and on the unintentionally (self-)destructive consequences of human sin.[64] When read in this context, Scheuchzer's *De portione* takes on a new and yet more urgent meaning. Although the manuscript stops short of presenting the catastrophe as an explicitly manmade event (the initial premise of a rise in temperatures is posited without any explanation of how such a phenomenon could occur), there is a deeper sense in which the entire process is in fact human in origin, to the extent that it would not have existed without the 'perverse reasoning' of those wishing to improve the Swiss climate. However subtle and phantasmic this agency may appear to us, it was a serious issue for Scheuchzer on account of the tight connection that he perceived between cognitive and moral states: thoughts and desires could be no less dangerous than actions, not least because they often led directly into them.

Although the idea of harmful environmental agency was certainly not foreign to Seneca (one can think of his famous critique of mining in *NQ* 5 and elsewhere),[65] human agency plays virtually no role in his account of the cataclysm, which deliberately portrays nature as an all-powerful agent while reducing humanity to the role of passive spectator and (in this instance) victim.[66] In this respect, his perspective is different from that of Scheuchzer, who instead recognises an element of active human agency at the very beginning of his imaginary catastrophe (only to emphasise humanity's powerlessness even more strongly once the inexorable process is set in motion). Still, one aspect of Seneca's natural philosophy may have played a part in Scheuchzer's thinking around these issues: namely, his belief (which he shared with many ancient philosophers) that the destruction of the world would come not from some external agent, but from causes immanent to the world itself.[67] For Scheuchzer too destruction comes from within: from a 'foolish' and 'perverse' humanity, which in destroying nature ends up destroying itself as well. This point is worthy of some attention because it challenges recent suggestions that Scheuchzer's view of humankind as immanent to nature made him oblivious to the many ways in which humans can endanger the environment.[68] The opposite, as we have seen, may very well be true. In this sense, it is interesting that while Seneca's cataclysm is recurrent and ultimately benign (insofar as it leads to the regeneration of nature), there is little sign of a happy tomorrow in Scheuchzer's *De portione*. In fact, Scheuchzer states quite unequivocally that the devastation caused by the disaster will be irreversible unless some further change occurs, such as an (equally unexplained) return to normal temperatures (§24). In adopting a linear narrative in deliberate contrast to Seneca's cyclical account, Scheuchzer further emphasises the urgency of thinking twice about initiating processes that cannot be undone.

Conclusion: Scheuchzer's floods

As I hope to have shown in this essay, Scheuchzer's *De portione* entertains a complex relationship with Seneca's *Natural Questions*. On the one hand, Scheuchzer found in Seneca a source of inspiration for the structure and initial concept of his paper, as well as literary material for some of his disaster scenes and a cosmically minded, anti-anthropocentric philosophy of nature that strongly resonated with his own physico-theological orientation. On the other hand, the two flood narratives diverge in crucial respects – from the causes of the disaster (climate change and melting glaciers for Scheuchzer, 'excessive rainfall' and the transmutation of earth into water for Seneca) to its scale and consequences (localised but irreparable devastation for Scheuchzer, regeneration and beginning of all things anew for Seneca) to the role of human agency in it. These discrepancies are meaningful because they help reveal Scheuchzer's intentions for his piece: to promote his particular brand of local natural history (patriotic yet outward-looking), by highlighting the crucial role of the Swiss environment in both the regional and continental ecosystem; and to nudge a particular readership – his Royal Society colleagues, the international Republic of Letters – towards a more humble, more holistic understanding of creation. By taking a stand against the 'perverse reasoning' that calls for greater heat in the Alpine region, Scheuchzer draws attention to the dangerous implications of the ideology of improvement, which presupposes a dissatisfaction with nature as created and a hubristic conviction that humans could do better through their imagination and ingenuity.[69] This attitude, besides being impious, is a slippery slope for beings who are blighted by sin and unable to anticipate the ultimate consequences of their actions or desires:[70] when it comes to tampering with nature, the line between utopia and dystopia – between 'improvement' and catastrophe – can be especially thin.

In comparing Seneca's universal cataclysm ('the entire earth is swimming') to the imaginary disaster in Scheuchzer's manuscript, one cannot help but think of another obvious intertext: the account of Noah's Flood in Genesis 7. Without attempting an exhaustive analysis, it is worth singling out here a few key differences between the cataclysm in the *De portione* and the biblical Deluge. Aside from the fact that one event is firmly in the past while the other takes place in an indistinct and tentative future, their scale is different (planetary destruction in Genesis, a more circumscribed European disaster in Scheuchzer) and so are also their origins and dynamics: whereas the biblical Flood is an embodiment of God's agency at work, Scheuchzer's disaster does not occur by divine decree – on the contrary, it clearly takes place *against* God's plans, disrupting His providential order in pursuit of humanity's 'foolish wishes'.[71] This point is important because it rules out an apocalyptic reading of Scheuchzer's manuscript: although the *De portione* is undoubtedly the child of an era imbibed with Christian eschatology and millenarian expectations,[72] the cataclysm that it portrays cannot be mistaken for a portentous phenomenon preluding to God's final destruction and new creation of the earth.

Scheuchzer's imaginary cataclysm does bear some resemblance to contemporary or near-contemporary accounts of the biblical Flood that described it in strongly naturalistic terms. One can think of the English cleric Thomas Burnet, whose controversial *Sacred Theory of the Earth* (first published in Latin in 1681 and well known to Scheuchzer) built on Cartesian principles to explain the Deluge as a mechanistic un-folding of natural causes and effects.[73] Burnet's account not only took extensive liberties with the text of Genesis, but it also downplayed God's agency to an extent considered unacceptable by many contemporary readers.[74] Though Burnet's attempt to 'mechanize' the Flood did inspire a generation of scholars to 'unite human and natural history' through the joint study of physics and Scripture,[75] those who followed in his footsteps tended to react to his perceived excesses by stressing the providential and supernatural origins of the Flood. This may help explain why Scheuchzer took such pains to keep his imaginary cataclysm well separate from a divinely ordained event such as the biblical Flood. If his *De portione* describes the disaster in strictly naturalistic terms – using Cartesian language (e.g., vortices) and accounting for its inexorable progress by a mechanistic chain of causes and effects – it can do so without danger because the phenomena that it portrays lie outside the scope of sacred history and belong instead to the realm of the scientific imagination.

By contrast, Scheuchzer's later account of Noah's Flood in the first volume of *Physica sacra* (1731) unequivocally places the Deluge under the sign of 'a divine and downright inscrutable miracle' (*divino et imperscrutabili prorsus miraculo*)[76] and acknowledges the cluelessness of philosophical reason in such mysterious matters (*philosophamur hic sicubi in obscuro*).[77] To be sure, echoes of the earlier manuscript can occasionally be detected here, for instance when Scheuchzer comments on how the Flood's 'fatal progress' (*fatalis progressus*) follows the natural 'laws of motion' (*leges motus*), or when he offers a glimpse of the advancing waters upon the 'Alpine yokes' (*Alpium juga*)[78] – two events beautifully captured in the accompanying copper plates (see Figures 4.1 and 4.2). But perhaps the most explicit gesture towards the *De portione* is his mention here of Seneca's NQ 3.27–30 among other ancient sources that 'graphically' depict 'diluvial catastrophes' past and future.[79] Still, there is no doubt that Noah's Flood – by its divine origins, miraculous nature and providential role in God's plan – sat in Scheuchzer's eyes in a wholly different category from the imaginary cataclysm narrated in his *De portione*. Here, Scheuchzer built on Seneca's seminal flood narrative not in order to comprehend a sacred event in the distant past (as others had done before him and as he himself would later do in his *Physica sacra*) but in order to craft a cautionary tale for the present and future. Written in the safe form of the thought experiment (which ensured a certain amount of philosophical and theological freedom), Scheuchzer's *De portione* stands to this very day as a powerful reflection on humankind's place in nature and on the precarious relationship between desires, actions and consequences.

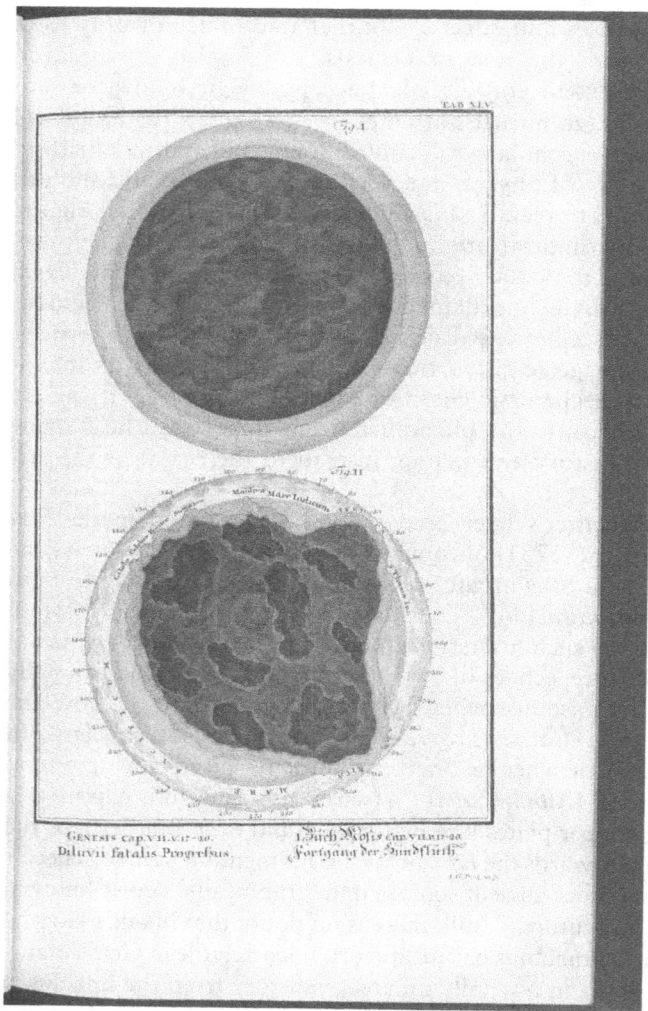

Figure 4.1 The 'fatal progress' of the universal deluge, as depicted in Plate 45 in Scheuchzer's *Physica sacra* (vol. 1, 1731).

Figure 4.2 The geomorphic impact of the Flood upon the Swiss Alps, as depicted in Plate 46 in Scheuchzer's *Physica sacra* (vol. 1, 1731).

Appendix. Text of Scheuchzer's 'De portione' (trans. William Barton and Sara Miglietti) [80]

[1] Just as God created all things according to fair weight, number and measure, he assigned a certain and proportionate amount of fire to both the Solar System in its entirety and to the individual planets within it: to the sun, indeed, as the central source and dispenser of heat; to the planets, as its receivers. [2] A fixed and perfectly fair share of this solar heat and fire reaches our terraqueous globe, and the proportion is the same for the whole as for its individual parts. [3] As for our Switzerland – this tiny corner of Europe from which so many great European countries receive the chief cause of their fertility – it receives just as much heat or fire, be it solar or subterranean, as is necessary. [4] This is *a priori* the will of God, in His wisdom and goodness; and it is proved *a posteriori* by the experience of so many, both natives and foreigners: but a perverse reasoning leads to a desire for greater heat in many parts of Switzerland, and especially in mountainous areas, so that crops could mature more quickly, perpetual snows could be melted, summers could be prolonged, winters could be shortened, and so that we would enjoy so many other comforts that the bitter cold now denies us. [5] To please those who make such foolish wishes, let us assume that Switzerland could receive more heat than we now actually perceive, and let us see what would happen as a consequence, provided all other conditions on Earth remained the same. [6] Those immense expanses of ice and snow on the highest Alpine summits will melt more abundantly. [7] Then what? [8] The number of Alpine meadows will grow, turning these bleak-looking places into a blooming, pleasant garden. [9] Hold your step, unfair judge. [10] As I shall show in greater detail at the right time, it is from these very masses of snow and ice that the Alpine meadows and the valleys below get their fertility; it is from them that great rivers, innumerable brooks and springs take their sources. [11] For as long as the snowy mountaintops melt more abundantly because of the greater heat, the brooks will overflow, rushing down the walls of the mountains so violently that the meadows will be completely covered by a huge amount of sandy mud and by large rocks, as we in fact experience whenever it rains heavily or a water reservoir pours out: then the streams grow and cause terrible damage to the valleys below, as I shall explain at the right time. [12] Then the rivers would break the banks that now contain them, with considerable damage for the Germans, French and Italians, through whose lands they flow. [13] But let us even assume that these waters roll down gently into the sea. [14] What is going to happen then? [15] More water will force its way into this wide reservoir than the latter can actually contain because more will flow into it than out of it. [16] Coastal plains will be flooded, and whole thriving nations driven away from their homes. [17]

Let me continue: were the Alpine tops to be stripped of their snow cap at any point, water levels in rivers would fall below normal, springs and brooks would dry up and thirsty plants would die along with animals, quickly turning this happy Switzerland of ours into the desert interior of Arabia or the sandy heart of Africa, and forcing us to leave. [18] Let us continue. [19] That our Switzerland is not only the fertile repository of Europe's waters for distribution but also its storehouse of clouds and winds, I will prove in time. [20] Suppose now, however, that the amount of fire in our homeland would be greater than the amount we currently experience, and observe closely how vapour will also be raised into the air in a greater quantity, as well as exhalations of every sort: greater clouds, which will terrify us and our neighbouring countries with their dark appearance, will gather in our lower atmosphere. Indeed, they will empty themselves in rains and storms as severe as they will be more frequent, and thunder will rumble out more often. Lightning will become more intense and hailstorms larger. [21] In a word, Switzerland would sadly become a theatre of weather phenomena of all sorts, particularly of the most frightening and harmful ones. [22] I now return to the plant life of our region and living animals to see what might remain of them in this proposed greater share of fiery heat. [23] As long as the melting of the Alpine snows and glaciers lasts, the plants will be overwhelmed by too much moisture and be rather hindered in their growth than helped along. [24] Once that storehouse of Alpine snows has been dissolved (and it will not be easily restored if the supposed greater degree of heat lasts) the pastures will dry up because of the lack of fluids. The meadows of the valleys too – which currently please the eye with their fresh greenery and are beneficial for both animals and mankind, indeed, with their most pleasing variety of colours and flowers exciting the eyes of onlookers in wonderful ways – would dry out to become parched and sterile fields, since they actually enjoy mild to cool air. [25] For in the valleys, one would feel a heat as intense as between the Tropics, because the warmth of the sun's reflected rays would be concentrated by the mountainsides, particularly once these are bare. [26] The Valais, the Val Telline and the other more fertile valleys of Switzerland would not only dry up in this excessive heat, but many other evils would also arise from it. [27] I had no reason to dwell on the inconveniences that would befall our homeland if the temperatures should become lower [rather than higher] than now, since that perverse reasoning already complains that the amount of heat presently assigned to Switzerland is too small; but it is not beside the point to see what would happen by further augmenting the subtraction of fire. [28] Those masses of snow and ice would endure, in fact they would keep growing enormously by daily accumulation, whereas now they continue to melt in the middle of winter because of the subterranean heat, for the sake of the continuous sustenance of the rivers. [29] Their courses would disappear and the streams would be filled up with glacial bulk, rather than with flowing water. The pores of the earth would shrink, and those vapours and exhalations which were designed for producing clouds would be held inside the bowels of the earth, thus causing earthquakes and other disasters. [30] Mountain plants in particular, which now barely manage to bear fruit, would stiffen up thoroughly and deprive animals of their food, and men of their benefits. [31] Let us praise with full mouths the infinite wisdom of God and His most powerful goodness, which blesses our regions, otherwise surrounded by thinner and colder air, with sufficient heat, well-proportioned for our land and the whole of Europe. [32] Do you see how? [33] The cold, in this highest point of Europe, lashed by winds from all sides and immersed in thinner air, is largely blocked by long chains of mountains, so that is less harmful to the valley dwellers; it is also moderated by the manifold reflection of the sun's rays from the mountainsides, and ultimately by that very subterranean heat, which reveals itself through many signs, and is fostered by those snowy masses of the Alps, lest it fly away easily in the breeze, so that it rarefies the plants, and it raises nourishing sap into them. For which reason, it is not surprising if we see glaciers alongside blossoming pastures in the middle of August and July: summer beside winter, indeed. [34] May we praise repeatedly this marvellous providence of our most holy God, along with the inhabitants of Iceland situated in the extreme North, for whom Mount Hecla serves as a sort of furnace, and takes up the office of kindly heat distributor through the whole island.

Notes

1 I am grateful to Ovanes Akopyan, Sandro La Barbera, David Lines, Lucy Nicholas, Gareth Williams, and an anonymous reviewer for their helpful comments on earlier drafts of this essay.

2 For a list of these papers, see William Barton and Sara Miglietti, 'An Eighteenth-Century Thought Experiment on Climate Change: Johann Jakob Scheuchzer's *De ignis seu caloris certa portione Heluetiae adsignata* (1708)', *Lias: Journal of Early Modern Intellectual Culture and Its Sources*, 42, 2 (2015), 135–66 (165–66). The manuscript discussed in this essay is held in the Archives of the Royal Society in London (Classified Papers, 14i/54). This is Scheuchzer's autograph, registered as read on 28 January 1708 (= 1707 old style). A fair transcription in an unknown hand, containing minor textual variations, is held in the Register Book Original, 9/68.

3 While this term can be perceived as anachronistic, we shall see that Scheuchzer had a strong sense of how climate, geography and animated life (plants, animals, humans) constitute an integrated and interdependent system. As this article will show, Scheuchzer's understanding of nature's interconnectedness largely derived from his physico-theological views.

4 For an English translation of Scheuchzer's manuscript, see below, Appendix (this quotation at §20). For the Latin original, see Barton and Miglietti, 'An

Eighteenth-Century Thought Experiment', 150. Scheuchzer also describes climate warming in terms of *caloris maior gradus* ('greater degree of heat') and *maior calidi ignei portio* ('greater portion of fiery heat'): see e.g., *De portione*, §22, §24 (p. 150).

5 Barton and Miglietti, 'An Eighteenth-Century Thought Experiment'. More on 'physico-theology' below.

6 Sara Miglietti, 'Environmental Ethics for a Fallen World: Johann Jakob Scheuchzer (1672–1733) and the Boundaries of Human Agency', *Earth Sciences History*, 39, 2 (2020), 447–73.

7 What is now the third book of *Natural Questions* was probably the opening book in Seneca's original ordering. See Piergiorgio Parroni, *Seneca: ricerche sulla natura* (Milan: Mondadori, 2002), pp. xlix–l.

8 For most Stoics, the theory of *ekpyrosis* was actually far more important than that of *kataklysmos*: see Jaap Mansfeld, 'Providence and the Destruction of the Universe in Early Stoic Thought', in *Studies in Hellenistic Religions*, ed. by Maarten Jozef Vermaseren (Leiden: Brill, 1979), pp. 129–88; Francesca Romana Berno, 'Apocalypses and the Sage: Different Endings of the World in Seneca', *Gerión*, 37, 1 (2019), 75–95 (82). Seneca himself discusses *ekpyrosis* in *NQ* 3.30 and in the final section of his *Consolatio ad Marciam*. Indeed, what he describes in *NQ* 3.27–30 seems not to be the end of the Stoic world-cycle in terms of *ekpyrosis* or *kataklysmos*, but the 'fated day' that takes place within a world-cycle ending in the Stoic conflagration (see esp. Anthony A. Long, 'The Stoics on World-Conflagration and Everlasting Recurrence', *Southern Journal of Philosophy*, 23, Supplement (1985), 13–37). I'm thankful to Gareth Williams for bringing this point to my attention.

9 Gareth Williams, *The Cosmic Viewpoint: A Study of Seneca's* Natural Questions (Oxford: Oxford University Press, 2012), pp. 110–32; Rita Degl'Innocenti Pierini, *Tra Ovidio e Seneca* (Bologna: Patron, 1990), pp. 177–210. While this essay will focus on Scheuchzer's intertextuality with *NQ* 3 (partly because his knowledge of this passage can be demonstrated through specific quotations: see below, n. 12), the *De portione* also clearly engages with Seneca's own sources, particularly Ovid and (to a lesser extent) Lucretius and Virgil, all of which Scheuchzer likely knew directly as well as through Seneca. Examples of this complex textual layering will be offered here when relevant.

10 Seneca, *Natural Questions* [henceforth: *NQ*], 3.29.9, trans. by Thomas H. Corcoran, 2 vols (Cambridge, MA; London: Harvard University Press; Heinemann, 1971), I, p. 293.

11 *NQ* 3.29.5, p. 289. See Williams, *The Cosmic Viewpoint*, pp. 110–32.

12 See, e.g., *Beschreibung der Natur-Geschichten des Schweizerlands*, 3 vols (Zurich: In Verlegung des Authoris, 1706–1708), I (1706), p. 187 (citing *NQ* 6.1); *Beschreibung*, III (1708), p. 80 (citing *NQ* 3.25), p. 104 (citing *NQ* 4); *Physica sacra … iconibus aeneis illustrata*, vol. 1 (Augsburg; Ulm: Johannes Andreas Pfeffel, 1731), p. 45 (citing *NQ* 3.27–30).

13 Williams, *The Cosmic Viewpoint*.

14 Fabio Nanni and Daniele Pellacani, 'Per una rassegna sulla fortuna delle *Naturales quaestiones*', in *Seneca e le scienze naturali*, ed. by Marco Beretta, Francesco Citti and Lucia Pasetti (Florence: Olschki, 2012), pp. 161–252.

15 Nanni and Pellacani, 'Per una rassegna', p. 172 (on Hélinand of Froidmont), p. 185 (on Niccolò Perotti), etc. See also (not included in their bibliography): Maria Susana Seguin, *Science et religion dans la pensée française du XVIIIe siècle: le mythe du deluge universel* (Paris: Champion, 2001), esp. pp. 44–47;

Jan Papy, 'Libertus Fromondus' Commentary on Seneca's *Naturales quaestiones*', *Lias: Journal of Early Modern Intellectual Culture and Its Sources*, 41, 1 (2014), 33–51.

16 On the importance of this text for early modern physico-theology, see Kerry V. Magruder, 'Thomas Burnet, Biblical Idiom, and Seventeenth-Century Theories of the Earth', in *Nature and Scripture in the Abrahamic Religions: Up to 1700*, ed. by Jitse M. van der Meer and Scott Mandelbrote, 2 vols (Leiden; Boston: Brill, 2008), II, pp. 451–90.

17 See Jill Kraye, 'Moral Philosophy', in *The Cambridge History of Renaissance Philosophy*, ed. by Charles B. Schmitt and Quentin Skinner (Cambridge: Cambridge University Press, 1988), pp. 303–86 (367–74). While this early modern revival focused on Seneca's *moral* rather than *natural* philosophy, speaking (as has occasionally been done) of an early modern 'marginalization' of Seneca's *NQ* seems too strong, as there is abundant evidence of *NQ*'s continuing influence in this period – for instance through its use as a textbook in seventeenth-century English universities (Yushi Ito, 'Hooke's Cyclic Theory of the Earth in the Context of Seventeenth-Century England', *The British Journal for the History of Science*, 21, 3 (1988), 295–314 [298n18]). This also means that Scheuchzer could have fully expected his readers to pick up the Senecan inspiration of his piece.

18 Nanni and Pellacani, 'Per una rassegna', pp. 206, 219–20.

19 Hiro Hirai, 'Seneca's *Naturales quaestiones* in Justus Lipsius' *Physiologia Stoicorum*: The World-Soul, Providence and Eschatology', in *Seneca e le scienze naturali*, pp. 119–41 (esp. pp. 129–30). See also *Justus Lipsius and Natural Philosophy*, ed. by Hiro Hirai and Jan Papy (Brussels: Koninklijke Vlaamse Academie van Belgie voor Wetenshappen en Kunsten, 2011).

20 The unrepeatable character of Noah's Flood is based on God's promise in Genesis 8:21 not to 'ever again strike down every living creature as I have done', which preludes to his subsequent covenant with Noah (Genesis 9:1–17). Scheuchzer comments on this passage in *Physica sacra*, vol. 1, pp. 59–60. The question of whether a similar catastrophic event could in fact occur again on a planetary scale began to be raised in the sixteenth century, partly in connection to astrological beliefs and millenarian expectations (see n. 19): see Lydia Barnett, *After the Flood: Imagining the Global Environment in Early Modern Europe* (Baltimore: Johns Hopkins University, 2019), pp. 27–31.

21 In this sense another important background for Scheuchzer's manuscript is the tradition (particularly strong in the German territories) of religious/astrological prophecies of catastrophic flooding events, circulated for instance via sermons, broadsheets, and cheap print. This type of literature often mixed classical disaster myths with Christian eschatological imagery, as noted for instance by Heike Talkenberger with respect to the 'Great Conjunction' of 1524 (*Sintflut: Prophetie und Zeitgeschehen in Texten und Holzschnitten astrologischer Flugschriften 1488–1528* (Tübingen: Max Niemeyer Verlag, 1990)). See also 'Astrologi hallucinati': Stars and the End of the World in Luther's Time*, ed. by Paola Zambelli (Berlin; New York: Walter de Gruyter, 1986).

22 Williams, *The Cosmic Viewpoint*, p. 113. As will become clear later in this essay, this 'cosmic viewpoint' does not merely consist in writing the disaster from the perspective of a detached onlooker (the classic trope of the 'shipwreck with spectator' studied by Hans Blumenberg) or from a divine vantage point (as in Genesis 7). It entails a radical shift from an immanent, fragmented perception of the natural world (typical of the human experience) to a

trascendent, holistic understanding of nature as an interconnected whole. This shift, very prominent in Seneca, is also recognisable in Scheuchzer's manuscript (see below, 'The cosmic viewpoint').

23 *NQ* 3.26.8 (p. 271). Interestingly, these are all 'anthropic' elements from which the sea is cleansing itself (I owe this point to Sandro La Barbera).

24 *NQ* 3.27.1 (p. 271), translation modified. Cf. Ovid, *Met.* 1.256 ff.

25 *NQ* 3.29.2–3 (p. 289).

26 See e.g., *NQ* 3.28.7 ('At the time of the deluge the tide, freed from its laws, advances without limit', p. 285); 3.29.8 ('winter will hold strange months, summer will be prohibited, and all the stars that dry up the earth will have their heat repressed and will cease', p. 291).

27 *NQ* 3.28.7 (pp. 285–87), my emphasis.

28 *NQ* 3.29.5 (p. 289).

29 *NQ* 3.30.7–8 (p. 297).

30 For an overview of this debate, see Franz Peter Waiblinger, *Senecas Naturales Quaestiones: griechische Wissenschaft und römische Form* (Munich: Beck, 1977), pp. 38–53. The tension between fate and moral responsibility in Seneca's (and more generally Stoic) thought is a leitmotiv in modern criticism. See, among others, Aldo Setaioli, 'Ethics III: Free Will and Autonomy', in *Brill's Companion to Seneca: Philosopher and Dramatist*, ed. by Gregor Damschen and Andreas Heil (Leiden; Boston: Brill, 2014), pp. 277–99.

31 On the tight alliance between physics and ethics in Seneca's thought, see Gareth Williams, '*Naturales quaestiones*', in *Brill's Companion to Seneca*, pp. 181–190 (188, with specific reference to *NQ* 3.27–30); Francesca Romana Berno, 'Exploring Appearances: Seneca's Scientific Works', in *The Cambridge Companion to Seneca*, ed. by Shadi Bartsch and Alessandro Schiesaro (Cambridge: Cambridge University Press, 2015), pp. 82–92.

32 *NQ* 3.27.2 (p. 273).

33 *NQ* 3.27.2 (p. 273).

34 *NQ* 3.27.7 (pp. 275–77).

35 *NQ* 3.27.5–6 (p. 275).

36 *NQ* 3.27.2 (p. 273).

37 *NQ* 3.27.4 (p. 273).

38 *NQ* 3.27.4–7 (pp. 273–75), my emphasis. The final line (27.4.7) contains a difficult textual problem, deriving from all the MSS transmitting the imperspicuous reading *in materia magna gentium clarus (h)onustus* (or *(h)onustusque) effunditur* (or *diffunditur* or *defunditur*). Corcoran's edition (on which the translation given here is based) reads: *novissime [in] materia magna gentium onustus diffunditur*. Parroni reads with Shackleton Bailey's conjecture: *novissime in maria magna gentium clade onustus effunditur* (*Seneca: ricerche sulla natura*, p. 228). Lipsius' edition, which was possibly the one read by Scheuchzer, gives this line as: *novissime ruina magna gentium clarus onustusque* (sc. *torrens*) *diffunditur* (in *L. Annaei Senecae Operum Tomus Secundus* (Amsterdam: Elsevier, 1672), p. 732). Whichever reading one follows, the overall meaning remains roughly the same.

39 *NQ* 3.27.8–9 (p. 277).

40 *NQ* 3.27.10–12 (pp. 277–79). For a similar crescendo, see Ovid, *Met.* 1.262 ff., culminating in the famous *omnia pontus erant* (1.292).

41 *NQ* 3.27.12–15 (pp. 279–81).

42 *NQ* 3.28.2 (p. 282).

43 *NQ* 3.27.2–3 (p. 273), translation modified.

44 This expression is used by Gregory Hutchinson to describe the relentless pace of Seneca's narrative in *NQ* 3.27–30 (cited in Williams, *The Cosmic Viewpoint*, p. 113).

45 Williams, *The Cosmic Viewpoint*, p. 113.

46 For an analysis of Scheuchzer's 'philosophy of limits' (as expounded in his later *Physica sacra*), see Miglietti, 'Environmental Ethics'.

47 Pierre Hadot, *Philosophy as a Way of Life: Spiritual Exercises from Socrates to Foucault*, trans. by Michael Chase (London: Wiley-Blackwell, 1995), p. 252. See also Gareth Williams, '*Naturales quaestiones*', p. 187.

48 Hadot, *Philosophy as a Way of Life*, pp. 238–50. The 'view from above' is a characteristic detachment technique pioneered by the Stoics and subsequently employed by philosophers of all orientations: two classic examples that would have likely been familiar to Scheuchzer are Cicero's *Somnium Scipionis* and Boethius's *De consolatione philosophiae*.

49 Miglietti, 'Environmental Ethics', 465. On the 'Little Ice Age' (a period of climatic cooling, peaking between the sixteenth and early eighteenth centuries, that deeply affected societies and economies both in Europe and worldwide), see Jean M. Grove, *The Little Ice Age* (London: Routledge, 1988). See William Barton's article in this collection on William Derham's account of the 'great frost' of 1708/1709.

50 This close intertwining of faith, science, and geopolitics also characterises other examples of 'global environmental imagination' in the early modern Republic of Letters: see Barnett, *After the Flood*, pp. 11–5.

51 For more on this point, see Miglietti, 'Environmental Ethics', 449–53. Assessing Scheuchzer's theological views is a difficult task, partly because of censorial pressures on his published work; his relationship with the Reformed church in Zurich was in any case far from straightforward: see Simona Boscani Leoni, 'Men of Exchange: Creation and Circulation of Knowledge in the Swiss Republics of the Eighteenth Century', in *Scholars in Action: The Practice of Knowledge and the Figure of the Savant in the 18th Century*, ed. by André Holenstein, Hubert Steinke, and Martin Stuber, 2 vols (Leiden; Boston: Brill, 2013), II, pp. 503–533. Comparing Scheuchzer's published and unpublished works would be a fruitful starting point to evaluate his stance on key tenets of Reformed theology.

52 It cannot be ruled out that the manuscript was initially conceived for a different and more local audience. Scheuchzer himself refers to 'both natives and foreigners' at §4, and the 'foolish wishes' for a warmer climate seems to refer primarily to local residents who had a direct stake in the issue (see discussion above). That said, Scheuchzer's very decision to send the manuscript to the Royal Society is proof that he saw its broader relevance for conversations that were taking place there as well as in the wider Republic of Letters.

53 Ann Blair and Kaspar von Greyerz, 'Introduction', in *Physico-theology: Religion and Science in Europe, 1650–1750*, ed. by Ann Blair and Kaspar von Greyerz (Baltimore: Johns Hopkins University, 2020), pp. 1–20.

54 Blair and von Greyerz, 'Introduction', p. 7.

55 John Ray, *The Wisdom of God Manifested in the Works of the Creation* (London: Printed for D. Williams in Fleet-Street, 1762), pp. 108–111. Ray's work was first printed in 1691; Scheuchzer knew at the very least the French translation of 1714, which he mentions in the 'Catalogus auctorum' at the beginning of his *Physica sacra* (vol. 1, sig. g2r). While this is too late for the *De portione*, it is likely that Scheuchzer also knew the English original, just as he

knew earlier works by Ray such as *Three Physico-Theological Discourses*, published in 1693 and also included in the 'Catalogus auctorum'.

56 Miglietti, 'Environmental Ethics', 455–69.

57 Miglietti, 'Environmental Ethics', 468–71. Unlike Scheuchzer's paper, Derham's account of the 'great frost' of 1708/1709 did make it into the *Philosophical Transactions*. The Society was far from uninterested in weather phenomena and perceived climate changes; weather records from different European locations – including those of Derham's from Upminster – were regularly published in the early 1700s, and in the 1720s another Fellow, James Jurin, developed the project of a collaborative 'Aëris historia ex undique collectis meteorologicis observationibus' for which he sought help from scientists all over Europe (see London, Royal Society, Classified Papers, 5/27). It was therefore not by a lack of interest in the topic that the Society chose not to publish Scheuchzer's paper.

58 NQ 3.27.1 (pp. 271–73); 3.29.4 (p. 289). On the transmutation of earth into water, see Francesca Romana Berno, 'Non solo acqua. Elementi per un diluvio universale nel terzo libro delle *Naturales quaestiones*', in *Seneca e le scienze naturali*, pp. 49–68.

59 NQ 3.27.7 ('After the clouds have massed more and more, and the accumulated snows of centuries have melted...', p. 275).

60 As recalled at the beginning of this essay, the *De portione* belongs in a series of papers on Swiss natural history that Scheuchzer sent to the Royal Society between 1703 and 1708. Around the same years Scheuchzer was also working on his *Beschreibung der Natur-Geschichten des Schweizerlands* (which appeared in three volumes between 1706 and 1708 and which contains some material that was later reused in his Royal Society papers, including *De portione*); and on his *Itinera alpina*, which includes detailed descriptions of the Alpine glaciers and which was published in London in 1708 with funding and sponsorship from the Royal Society.

61 On Scheuchzer's strong sense of national identity and its impact on his scientific thought, see the essays in *Wissenschaft – Berge – Ideologien. Johann Jakob Scheuchzer (1672–1733) und die frühneuzeitliche Naturforschung*, ed. by Simona Boscani Leoni (Basel: Schwabe, 2010), particularly those by Ezio Vaccari, Thomas Maissen and Guy P. Marchal.

62 Williams, *The Cosmic Viewpoint*, p. 114, my emphasis.

63 See Bernard C. Schär, 'On the Tropical Origins of the Alps: Science and the Colonial Imagination of Switzerland, 1700–1900', in *Colonial Switzerland: Rethinking Colonialism from the Margins*, ed. by Patricia Purtschert and Harald Fischer-Tiné (Basingstoke: Palgrave Macmillan, 2015), pp. 29–49.

64 Barnett, *After the Flood*, p. 17.

65 Indeed, another hint at mining occurs towards the end of the flood narrative (3.30.3, p. 295). It is worth noting that the focus of Seneca's critique is generally more on mining's nefarious moral connotations – particularly its connection to greed and luxury – than on its negative impact on nature per se.

66 Williams, *The Cosmic Viewpoint*, p. 113.

67 Mansfeld, 'Providence and the Destruction', p. 144.

68 Michael Kempe, 'Noah's Flood: The Genesis Story and Natural Disasters in Early Modern Times', *Environment and History*, 9, 2 (2003), 151–71 (165–66).

69 As I have explained elsewhere, one major point of disagreement between Scheuchzer and other physico-theologians like Burnet or Woodward was whether the postdiluvian earth still reflected the original perfection of God's

creation or whether the Flood had disfigured it to such an extent that artificial intervention on it could be presented as legitimate, indeed indispensable, restorative work. See Miglietti, 'Environmental Ethics', pp. 455–466.

70 This is a point that Scheuchzer will develop further in his *Physica sacra*: see Miglietti, 'Environmental Ethics'.

71 Scheuchzer does not venture here into the complex theological issue of the relationship between God's providence and foreknowledge and human freedom. One could argue that the imaginative form of the manuscript, written as *mythos* rather than *logos*, spares him from having to explain in detail how humans might have the power to disrupt God's providential plan to such a cosmic extent.

72 Magruder, 'Thomas Burnet'.

73 Kerry V. Magruder, 'Global Visions and the Establishment of Theories of the Earth', *Centaurus*, 48, 4 (2006), 234–57 (247–51).

74 For an overview of Burnet's reception in late seventeenth-century England (including among Royal Society Fellows), see Barnett, *After the Flood*, pp. 107–16.

75 Barnett, *After the Flood*, p. 107.

76 Scheuchzer, *Physica sacra*, vol. 1, p. 42 (ad Gen. 7, 8–9).

77 Scheuchzer, *Physica sacra*, vol. 1, p. 44 (Tab. 44, ad Gen. 7, 11).

78 Scheuchzer, *Physica sacra*, vol. 1, pp. 44–46 (Tab. 44, ad Gen. 7, 11; Tab. 45, ad Gen. 7, 17–20).

79 Scheuchzer, *Physica sacra*, vol. 1, p. 45 ('graphice Catastrophen Diluvianam describit Philo p. 335 et futuram quondam inundationem Seneca *Nat. Quaest.* III. 27. 30 qui videantur omnino').

80 The English translation of Scheuchzer's *De portione* is reproduced here by courtesy of the editorial board of *Lias*. For the Latin original of this text, see Barton and Miglietti, 'An Eighteenth-Century Thought Experiment'.

5 Communicating research on the Great Frost in the republic of letters

From Halle to London

William M. Barton

Introduction

The winter of 1708/1709 has long been acknowledged as having been a period of extraordinary cold in Europe. Contemporaries asserted repeatedly that 'the Great Frost' (*Der Jahrtausend Winter; Le Grand Hiver*) – as it came to be known – was the coldest winter they had ever known.[1] Modern climatologists have confirmed that 1708/1709 indeed exhibited the lowest temperatures on record in the last 500 years.[2] This period had extensive and often disastrous results for people, land and animals across the continent: the snow that fell in Britain did not begin to thaw until well into the following spring; France saw widespread crop failure and suffered an enduring famine; across Germany and Eastern Europe livestock succumbed to the cold in their stables; and the Baltic froze so deeply that people could make the journey over the sound between Zealand and Scania on foot.[3] Whilst science today places the winter of 1708/09 within the Little Ice Age (ca. 1450–1850) and then within these centuries of colder temperatures again inside the period of low sunspot activity known as the Maunder Minimum (1645–1715), a precise cause for the extraordinary bitterness of the Great Frost has yet to be identified. Unsurprisingly, the mystery behind this harsh winter spurred intensive inquiry amongst contemporary philosophers, who rigorously measured, recorded and deliberated upon the conditions they were experiencing.

Among the myriad early modern texts that preserve records and analysis of the Frost, historical climatologists have highlighted two sources of particular value for their contribution to our understanding of the phenomenon. Of these, the most prominent has been William Derham's (1657–1735) paper for the Royal Society's *Philosophical Transactions*, 'The history of the great frost in the last winter 1703 and 1708/09'.[4] The English ecclesiastic, perhaps best known today for his remarkably precise measurement of the speed of sound in 1709, similarly published a wider selection of meteorological reports in the years before and after 1708/09.[5]

DOI: 10.4324/9781003029823-7

Derham's meticulous daily records, his descriptions of conditions, methodological reflections on the instruments he used and his theorising about the causes of the cold have led climatologists to express their hopes for the discovery of further data from Derham in particular.[6] As this contribution demonstrates, however, Derham's 'History of the great frost' relied heavily on a Latin dissertation defended by Georg Remus (1687/ 1688–1756) under Christian Wolff (1679–1754) at the University of Halle, Germany in the same year. Remus' and Wolff's *Consideratio physico-mathematica hiemis proxime praeterlapsae*, 'A Mathematical and Physical Study of Last Winter' at 59 pages of dense Latin text and tables, ranks easily amongst the most sustained accounts of winter 1708/09.[7] Already in 1964, Walter Lenke hailed this text as the most detailed surviving interpretation of the phenomenon, and while the dissertation's temperature data is of questionable value today, Remus' and Wolff's rich descriptions of conditions throughout the winter represent a singular source for our knowledge of events.[8] Several other contemporary records – above all perhaps the *Observations météorologiques depuis l'année 1665 jusqu'à l'année 1709 inclusivement* of Louis Morin (1635–1715)[9] – have also been acknowledged as important sources for the history of the late-seventeenth and early eighteenth-century climate in general. However, Remus'/Wolff's and Derham's special attention to the winter of 1708/09 itself and their thorough, contemporary descriptions of the season's conditions clearly make them privileged sources for the Great Frost.

The work of extracting and converting the climatological data from this body of texts on the Great Frost in order to gain access to a long-term perspective on the changes in European climate has already been completed by climatologists.[10] This chapter now focuses on the place of Remus and Wolff's *Consideratio physico-mathematica* in communicating information about the season's conditions among the early scientific community, and the precise form of this transfer of knowledge. In a first step, I will provide a brief description and overview of Remus and Wolff's dissertation, considering in particular the context of its production and the work's distinguishing features. A second section will consider how the *Consideratio*, already recognised as one of the earliest and most extensive reports on the event,[11] acted as a pivotal text for the transmission of research on the Great Frost more generally and its place in wider contemporary writing on the event. The third and final step of the present piece will be a closer study of this dissertation's influence on William Derham's 'History of the Great Frost'. In doing so, I intend to shed fresh light on the close relationship between the two crucial sources of historical evidence for the winter of 1708/09.

This close reading of Remus and Wolff's *Consideratio physico-mathematica* and the subsequent analysis of its influence aims to go some way to answering the call for reflection on the '"intermedial" [multimedia] connections' which formed the basis for the transmission of knowledge in early modern intellectual culture, voiced most recently by Marian Füsserl in 2016.[12] The place here of a Latin dissertation – a genre of early scientific literature largely overlooked in earlier approaches to the field – defended by a forgotten name under a very well-known figure of Enlightenment science is of special interest for its potential to corroborate the increasing awareness among scholars of the precise character of these works and their often surprising intellectual reach. In thus selecting a disputation piece as the focus of the present case study, I hope to contribute to wider discussion over the genre and its place in scholarly communication within the *res publica literaria* of eighteenth-century Europe.[13]

Remus' and Wolff's *Consideratio physico-mathematica hiemis proxime praeterlapsae*

The *Consideratio physico-mathematica* was published as a 59-page booklet in the form of a dissertation following the typical format of numerous German universities of its time. The detailed information on the work's title page accordingly gives, in descending order of importance, the name of the rector at the University of Halle (Philipp Wilhelm von Brandenburg-Schwedt), the name of the *praeses* under whom the dissertation was defended (Christian Wolff, then Professor *ordinarius* of Mathematics), the date and time of the event (the morning of the 13th June 1709) and the name of the respondent (Georg Remus from Gdánsk): *Consideratio physico-mathematica hiemis proxime praeterlapsae quam rectore magnicientissimo serenissimo principe ac domino Dn. Philippo Wilhelmo, principe Borussiae, marchione Brandenburgico, caetera, praeside Christiano Wolfio, mathematum Professore Publico Ordinario, ad diem XIII Junii anno MDCCIX. horis antemeridianis in auditorio majori publico eruditorum examini submittet Georgius Remus, Gedansis.*

As well-recognised figures of early eighteenth-century German political and intellectual life, the university's rector and the disputation's *praeses* require little introduction. About Georg Remus, however, the day's *respondens*, rather less is known. In addition to the information about his origins in Gdánsk and his studentship at the University of Halle stated on the dissertation's title page, among the earliest and most detailed account of Remus' career appeared in Richter's *Geschichte der Medicin in Russland*.[14] Of special relevance for our present interests is the information that the *Consideratio* was the topic Remus' disputation for his doctorate.[15] We can thus confirm that the *Consideratio physico-mathematica* was the result of a *disputatio pro gradu* and represented the inaugural dissertation for Remus' inception of his degree. Remus was the first of Wolff's doctoral students in the course of the philosopher's long career.[16]

The dissertation is accompanied by a preface written by the respondent and a congratulatory letter from Wolff.[17] Among the expressions of modesty, thanks and awareness of the work's limitations on the one hand, praise and well wishes on the other, a number of passages in these two paratexts bring us to the question of the dissertation's authorship. It is now common knowledge that inaugural dissertations of this type were by and large written by the *praeses* in German universities until well into the eighteenth century and beyond.[18] The respondent's task at the disputation event was to demonstrate his knowledge by defending the dissertation against *opponentes,* who tried to find holes in the work's argument or the candidate's understanding of the field. Over the course of the eighteenth century, the number of respondents who also wrote the dissertation that they defended (the system we recognise at universities today) gradually grew, and the *respondens* might occasionally be indicated as *auctor* on the dissertation's title page.[19] This is not the case on the title page of our *Consideratio,* but several features of the work suggest that Remus was probably the primary author of the text. The first appears in the work's *praefatio,* where the author's voice is that of the student about to graduate. On the first page of the work, he describes his choice of topic for the dissertation and explains that he has chosen to combine his studies in physics, mathematics and medicine to ensure the usefulness of his text for the understanding of the severe winter.[20] In the main text of the dissertation, the first-person voice of the author then references the work of Wolff in the third person – *dux Dominus Praeses* – throughout, and even occasionally addresses smaller differences in approach between the student and the professor.[21] In his letter to the respondent appended to the text, Wolff then confirms his student's (apparently) independent work on the dissertation in writing that Remus was able to defend the piece alone (*solus defendere poteras*) and that it will stand as a testament to his erudition and virtue for the outside world.[22]

After the work's title page and the author's preface discussed above, the body of the dissertation begins. The text is split into two larger sections, the first (pp. 1–29) offers a descriptive account of the winter of 1708/09 before the second (pp. 29–51) examines the causes of the season's extraordinary cold. Each of these sections is further divided into numbered paragraphs – in line with the widespread practice in German dissertations of the period – which deal with individual issues of the study. Section one of the *Consideratio* opens with a series of methodological and theoretical considerations: the author specifies what he means with *hiems,* 'winter' (p. 1); he discusses the limitations of temperature and wind measurements generally and the restraints of his instrumentation in particular (pp. 2–9).

In passages remarkably reminiscent of modern doctoral dissertations, the author also attempts to set his work against a background of scholarly writing on similar subjects (pp. 2, 4–6) and makes an explicit statement of his goals in the work. In the programmatic paragraph §7 of *Sectio I* (p. 5), he writes:

Neque enim nudam hiemis historiam tradere intendimus, sed phoeno-menorum quoque rationem superaddere constituimus: quod posterius fieri non posse, nisi modo enumerata satis considerentur ex subsequen-tibus patebit.

Nor do I aim to recount a plain history of the winter, rather I am resolved to include also a reasoning [explanation] of the phenomena: it will be clear from the following that the latter could not be done unless the things just mentioned be examined with sufficient care.

After these preliminary paragraphs, there next follows a description of the winter's conditions. It is here that we meet the first of the *Consideratio*'s primary distinguishing features, which allow us to trace the work's influence on later studies of the Frost: Remus/Wolff divide the winter into five distinct periods, each marked respectively by the freezing of water at its beginning and the following thaw at its end.[23] The subsequent chapters §15–§22 (pp. 9–20) describe in detail the weather conditions in each individual period with special attention to temperature readings from the thermometer (*thermoscopium*), precipitation in the form of rain or snow, cloud cover, frosts and daylight conditions, as well as the direction of the prevailing winds.

Remus'/Wolff's first *periodus* ran from 19 October 1708 to the beginning of November and was characterised by an extraordinary cold snap for the *Weinmonat*, which saw icicles develop on window frames in Halle and a good deal of rain.[24] In the second period, the month of November passed in relative comfort (*nihil extraordinarii observatum et frigus adeo tolerabile extitit*, p. 10) before strong winds, rain and snow in December brought another bout of cold temperatures, which were not however especially unusual for the time of year. At the turn of the new year, Remus/Wolff mark a pause (*intervallum*) in the succession of their periodisation. This short hiatus between 1 and 4 January 1709 was characterised by the arrival of south-westerly wind bringing heavy, warmer rain to central Germany. The temperatures thus briefly rose and December's ice and snow melted. The 5 January, however, brought an unexpected change:

Enimvero scena subito mutabatur et cum universae Europae admir-atione coepit periodus tertia, solito prorsus frigore notabilis.[25]

Indeed, the scene suddenly shifted and, to the astonishment of the whole of Europe, the third period began, utterly remarkable for its unusual cold.

This third period was marked above all by its low air temperatures and the ice that formed across the landscape. The weather conditions were changeable with a mixture of cloud, precipitation and clear skies, but it was the low thermometer readings which particularly interested our author. The remarkable conditions also brought him to record moments of reflection on the unusual features of the winter. Remus/Wolff comment, for example, on the particularly painful experience of exposing one's face to the cold and on the pleasant sight of the midday sun reflecting off ice crystals in the air.[26] Period four of the Frost began on 31 January and lasted until 9 February. Notable here was the persistent cloud cover, frequent snow and the very variable winds. Together these factors made for temperatures almost as cold as those experienced in January. The arrival of a westerly wind and calmer skies marked the end of this phase. The fifth and final *periodus* of the winter then began already on 19 February and brought once again a particularly deep and lasting cold. In this, the longest subdivision of the Frost and the most extensive individual chapter of his description, the author describes the dominant northerly winds and the resulting mixture of cloud, precipitation with only occasional clear skies. The fifth period ended on 17 March and whilst the temperatures rose considerably, our author describes further rain and cloudy conditions until 8 April. As Remus'/Wolff's description of the five periods ends and warmer temperatures return, a sense of relief and renewed appreciation of the natural surroundings are tangible in the text.[27]

After a chapter (§23) on the use of the Torricellian barometer tube and tables of observations made using the instrument, the *Consideratio*'s following chapters turn to the effects of the Frost. These describe the consequences of the season's extraordinary temperatures on liquids and bodies of water (§24), on plant life (§25) and on both animals and humans (§26). Remus'/Wolff's attention to the effects of the phenomenon they have described stands out as a second distinguishing feature of their *Consideratio* – one which will allow us to identify their work's specific influence on later writing about the Frost – and which allows us already to note the 'intermedial' (*multimedia*) references visible in the text.[28] Here, in his chapter on the effects of the frost on bodies of water, the author calls on the authority of his correspondence with the fellow of the Royal Society Johann Philipp Breyne (1680–1764), a fellow *gdańszczanin*, who reports that the city's gulf onto the Baltic Sea remained frozen solid well into April.[29] Similarly, in the following descriptions of the season's consequences for plant and animal life, the author references the reports of international newspapers and reviews for their accounts of the loss of life and crop failures.[30]

The second section of the *Consideratio* begins on page 29 and offers a study of the potential causes of the harsh winter. As in section one, the author begins with a series of methodological reflections on his approach. These include a statement of the limitations of the meteorological sciences (§1), a justification of his study for science (§2) and a short explanation of the causes of warmth and cold on the earth in general (§4). In the following chapters §5–§13 (pp. 31–36), the *Consideratio* moves to its first set of assessments of potential causes for the cold. Remus/Wolff deal with the effect of the earth's yearly aphelion and perihelion on winter temperatures in Europe, the idea that the earth's tilt may have changed (affecting the position of the sun and the warming effects of its rays on the earth's surface) and the effect of the sun's angle in respect of the earth on climate. Until this point, Remus/Wolff emphasise, the dissertation's explanations of meteorological and cosmological phenomena are in line with contemporary accepted opinion. In chapter §14 (pp. 36–38), Remus/Wolff go a step further and deal with the potential effect of sunspots – one of the previous century's 'hot topics' among cosmologists[31] – on the sun's warmth. Remus/Wolff accept here the idea that the sun's *maculae* can have a cooling effect, but they calculate its influence as so minor that they dismiss this theory for the winter of 1708/09.

Having discussed the general astronomy of winter, and having rejected the role of sunspots on the earth's temperature, Remus/Wolff go on to propose a series of more localised weather phenomena that might affect the temperatures in a given place. These chapters deal with the effect of clouds in blocking sunlight (§16, pp. 38–39), the effect of humid air on the solar rays (§17, pp. 39–40) and the role of the winds in contributing to cool temperatures (§19, pp. 40–41). In chapter §20, Remus/Wolff offer a concise summary of their explanation of the winter conditions. This summary acts as a basis for the following detailed discussion of the conditions in their five periods of 1708/09, and it will also be relevant later in this chapter. It is worth quoting Remus/Wolff at length on this point:

> *Ex hactenus dictis apparent quaenam ad frigus hibernum producendum concurrere possint. Nimirum parte solis requiritur ingens a vertice distantia et exigua supra horizonte mora, per §9 et 11; ex parte Telluris vero atmosphaera exhalarionibus plena et nubibus gravida per §16 et 17, ventique orientales et septentrionales, praesertim impetuosi per §19 requiruntur. Omnium autem maxime necessarius, ut actiones solis et diu, et tum inprimis impediantur, quando causae frigoris concurrunt.*[32]

From the above-mentioned, the factors which can come together to bring about the cold in winter become clear. Obviously, for the sun's part, a great distance from the pole and a short period above the horizon, as per chapters 9 and 11, are needed. From the earth's side, on the other hand, an atmosphere full of moisture and heavy with clouds, as per chapters 16 and 17, and also easterly and northerly winds, especially fierce ones, as per chapter 19, are required. The most necessary thing of all, however, is that the power of the sun be impeded both long- and short-term when these causes of the cold come together.

After a further methodological reflection on their thermometer's results (§21), Remus/Wolff offer in the following chapters their explanation of the winter's extraordinary cold. This explanation builds straightforwardly on the standard conditions of winter in the northern hemisphere described in §§5–14 (shorter days, the earth's tilt on its axis and the earth's variable distance to the sun between its ap- and perihelion) and then the relevant local weather conditions dealt with in §§16–19 (particularly cloud cover, atmospheric humidity and wind conditions). In this way, in chapter §22 (pp. 42–43), Remus/Wolff describe the particular factors contributing to their first period's cold weather (a very wet summer of 1708 and then widespread cloud cover in the autumn) and then explain that the same manner of reasoning can be applied to their other four periods (§23, p. 43). They next detail particularly noteworthy features and causes of the first, third and fifth periods in particular (§§24, 27 and 28). Representative of their miscellaneous approach throughout these chapters of explanation is the first part of their treatment of January's biting cold, for example:

Frigus extrordinarium cum mense Januario fere coepit. Erat vero tum maxima Solis a vertice distantia et exigua eius supra Horizonte mora, quod nemini dubium. Pauculum caloris quod in Tellure et atmosphaera a Zephyro (W.), Noto (S.) et Euronoto (S.O.) atq[ue] exigua Solis per spissas nubes et aerem exhalationibus plenam actione superesse poterat, ventus suo impetu die quarta Januarii maxima ex parte extinxerat.[33]

The extraordinary cold began roughly with the month of January. The sun was then at its greatest distance from the pole, and its time above the horizon was at its shortest. This no one doubts. The minimal warmth that was able to endure on the earth and in the atmosphere after the westerly, southerly and south-easterly winds and the minimal effect of the sun through the dense clouds and moist air was removed almost completely by the force of the wind on 4 January.

Having expanded their explanatory approach to the winter as they experienced it, our authors turn to their penultimate topic, namely, an explanation of the unusual effects of the Frost. Remus/Wolff split this theme between two chapters: the first dealing with the bodies of water (§31, pp. 48–49), the second with the effects of the cold on plants, animals and humans (§32, pp. 49–50). In §31, Remus/Wolff call on the evidence of the waterbodies which froze during the Frost to show that the winter's cold in 1708/09 was indeed deeper than usual. Here they cite a series of well-known local examples where common experience held springs and lakes for warmer than others, and they underline that these also froze. For the question of the winter's effect on plants, animal and human bodies (§32), Remus/Wolff cite, for instance, the early modern period's anatomical discoveries and the art of the microscope to reveal that all living organisms are dependent on veins, or 'tubes', *tubuli*, for the transmission of vital fluids throughout their bodies. Should these tubes, or the liquids they carry, be adversely affected by the cold, or even freeze, the resulting damage to the whole organism is only to be expected. The dissertation's final chapter (§33, pp. 50–51) deals briefly with a remarkable storm in May 1709, which Remus/Wolff connect to the unusual winter that preceded it.

This close reading of the most elaborate and earliest contemporary reports on the Frost of 1708/09 reveals several significant features of the work's structure and approach. As we have seen, the first of these features is Remus'/Wolff's division of the winter into five distinct periods of unusual cold. Both the *Consideratio*'s descriptive part one and its second analytical section follow this structural framework. The second noteworthy characteristic of the piece is the place it gives to description and explanation of the Frost's effects. The final two features we can usefully underline are the *Consideratio*'s rejection of the changes in sunspot activity as a cause for the cold, and their explanation of the harsh conditions in terrestrial and atmospheric weather conditions. As we now turn to a wider selection of literary responses to the Frost, these features will function as dots on the map of the *Consideratio*'s diffusion and reception.

The wider contemporary reception of the *Consideratio*

This section of the present chapter will consider several groups of textual genres and forms where the reception of the *Consideratio* allows for closer study of the dissertation's communication of knowledge. Whilst not the first, temporally, the most obvious and most direct moment of the *Consideratio*'s later 'intermedial' influence was its inclusion within later collections of works by the disputation event's *praeses*, Wolff. Although the evidence for Remus' authorship of the piece discussed above seemed straightforward,[34] the work's subsequent appearance among Wolff's *œuvre* offers a significant perspective on the *dissertatio*'s role in transferring information about the Frost.

The Latin dissertation was first reprinted in 1727 as an annex to the *Melemata* of Ludwig Thümmig (1697–1728), a student and follower of Wolff who would later become professor of philosophy himself in Kassel.[35] This reprint was made as part of Thümmig's programmatic efforts to organise and disseminate Wolff's teachings, which included the earlier two-volume *Institutiones philosophiae Wolfianae* (1726–1727).[36] Next, a redacted German translation based on Thümmig's reprint later appeared in Wolff's *Gesammelte kleine philosophische Schriften* in 1736,[37] before the Latin text appeared once again in Wolff's own *Melemata mathematico-philosophica* in 1755.[38] As the translator's note to the 1736 German version of the work tells us, the text of the dissertation had already been altered by the time of its first reprint in Latin in 1727:

> Herr Thümmig hat sie [die Untersuchung = *consideratio*] nachmals als einen Anhang an seine *Melemata* drucken lassen [...] Es ist daselbst sowohl die vorgesetzte Vorrede weg gelassen, als auch mancher Ausdruck geändert worden, der sich auf die damahligen Umstände, besonders hrn Rem bezogen, und ich habe ihm darinnen gefolget.[39]

> Mr Thümmig later had it [the study = *consideratio*] printed as an appendix to his *Melemata*. There, both the preceeding prefatory speech is left out and some expressions changed with respect to the contemporary circumstances (of the text), particularly those related to Mr Remus. In this, I have followed him [Mr Thümmig].

A closer look at the later Latin and German texts reveals the precise character of these changes. The passages, which in the original print of 1709 saw the authorial first person (Remus?) refer to the work and ideas of the *praeses* in the third person, are now edited so as to come from a new author's voice, aligned with Wolff. Where Remus had referenced his teacher's work in the original, for example, *Ostendit Dn. Praeses in suis* Aerometriae Elementis, the phrase is now expressed in the first person, *Ostendi in* Aerometriae Elementis, and 'Ich habe in den *Gründen der Wett-Meß-Kunst* [...] gewiesen'.[40] The passage cited above (see note 20) on the occasionally varying methods of observation between Remus and Wolff is now printed as:

> *Hactenus ex diario meteorologico nostro excerpsimus, quae Halae annotata sunt. Cl. Respondens iter per Misniam tum facie[n]s campos nive aliquot pedes alta tectos...conspexit.*[41]

Bisher habe ich aus meinem über die Witterung gehaltenen Tage-Buch ausgezogen, was in Halle bemerket wurde. Der herr Vertheidiger dieser Schrift, welcher dazumahl eine Reise durch Meisen that, hat die Felder einige Fuß hoch mit Schnee bedeket [...] sehen.[42]

So far, I have been excerpting from my weather diary on the things recorded from Halle. The respondent saw fields covered in several feet of snow as he was making a trip through Meisen at the time.

The question over the authorship of a single university dissertation, produced in a period when collaborative writing and collective composition were the norm throughout the European academy and the modern emphasis on individual authorship far less pronounced,[43] is admittedly less urgent than this discussion would make it appear. After all, the concepts discussed during a disputation event in early modern Europe were customarily considered to be those of the *praeses*, regardless of who actually penned the *dissertatio* which appeared later. Dissertations were thus systematically attributed to the professor irrespective of authorship throughout the period. That the *Consideratio* was attributed to Wolff is, in this respect, simply standard for the period.[44] However, the extensive changes to the document in later editions of the work are unusual. That Thümmig's reprint contained these changes as part of his commitment to the promotion of Wolff's ideas (and the associated commendation of his teacher) already indicates the significance of Wolff's new prominence as the authorial voice in the text. The fact that a contemporary translator should comment at length on the modifications he found further emphasises the peculiar situation of our text. Indeed, a historical overview of German intellectual culture from 1813 underlined the importance of precisely the *Consideratio* for Wolff's early international renown and explicitly credits the dissertation, alongside his *Aerometriae elementa*, as the basis for his acceptance as a fellow of England's Royal Society.[45] The efforts made to have the text appear as truly the product of the Wolffian pen thus perhaps had more weight for the image of the man's career than the typical fluidity of authorial attribution among Germany's eighteenth-century dissertations.

Outside of Wolff's *œuvre*, the next obvious group of texts in which the *Consideratio*'s and its ideas were quickly taken up in two dissertations which appeared very soon after Remus'/Wolff's own.[46] The first of these was the *De tempestatum apparenter et vere extraordinariarum ac speciatim Frigoris quod hyeme superiori sensimus intensissimi causis dissertatio* published in 1710.[47] In this text, the student (specifically indicated as the author in the work's full title), Johannes Nicolaus Sybellius, opens by giving a standardised 12-page explanation of the seasons and temperature changes.[48] He then proceeds to more inconsistent potential causes of cold, which he lists as solar eclipses (§15, p. 13), comets (§16, pp. 13–15) and sunspot activity (§17, pp. 15–17). These, he writes, certainly produce colder weather conditions on Earth and may contribute to the prevailing circumstances in any given season, but they cannot be reliably connected to either the rhythm of the seasons or to particular cold snaps. For these, Sybellius argues, we must look to the *causae intermediae*, 'intermediate causes', which are the clouds, fogs, humidity, winds and rains (§18, pp. 17–18). Here, we already hear an echo of Remus'/Wolff's explanation of the cold.

Sybellius now continues with general meteorological explanations for his 'intermediate causes' (§§19–24, pp. 18–22), before turning to a theological argument which ultimately has God as the principal cause for conditions on earth (§§25–30, pp. 22–30). Only now in the dissertation's penultimate chapter (§35, pp. 30–35) do readers learn the causes for the previous winter advertised in the work's title. It comes as no surprise to find that the local weather conditions (the clouds, fog, humidity, winds and rains) are here proposed as principal reasons for the Frost by Sybellius, who now also cites Remus'/Wolff's *Consideratio* on the question of sunspots where his own conclusions are similarly ambiguous. The prominent place of our *Consideratio* in this explanatory chapter and the clear overlap of the two dissertations' results reveal the influence of Remus/Wolff on Sybellius' work.

The same is true of the *Disputatio physica de frigore anni MDCCIX memorabili* defended by Christian F. Rast under Heinrich von Sanden at the University of Königsberg (now Kaliningrad) in 1712.[49] Profiting from its publication now three years after the event with which it deals, Rast's/Sanden's *Disputatio* has as a defining characteristic its extensive engagement with earlier reports on the Frost of 1708/09 from wider Europe. After the introductory chapter §1 and §2, Rast/Sanden take the reader through accounts from France, Portugal, Spain, Italy, Switzerland, Britain, Belgium and the Netherlands, Hungary, Poland and the German-speaking area in a series of multilingual chapters (§§3–12, pp. 3–20). Indicative of the repetitions and commonalities characteristic of dissertation-writing in this period, there next follows a series of chapters on the effects of the Frost, which Rast/Sanden identified as particularly harsh in their contemporary Prussia.[50] These sections, as we saw in Remus'/Wolff's *Consideratio*, deal with the damage to human, animal and vegetable life, as well as frozen water-bodies, where the Baltic Sea is particularly prominent for this *Dispuatio* (§18, pp. 27–28).

Having worked through the descriptive parts of their account, Rast and Sanden next turn to their explanatory chapters. These begin with a consideration of cold in general (§25, pp. 31–32) and the effect of the different qualities of air (§26, pp. 32–33) before dealing with various sorts of precipitation and the effects of frozen liquids (§§27–33, pp. 33–40). There then follow a series of chapters where Rast's/Sanden's explanations of these phenomena are applied to the specific cases witnessed during the Frost of 1708/09, including the effect of the cold on man's blood circulation (§31, p. 46); fish in frozen water (§32, p. 47) and ruined plant life (§38, pp. 48–49). Hearing already the echoes of Remus/Wolff in this account, the reader is now ultimately – as already in Sybellius' study – directed to the *Consideratio* in the *Disputatio*'s final chapter (§43):

> *Quales meteorologicas observationes celeberrimus Mathematum Professor Halensis Wolfius in Dissertatione ann.* MDCCIX *habita, quae inscribitur* Consideratio Physico-Mathematica hyemis proxime praeterlapsae, *orbi erudito communicavit, in qua praeter observationes* Cizensis [...] atque Jenenses [...] et proprias concessit, illiusque in quinque hyemis periodos [...] ann. MDCCIX digessit, simulque harmoniam factarum observationum indicavit.*[51]

The very famous professor of mathematics at Halle, Wolff, imparted meteorological observations of this kind to scholarship a dissertation defended in 1709 entitled *Consideratio Physico-Mathematica hyemis*. In this work he recorded his own observations alongside those from Zeitz [...] and those from Jena [...], and divided 1709 into five periods of winter [...] and simultaneously revealed the agreement of the observations he made.

Having now traced the uptake of the *Consideratio* in the long tradition of Wolff's own *opera* and within dissertations on the topic of the Great Frost produced by contemporaries, we turn to a larger-scale monograph on cold winters in Germany. The *Observationes Meteorologicae* by the Bavarian polymath Johann Alexander Döderlein appeared in 1740.[52] The book makes a systematic study of the harsh winters of 1708/09 and 1740 with the aim of drawing conclusions from the comparison. Though rather different in form, approach and content to the works discussed thus far, Döderlein's *Observationes* demonstrate their dependence on the *Consideratio* in particular for their account of the conditions of 1708/09. The most obvious of these is Döderlein's reliance on Remus'/Wolff's division of the earlier winter into five *periodi* (Döderlein uses the Latin word throughout his German study) for his own account of 1708/09.[53] Döderlein then goes on to make his own description of the harsh winter of 1740 using similar, corresponding periods to structure his account.[54] That the five-period scheme was considered characteristic of the *Consideratio* already among contemporaries is underlined in the passage from Rast/Sanden cited above.[55] The indication of Döderlein's overwhelming reliance on *Consideratio* for his account of 1708/09 is the striking resemblance between his own summary of the winter season and that of the digest in Remus/Wolff cited above (see note 32). Döderlein's summary reads:

> Kurz endlich von den Ursachen eines ausserordentlichen kalten Winters und also des dißjährigen und die noch folgen könnten, zu sagen; so förderen vornehmlich eine aus nehmende Kälte, viele und kalte vorhergehende Regen, scharffe und lang anhaltende stürmische Schnee und Eiß, so die von dem Sommer übrige warme Dünste der Erden vor der Zeit dämpfen und consumiren; die Luft mit kalten Dünsten angefüllet; und sonderlich auch wann die liebe Sonne, besagter massen, an ihren Würkungen mehr als sonsten, gehindert wird.[56]

> To finally say something brief about the causes of an extraordinarily cold winter, and therefore also about that of this year and those that still might follow; they primarily promote an increasing cold, many cold and earlier rains, fierce and long-lasting stormy snow and ice, thus dampening and consuming before the appropriate time the warm fumes of the earth remaining from the summer; the air filled with cold vapors; and especially when the dear sun, as explained, is prevented from doing its work more than usual.

Moreover, though the *Consideratio* is never cited explicitly in the *Observationes* – nor would we expect necessarily it to be so in a piece of mid-eighteenth-century vernacular scholarship – Döderlein nonetheless confirms his thoroughgoing knowledge of Wolff's work in his frequent references to the professor's name.[57]

Derham's use of the *Consideratio*

Whilst attention to the *Consideratio* was, then, intensive within Germany in the years after its publication, its most substantial and widest-reaching response came in the form of the journal article by William Derham. 'The History of the Great Frost in the Last Winter 1703 and 1708/09' appeared in the Royal Society's *Philosophical Transactions* just months after the Remus'/Wolff's dissertation was printed. After reporting observations from England, Scotland and Ireland, Derham makes his dependence on the *Consideratio* explicit when he turns to the effects of the Frost in wider Europe:

> I have an Ingenious printed Account put into my Hands by the fore-mention'd Dr. Woodward. The Title of the Book is, *Consideratio Physico-mathematicae Hyemis proxime praeterlapsae [...]*. This Dissertation

relating directly to our Subject, and being I suppose in but few Hands with us, a short Account thereof may not be unacceptable.[58]

The following English account of the *Consideratio* extends to four pages. It follows Remus'/Wolff's five-period framework and reports on the details of temperature, wind and the extent of the frosts. Derham finishes the summary with the promise to reproduce the *Consideratio*'s barometrical data in a separate article.[59] The English author then turns to the effects of the Frost.

Here, though no longer reporting explicitly on Remus'/Wolff's dissertation, Derham nonetheless follows closely the approach we saw in the *Consideratio*'s first section. He deals first with the Frost's effect on fluids (*cf. Consideratio* I.§24), then on animals (*cf. Consideratio* I.§26) and finally plant life (*cf. Consideratio* I.§25).[60] In these sections, which make up the greater part of the article, Derham mixes his own observations of the effects he witnessed in London and the surrounding areas with reports he had received from friends and colleagues both within England and abroad. Here, readers are offered an insight on the transmission of information within the circles of the Royal Society, where the interchange of letters, the loan of research literature and materials between members, as well as privileged access to collections of objects were central to intellectual exchange.[61] We hear, for example, excerpts from a letter from then-editor of the *Philosophical Transactions*, John Lowthorp, on the frozen Thames (p. 463). Derham reproduces from a letter passed on to him by John Woodward a list of birds, which had succumbed to the cold on the Essex Marshes (pp. 465–66). And he reports on a letter from Jacob Bobart the Younger, superintendent of the Oxford 'Physic Garden', on the state of the plants (pp. 469–71). Dominant throughout these paragraphs on the effects of the Frost once again, however, is Remus'/Wolff's *Consideratio*, which Derham cites at length – both in Latin and English – on frozen waterbodies (p. 464), on harm to animals (pp. 466–67) and on damage to plant life (pp. 474–75). Derham owed his access to the *Consideratio*, as we learned in the passage quoted above (see note 58), to geologist and antiquarian John Woodward (1665–1728). Woodward had become a Fellow of the Royal Society in 1693, one year after having taken the position of Professor of Physic at Gresham College.[62] Woodward's numerous and productive connections to the German-speaking world, through which he conceivably gained initial access to the print of Remus/Wolff's *Consideratio*, is perhaps best represented in his relationship with Swiss natural philosopher Johann Jakob Scheuchzer (Zurich, 1672–1733): Scheuchzer translated Woodward's physico-theological *Essay toward a Natural History of the Earth* (1695) into Latin (1704) before the expanded Latin edition of 1714 appeared with Woodward's responses to comments by Elias Camerarius (Tübingen, 1673–1734).[63] Scheuchzer sent his son, Johann Caspar (1702–1729), to work with Woodward in London where the young man eventually also became a Fellow of the Royal Society.

Continuing its structural dependence on the *Consideratio*, the final section of the article now turns to the causes of the Frost. Here Derham also relies explicitly on Remus'/Wolff's explanation for the content of his study:

> Having dispatched the two things proposed, the Degree and Effect of the Frost I intended here to have put an end to my History: But upon a review of the forementioned Dissertation, I cannot easily forbear saying something to The Causes of this Great Frost. These are to me, I confess so very much hidden that upon that Account I intended wholly to have passed over this Matter; but the last commended Author having ingeniously enquired thereinto, I shall as briefly as may be shew his Opinion.[64]

After a paragraph-length digest of the *Consideratio*'s basic list of causes, Derham cites Remus'/Wolff's own summary chapter II.§20 (see note 20 above) which, as we have seen, would go on also to echo in Döderlein's later work.[65]

As Derham's study now reaches its final two pages, the reader finds, for the first time, a critical engagement with Remus'/Wolff's *Consideratio*, which has informed the article's form, approach and basic results thus far. Derham begins his short critical discussion of the dissertation with a statement of his agreement with its standardised account of the causes of the winter cold. He then turns to one specific point of contention and two suggestions for further research: Derham's point of disagreement with the *Consideratio* concerns the effect of atmospheric humidity on temperature. Derham naturally agrees that mist, cloud cover and 'dark Weather' hinder the warming effects of the sun, but emphasises that 'the warm Vapours of the Sea' have a considerable warming effect 'in our Island-places'. He backs this up by reminding readers that clear conditions bring colder temperatures at night, but he does not go so far as to explain the insulating effect of humid air, nor the differences between maritime and continental weather conditions more generally.[66] Turning to his suggestions for further enquiry, Derham immediately acknowledges the vague character of the two proposed causes: 'it may be expected I should subjoyn others [causes]. But as I have declared my Ignorance of them, little can be expected'. In a short paragraph, Derham then points to the potential effects of meteors on the earth's temperatures[67] and makes a plea for closer attention to the 'Dispositions of our Globe, at least to the greater or less Plenty of Vapours and Exhalations'.[68]

Whilst it would be anachronistic and quite unproductive to hold Derham's work to the requirements for innovative results and well-evidenced claims in modern research, it is nonetheless clear from our close reading that his article's primary contribution to the study of the Great Frost was its role in the transmission of Remus'/Wolff's account to a wider scholarly community. This was not primarily achieved through Derham's use of the vernacular (English was used by few outside the islands in the early eighteenth century, and Derham cites Remus/Wolff extensively in Latin throughout his piece), but rather a question of distribution: Derham himself notes the apparent scarcity of the *Consideratio* in his introduction to the article: 'this being I suppose in but few Hands with us' (see note 58 above). Indeed, recent work on the genre has shown that German dissertations of the period were published in very restricted numbers: print-runs of between only 200 and 300 copies were standard.[69] In comparison, individual numbers of the early Royal Society's *Philosophical Transactions* were usually printed in sets of between 750 and 1000 copies.[70] Moreover, early modern dissertations, as a non-commercial form of publication, served primarily a relatively small audience of readers, a large number of which were connected to the actors in the related disputation event and who received copies as gifts or tokens of friendship.[71] On the other hand, early scientific journals already had the wider goal of general research dissemination (for which they are best known today) as one of the motivating factors for their creation. The founder of the Royal Society's *Philosophical Transactions* famously included in the title of the journal's early numbers the statement that the journal should give 'some accompt of the present undertakings, studies and labours of the Ingenious in many considerable parts of the World'.[72]

The success of Derham's journal article as an intermediary for the transmission of Remus'/Wolff's account, and as a refinement of the earlier text, is clear from the later scholarship concerning the Frost and historical temperature measurements in Europe. Here, it is Derham's article which is routinely mentioned, and later cited, as a standard account of the Frost in 1708/09 from the mid-eighteenth century onwards.[73] With the obvious exceptions of Remus' and Wolff's countrymen Döderlein and Lenke, the *Consideratio*, on the other hand, finds mention only in later *repertoria* of German dissertations before disappearing from view.[74] The only significant exception was its appearance in the article by Dutch physician Jean Henri van Swinden (1746–1823) for the French *Journal de physique, de chimie, d'histoire naturelle et des arts*, in which he compared Derham's and Remus'/Wolff's studies, apparently without recognising their interdependence.[75] It goes almost without saying that in popular scientific accounts of the Great Frost intended for general audiences today, the English language account of Derham dominates almost entirely.[76]

Conclusion

Based on a close reading of Remus'/Wolff's early study of the 'Great Frost' of 1708/09, this chapter has attempted to illustrate the role of a Latin dissertation in communicating scientific information about a natural disaster which had a widespread and damaging impact on life across Europe. The role of the genre of the dissertation in fulfilling this function for the early modern scientific community has been generally acknowledged in scholarship since dedicated study of the genre began. But the present inquiry into the precise mechanisms of knowledge transfer between early modern dissertations and other contemporary genres hopes to answer to the call for greater attention to the 'intermedial connections' of the genre previously voiced.[77]

In this chapter's case study, the *Consideratio physico-mathematica hiemis proxime praeterlapsae* found a significant place within the *opera omnia* of the disputation event's prominent *praeses*, Christian Wolff. The work was lightly edited, translated and reprinted several times into the mid-eighteenth century and later contributed to Wolff's incorporation as a fellow of England's Royal Society. The work's approach, structure and primary results were then demonstrably influential for later German studies of the Frost, which included two further dissertations and a comparative study of the cold winter conditions in 1740. The best-known response to Remus' and Wolff's *Consideratio* was, however, William Derham's article on the Great Frost for the *Philosophical Transactions*. Amongst a collection of Derham's own data from London and reports from a series of correspondents across Europe, the *Consideratio* was the single most significant source for Derham's article. The present chapter has shown that Remus' and Wolff's data, results and approach were fundamental to Derham's study. Thanks to the innovative and very effective dissemination of scientific journals when compared to the earlier tradition of academic dissertations, it was then primarily through Derham's article that Remus and Wolff's report could contribute to our historical understanding of the Frost: it is, indeed, Derham's article that remains most prominent in modern accounts of the event. Though the title of Remus' and Wolff's work has thus disappeared almost entirely in later writing about the disaster, then, its influence has certainly not.

Notes

1 Walter Lenke, 'Untersuchung der ältesten Temperaturmessungen mit Hilfe des strengen Winters 1708–1709', *Berichte des Deutschen Wetterdienstes*, 92, 13 (1964), 3–49. Lenke's overview of contemporary reports on the period from all over western Europe (pp. 43–5) remains the single most useful bibliographic guide for historians.

2 Jürg Luterbacher et al., 'European Seasonal and Annual Temperature Variability, Trends, and Extremes Since 1500', *Science*, 303 (2004), 1499–1503.

3 Contemporary observations on the effect of the cold on life in Europe can be found throughout the two texts at the heart of this article: Georgius (Georg) Remus (rs.) and Christian Wolff (pr.), *Consideratio physico-mathematica proxime prae-terlapsae* (Halle: Zeidler, 1709) and William Derham, 'The History of the Great Frost in the Last Winter 1703 and 1708/09', *Philosophical Transactions of the Royal Society of London*, 26, 324 (1709), 453–78. We will return to their record of events in the detailed discussion of these works that follows. Modern accounts of this brutal season's conditions are numerous. Among the most useful in pre-paring this study have been, see Francis Asaf, 'L'Hiver de 1709', *Cahiers du dix-septième*, 12, 2 (2009), 1–29; Stephanie Pain, 'The Year that Europe Froze Solid', *New Scientist*, 201, 2694 (February 4, 2009), 46–47; Hans von Rudloff and Hermann Flohn, *Die Schwankungen und Pendelungen des Klimas in Europa seit dem Beginn der regelmäßigen Instrumenten-Beobachtungen (1670), Mit einem Beitrag über die Klimaschwankungen in historischer Zeit von Hermann Flohn* (Braunschweig: Vieweg, 1967); Émile Bouant, *Les grands froids*. Bibliothèque des Merveilles (Paris: Librarie Hachette, 1888).

4 Derham, 'The History of the Great Frost in the Last Winter 1703 and 1708/09'. See note 3 above.

5 The most relevant of these for the represent study is William Derham, 'Tables of the Barometical Altitudes at Zurich in Switzerland in the Year 1708. Observed by Dr. Joh. Ja. Scheuchzer, F. R. S. and at Upminster in England, Observed at the Same Time by Mr. W. Derham, F. R. S. as Also the Rain at Pisa in Italy in 1707. and 1708. Observed There by Dr. Michael Angelo Tilli, F. R. S. and at Zurich in 1708. and at Upminster in All That Time: With Remarks on the Same Tables, as Also on the Winds, Heat and Cold, and Divers Other Matters Occurring in Those Three Different Parts of Europe', *Philosophical Transactions of the Royal Society of London*, 26, 321 (1709), 334–66.

6 V. C. Slonosky, P. D. Jones, and T. D. Davies, 'Instrumental Pressure Observations and Atmospheric Circulation from the 17th and 18th Centuries: London and Paris', *International Journal of Climatology*, 21, 3 (2001): 285–98. The authors formulate their desire for more from Derham in their conclusion pp. 296–97.

7 Remus/Wolff, *Consideratio mathematica-physica*.

8 Lenke, 'Untersuchung der ältesten Temperaturmessungen mit Hilfe des strengen Winters 1708–1709', 17. Remus and Wolff took their temperature readings indoors (*cf. Consideratio physico-mathematica*, p. 4) and thus com-promised their value for comparison with other scientists' data made outdoors.

9 Jean-Pierre Legrand and Maxime Le Goff, 'Louis Morin et les observations métérologiques sous Louis XIV', *La Vie des Sciences*, 4, 3 (1987), 251–81. The modern edition of Morin's manuscript record is Jean-Pierre Legrand and Maxime Le Goff, *Les observations météorologiques de Louis Morin* (Boulogne-Billancourt: Direction de la météorologie nationale, 1992).

10 See in particular Luterbacher et al., 'European Seasonal and Annual Temperature Variability, Trends, and Extremes Since 1500'; Lenke, 'Untersuchung der ältesten Temperaturmessungen mit Hilfe des strengen Winters 1708–1709'; Slonosky, Jones, and Davies, 'Instrumental Pressure Observations and Atmospheric Circulation from the 17th and 18th Centuries'; Rudloff and Flohn, *Die Schwankungen und Pendelungen des Klimas in Europa seit dem Beginn der re-gelmäßigen Instrumenten-Beobachtungen (1670)*.

11 See note 8 above.

12 Marian Füssel, 'Die Praxis der Disputation. Heuristische Zugänge und theoretische Deutungsangebote', in *Frühneuzeitliche Disputationen: Polyvalente Produktionsapparate gelehrten Wissens*, ed. by Marion Gindhart, Hanspeter Marti, and Robert Seidel (Cologne; Weimar; Vienna: Böhlau Verlag, 2016), pp. 27–48 (p. 46): 'Stärker als bislang sollten auch die intermedialen Zusammenhänge zwischen Disputationen, Briefkorrespondenzen und Zeitschriften in den Blick genommen werden, gewann die gelehrte Wissensorganisation ihre Dynamik doch gerade aus dem Zusammenspiel unterschiedlicher Medien und Praktiken'. An earlier contribution to this field of study was Martin Gierl, 'Korrespondenzen, Disputationen, Zeitschriften', in *Macht des Wissens* (Vienna; Cologne: Böhlau Verlag, 2004), pp. 417–38.

13 Particularly useful for the present paper in this wider field have been Ku-ming (Kevin) Chang's series of articles 'From Oral Disputation to Written Text: The Transformation of the Dissertation in Early Modern Europe', *History of Universities*, 19 (2004), 129–87; 'Kant's Disputation of 1770: The Dissertation and the Communication of Knowledge in Early Modern Europe', *Endeavour*, 31, 2 (2007), 45–49; 'Collaborative Production and Experimental Labor: Two Models of Dissertation Authorship in the Eighteenth Century', *Studies in History and Philosophy of Science*, 41, 4 (2010), 347–55; 'For the Love of Truth: The Dissertation as a Genre of Scholarly Publication in Early Modern Europe', *KNOW: A Journal on the Formation of Knowledge*, 5, 1 (2021), 113–66; and Olga Weijers, *In Search of the Truth: A History of Disputation Techniques from Antiquity to Early Modern Times* (Turnhout: Brepols, 2013).

14 Wilhelm Michael von Richter, *Geschichte der Medicin in Russland*, vol. 3 (Moscow: Tipografiya Vsevolozhskogo, 1817), pp. 160–61. The modern *GND* information on Remus from the *Deutsche National Bibliothek* goes no further than Richter, see http://d-nb.info/gnd/120603381.

15 He would leave for Russia to serve as a field surgeon for the country's army in 1719 and later as a private doctor for a local dignitary before returning to Prussia in 1723, Richter, *Geschichte der Medicin in Russland*, pp. 160–61.

16 Günther Mühlpfördt, 'Christian Wolffs Lehre in östlichen Europa', *Aufkläung*, 12, 2 (special issue: *Christian Wolff—seine Schule und seine Gegner*) (1997), 77–100 (79).

17 Remus/Wolff, *Consideratio physico-mathematica*, a2^{r-v} and p. 52, respectively.

18 For the extensive discussion on this question *cf.* Ewald Horn, *Die Disputationen und Promotionen an den Deutschen Universitäten vornehmlich seit dem 16. Jahrhundert* (Leipzig: Harrassowitz, 1893); Georg Kaufmann, 'Zur Geschichte der akademischen Graden und Disputationen', *Centralblatt für Bibliothekswesen*, 11 (1894), 201–25; Chang, 'Collaborative Production and Experimental Labor', 349; Marion Gindhart, Hanspeter Marti, and Robert Seidel, 'Einleitung', in *Frühneuzeitliche Disputationen*, p. 8.

19 Gindhart, Marti, and Seidel, 'Einleitung', p. 8.

20 Remus/Wolff, *Consideratio physico-mathematica*, a2^{r}: 'Quare cum ante abitum ex illustri hac Musarum sede specimen mihi edendum sit studiorum meorum publicum, atque cum medico physicum et mathematicum constanter coniugere libuerit, ut tanto certiores in illo facerem progressus, non utiliorem a me operam collocari posse ratus sum, quam si in causas hiemis tam notabilis inquirerem'.

21 The authorial first-person references Wolff's published work already on p. 2, for example, and again on p. 6 of the text. This practice continues throughout the work. At the end of the first section of the *dissertatio*, the author clearly distinguishes between his activity and that of Wolff when he writes (p. 29): 'Hactenus ex diario meteorologico DN. Praesidis excerpsi, quae Halae annotata sunt. Ego iter per Misniam tum faciens campos nive aliquot pedes alta tectos...conspexi' (Until this point I have taken excerpts from the Master *Praeses*' weather diary, which observations were made in Halle. I then saw the fields covered in snow several feet high whilst making a trip through the countryside around Meissen).

22 Remus/Wolff, *Consideratio physico-mathematica*, p. 52: 'Satis ingenuii tui vim et tua, quam me Praeside nunc publice defendis, solus defendere poteras, dissertatione perspicient talium rerum gnari, et quemadmodum huc usque mihi in delitiis extitisti, ita nillus dubitus fore, ut etiam apud exteros, ad quos excurrere visum est, virtutem tuam et eruditioni junctam modestiam suspiciant omnes boni'.

23 See §13 *Divisio hiemis in periodos*, pp. 8–9 for a description of this system.

24 Remus/Wolff, *Consideratio physico-mathematica*, p. 10.

25 Remus/Wolff, *Consideratio physico-mathematica*, p. 13.

26 Remus/Wolff, *Consideratio physico-mathematica*, p. 15.

27 Remus/Wolff, *Consideratio physico-mathematica*, p. 20.

28 For the interest of this topic to this article and the study of early modern dissertations more generally see Füssel, 'Die Praxis der Disputation' and Gierl, 'Korrespondenzen, Disputationen, Zeitschriften' in note 12 above.

29 Remus/Wolff, *Consideratio physico-mathematica*, p. 24.

30 Remus/Wolff, *Consideratio physico-mathematica*, pp. 25–7. An example of such a review that offers a vivid account of the effects of the frost would be *La Clef du Cabinet des princes de l'Europe, ou Journal historique sur les matières du tems: contenant aussi quelques nouvelles de litterature, & autres remarques curieuses* initally edited by Claude Jordan out of Verdun and Paris. The first issue of 1709 (Janvier) is available under ark:/12148/bpt6k1050076 from la Bibliothèque nationale de France and has a report on the Frost on pp. 158–62.

31 Jean Dietz Moss, *Novelties in the Heavens: Rhetoric and Science in the Copernican Controversy* (Chicago: University of Chicago Press, 1993), pp. 97–125.

32 Remus/Wolff, *Consideratio physico-mathematica*, p. 42.

33 Remus/Wolff, *Consideratio physico-mathematica*, p. 43.

34 See note 19 and following above.

35 For the standard *GND* data on Thümmig see http://d-nb.info/gnd/120076276. The *Consideratio* was reprinted in Ludwig Philipp Thümmig, *Meletemata varii et rariores argumenti in vnum volumen collecta* (Leipzig: Sumptibus Simonis Iacobi Rengeri), p. 265ff.

36 Ludwig Philipp Thümmig, *Institutiones philosophicae Wolfianae in usus academicos adornatae*. 2 vols. (Leipzig; Frankfurt: In officina libraria Regneriana, 1726–727).

37 Christian Wolff, *Gesammelte kleine philosophische Schriften welche besonders zu der Natur-lehre und den damit verwandten Wissenschaften nehmlich der Mees- und Arzney-kunst gehoren* (Halle: Renger, 1736), pp. 11–107.

38 Christian Wolff, *Meletemata mathematico-philosophica cum erudito orbe literarum commerecio communicata* (Halle: Renger, 1755), pp. 319–63.

39 Wolff, *Gesammelte kleine philosophische Schriften*, pp. 11–12. I have followed
    the orthography and punctuation of the early modern German. My additions
    to and subtractions from the original text are marked in square brackets.
40 Wolff, *Meletemata mathematico-philosophica*, p. 320 and Wolff, *Gesammelte
    kleine philosophische Schriften*, p. 13. The work on Meteorology referenced
    repeated by Remus/Wolff in the *Consideratio* is Christian Wolff, *Aerometriae
    Elementa, in quibus aliquot Aeris vires ac proprietates juxta methodum
    Geometrarum demonstrantur* (Leipzig: Heredes Lankisianorum, 1709). The
    clear overlap between the subject, methodology and instrumentation of this
    work and that of the *Consideratio* make clear the collaborative nature of
    Remus' and Wolff's work in 1709.
41 Wolff, *Meletemata mathematico-philosophica*, p. 343. The original's misprint
    facies for *faciens* is noted here in square brackets.
42 Wolff, *Gesammelte kleine philosophische Schriften*, p. 60.
43 Chang, 'Collaborative Production and Experimental Labor', 347.
44 Gindhart, Marti, and Seidel, 'Einleitung', p. 8.
45 Friedrich Carl Gottlob Hirsching, *Historisch-literarisches Handbuch
    berühmter und denkwürdiger Personen welche in dem XVIIIen Jahrhundert
    gelebt haben* (Leipzig: Schwickert, 1813), p. 242. We will be turning to this
    English context in the later part 3 of this study.
46 I know of one short (8-page) *dissertatiuncula* on the Frost from 1709, which
    does not mention Remus'/Wolff's *Consideratio*. This is likely because the two
    works were produced at almost precisely the same time. Gabriel Erasmus
    Krafft (rs.) and Johannes Guilielmus Baierus (pr.), *Dissertatiuncula physica De
    frigore proximi mensis Ianuarii insolito, in ordine, ut vocant, circulari sub
    praesidio Johannis Guilielmi Baieri philos. natur. et mathem. prof. publ. pla-
    cidae commilitionum disquisitioni exposita per Gabrielem Erasmum Krafft
    Norimbergensem.* (Altdorf: Literis Kohlesii, 1709).
47 Johannes Nicolaus Sybellius (rs.) and Johannes Georgius Liebknecht (pr.), *De
    tempestatum apparenter et vere extraordinariarum ac speciatim Frigoris quod
    hyeme superiori sensimus intensissimi causis dissertatio, quam sub praeside
    domino Io. Georgio Liebknecht Mathematum in alma Ludouica Professore
    Publ. Ordinario, Patrono et Praeceptore suo omni honoris cultu prosequendo
    ad diem* [not provided] *A. MDCCX examinando proponit Io. Nic. Sybellius,
    sustat. Westph. Autor.* (Giessen: Vulpius, 1710).
48 There is no further information on J. N. Sybellius in the standard German
    GND, nor in the *VD18*.
49 Christiano Friderico Rast (rs.) Heinrich von Sanden (pr.), *Annuente Divina Gratia
    Rectore Magnificentissimo Serenissimo ac Excelsissimo Principe ac Domino,
    Domino Friderico Wilhelmo Regni Prussiae et Electoratus Brandenburgici
    Haerede etc. etc. etc. Consensu Amplissimae Facultatis Philosophicae, Praeses
    Henricus von Sanden, D. Physices Prof. ordinar. Disputationem Physicam De
    Frigore Anni M. DCC. IX. Memorabili Publicae submittit ventilationi
    Respondente Christiano Friderico Rast, Regiom. Medic. Stud. In Auditorio
    Maximo D.* [not given] *Febr. Ann M. DCC. XII. hor. antemeridianis* (Königsberg:
    Typis Zaenckerianis: 1712). For the standard *GND* data on Rast see: http://d-nb.
    info/gnd/13979011X; and for Sanden: http://d-nb.info/gnd/12036056X.
50 Rast/Sanden, *Disputatio physica de frigore*, p. 20: 'Adductis noxiis praegelidae
    hyemis effectibus, quibus memorata Europae loca fuere subjecta, superest ut et
    mentionem damnorum faciamus, quae dulcissimam nostram Patriam ubique
    locorum afflixerunt'.

51 Rast/Sanden, *Disputatio physica de frigore*, p. 52.
52 Johann Alexander Döderlein, *Observationes Meteorologicae, oder Historisch-Physicalische Nachrichten Von dem Strengen Winter An.* 1740: *Nach seiner eigentlichen Beschaffenheit, wahren Ursachen, besorglichen Folgen, und einigen merckwürdigen Umständen und Vorfälligkeiten, In einem ordentlichen Parallelismo, oder Vergleich, Mit dem durch gleiche Kälte bekandten Winter An.* 1709. *Nach den neuesten Philosophischen Principiis zum Angedencken vorgestellet von einem Mitglied der Kaysel. Reichs-Academie Nat. Curiosorum, Clitomachus beigenannt.* (Frankfurt; Leipzig: [n.d.], 1740).
53 See, example, Döderlein, *Observationes Meteorologicae*, pp. 20, 34, 64.
54 See, example, Döderlein, *Observationes Meteorologicae*, p. 33, where his periodised comparison of the two winters begins.
55 See note 49.
56 For Remus'/Wolff's summary see note 32 above. Döderlein, *Observationes Meteorologicae*, pp. 67–68. The linguistic resemblance of this summary to that of Remus/Wolff is further emphasised in Döderlein's echoes of the German translation of the *Consideratio*, where the relevant passage appears on p. 88.
57 See, for example, Döderlein, *Observationes Meteorologicae*, pp. 98, 107, 110.
58 Derham, 'The History of the Great Frost', 458–59.
59 To the best of my knowledge, this article never made it to publication.
60 Derham, 'The History of the Great Frost', 463–75.
61 On the central place of letters in scientific exchange in the late-seventeenth and early eighteenth centuries, as well as on letter writing as the dominant form selected for the presentation of information in the Royal Society's *Philosophical Transactions*, see, for example, Dwight Atkinson, 'The 'Philosophical Transactions of the Royal Society of London, 1675–1975: A Sociohistorical Discourse Analysis', *Language in Society*, 25, 3 (1996), 333–71. Derham also makes the established methods of exchange explicit at the beginning of his article, where he writes: 'THIS Famous Society having done me the Honour, to put into my Hands their Papers relating to the late Great Frost, and having also my self received divers Relations thereof from my Friends at Home and Abroad, as well as made Observations my self, I shall endeavour to give an Account [...]', (Derham, 'The History of the Great Frost', 454.
62 For an overview of Woodward's life and career see Joseph M. Levine, 'Woodward, John (1665/1668–1728)', in *Oxford Dictionary of National Biography* (Oxford: Oxford University Press, 2004): doi:10.1093/ref:odnb/29946.
63 Woodward's *Essay* appeared with the full title *An Essay toward a Natural History of the Earth and Terrestrial Bodies, especially minerals, also of the sea, rivers, and springs: with an account of the universal deluge: and of the effects that it had upon the earth* (London: Printed for R. Wilkin, 1695). Scheuchzer's Latin translation of his friend's work was *Specimen geographiæ physicae quo agitur de terra, et corporibus terrestribus speciatim mineralibus: nec non mari, fluminibus, & fontibus. Accedit diluvii universalis effectuumque ejus in terra descriptio* (Zurich: Typis Davidis Gessneri, 1704). The expanded Latin edition with responses to Camerarius appeared as *Naturalis historia telluris illustrata et aucta: una cum ejusdem defensione; praesertim contra nuperas objectiones D. El. Camerarii* (London: Typis J. M. Printed for R. Wilkin, 1714).
64 Derham, 'The History of the Great Frost', 475–76.
65 Derham, 'The History of the Great Frost', 476. Derham cites the *Consideratio* here again in Latin. For the echo of Remus'/Wolff's II.§20 in Döderlein's German work, see note 56 above.

66　Derham's argument on this point are to be found at 'The History of the Great Frost', 476–77.

67　For contemporary interest in the effects of meteors on conditions on earth, particularly concerning comets as the source of ice and water for the Biblical flood compare for example, William Whitson's *A New Theory of the Earth, From its Original, to the Consummation of All Things, Where the Creation of the World in Six Days, the Universal Deluge, And the General Conflagration, As laid down in the Holy Scriptures, Are Shewn to be perfectly agreeable to Reason and Philosophy* (London: Benjamin Tooke, 1696). For a comparative study of Whitson's *Theory* within contemporary English natural philosophy see William Poole, *The Earth Makers: Scientists of the Restoration and the Search for the Origins of the Earth* (Oxford; Bern: Peter Lang, 2010), pp. 55–75.

68　By this can be taken to mean the study of the underlying reasons for the occurrence of atmospheric humidity on earth.

69　Chang, 'For the Love of Truth', 134–35.

70　David A Kronick, 'Notes on the Printing History of the Early 'Philosophical Transactions'', *Libraries and Culture*, 25, 2 (1990), 243–68.

71　Chang, 'For the Love of Truth', 137.

72　For this point see Kronick 'Notes on the Printing History of the Early 'Philosophical Transactions'', 244. For the original title of the *Philosophical Transactions* see the first issue in 1665.

73　Derham's article saw a lively uptake, especially in Britain. See, for example, Thomas Hale, *A Compleat Body of Husbandry* (London: Thomas Osborne, Thomas Trye, Simon Crowder, 1758); *The Universal Magazine of Knowledge and Pleasure: Containing News, Letters, Debates, Poetry, Musick, Biography, History, Geography…and Other Arts and Sciences*, ed. by John Hinton and William Bent, vol. 32. (London: J. Hinton, 1763), Royal Statistical Society, *Journal of the Royal Statistical Society: Series A (General)*, 41 (1878) (London: RSS). But later also across Europe within the fields of mathematics, physics and historical meteorology: Kirstine Bjerrum Meyer, *Die Entwickelung des Temperaturbegriffs im Laufe der Zeiten sowie dessen Zusammenhang mit den wechselnden Vorstellungen über die Natur der Wärme* (Braunschweig: Vieweg, 1913); Kirstine Bjerrum Meyer, 'De quelques manuscrits d'Ole Rœmer', *Bulletin des sciences mathématiques*, 34 (1910), 73–96; Arthur de Boislisle, 'Le Grand Hiver et la disette de 1709', *Revue des questions historiques*, 72 (1903), 486–542. For more recent studies which also call on Derham see note 3 above.

74　The *Consideratio* appears in Johann Carl Heffter, *Museum Disputatorium Physico-Medicum Tripartitum, vol. 1, pars I. Editio nova* (Zittau: Sumptibus Schoepsianis, 1763). Mention of the work is also later made in a similar compilation: Friedrich Wilhelm Strieder, *Friedrich Wilhelm Strieders Grundlage zu einer Hessischen Gelehrten und Schriftsteller Geschichte Von der Reformation bis 1806*, ed. by Karl Wilhelm Justi, vol. 17 (Marburg: Gedruckt mit Bayrhoffer'schen Schriften, 1819).

75　Jean Henri van Swinden, 'Seconde lettre […] sur les grands hivers', *Journal de physique, de chimie et d'histoire naturelle et des arts*, Nivose an VIII de la République, Tome I (1800), 348–57.

76　Exemplary for this genre are Stephanie Pain, 'The Year that Europe Froze Solid', *New Scientist*, 201, 2694 (February 4, 2009), 46–47, and Juan José Sánchez Arreseigor, 'Winter Is Coming: Europe's Deep Freeze of 1709', *The National Geographic History* (January/February 2017) accessed online: https://www.nationalgeographic.com/history/history-magazine/article/1709-deep-freeze-europe-winter.

77　See note 12 above for this call in Füssel, 'Die Praxis der Disputation', 46.

Part 2
Representations

6   What was an avalanche?

Death in the snow from antiquity to early modern times[1]

Martin Korenjak

Like many natural disasters discussed in this volume, avalanches are a phenomenon as impressive as they are dangerous. Unpredictable and mysterious, vehement and powerful, they constitute a source of horror, but also of awe and fascination. At the same time, they represent a peculiar kind of disaster in a number of regards. They are geographically restricted to a small fraction of the earth's surface, high mountain regions. There, they occur quite frequently, but the damage they do, terrible as it may appear to those affected, is usually small-scale in comparison to most other disasters. Even the most catastrophic avalanche descent kills few people compared with the number of casualties caused by earthquakes, plagues or droughts. Because of their geographical limitation and their restricted effect, the information provided about them by historical sources is comparatively late and fragmentary. For instance, no full-length treatise seems to have been devoted to avalanches before the late nineteenth century.[2] Moreover, although the early modern interest in avalanches is indirectly linked to the rise of patriotism, as will become clear later, avalanches were not, like other disasters, integrated into the framework of an overarching theoretical discourse, be it theology, Aristotelian natural philosophy or the new science.

A century ago, however, avalanches did indeed become an object of scientific research. Today, the physical, geographical and meteorological causes leading to their descent are being intensely investigated, and a large body of knowledge and understanding has been accumulated.[3] Moreover, in mountainous regions, where avalanches are no rare sight and a factor to be reckoned with, even many lay people have a basic understanding of the phenomenon.[4]

Historical avalanche research has so far been conducted mainly from the perspective of the history of events: attention has focussed on specific disastrous avalanche descents, which form popular topics of local history.[5] By contrast, historians of mentalities have been slow to devote any attention to the phenomenon, and its perceptual history largely remains to be written. A start in this direction has been made a few years ago by Raphael Rabusseau,

DOI: 10.4324/9781003029823-9

who analysed how people felt, thought and spoke about avalanches in the eighteenth century, and Peter Utz, who examined the avalanche as a representational challenge from the sixteenth century to the present.[6]

The present contribution is intended to make one further step in the direction indicated by the studies just mentioned. It will focus on two basic questions that to the best of my knowledge have never been asked or answered so far: when, where and how did the very notion of an avalanche emerge in the first place? And once emerged, how did it evolve over time? In seeking answers to these questions, I will start from a brief sketch of what may be called the notion's ancient and medieval prehistory. After that, I will focus on the period around and after 1500, when a full-fledged concept of an avalanche appeared for the first time – or, to be more precise, two concepts. The first, a precursor of the modern view of the subject, emerged and spread in the Swiss Confederation, to which it remained confined for quite some time. The origins of the second notion, which strikes a modern observer as bizarre, are less clear, but despite its incompatibility with the first, it too was incorporated in the early modern avalanche discourse of Switzerland, creating a paradox for which some tentative explanations will be offered. The article will conclude with a few remarks on the long-term consequences of the story told, hinting at how the developments of the early modern era are still reflected in the present state of things.

Antecedents: Antiquity to late middle ages

That the development of the notion of an avalanche was long, complex and correspondingly interesting is already suggested by some elementary considerations regarding the sociology of knowledge. Those people who had first-hand experience of the phenomenon, namely, the rural mountain population, were not inclined to theoretical reflection and even less to leave written records of their knowledge. On the other hand, there were educated Greeks and Romans, medieval clerics and early modern humanists: they were used to putting pen to papyrus, parchment or paper, but knew avalanches from hearsay at best. Factual knowledge, theoretical interest and writing did not come together easily. Before this happened, a considerable number of misconceptions, discrepancies and confusion were produced.

Although the ancient Near Eastern and Mediterranean cultures had their experiences with high mountains, avalanches apparently left no traces in the written record over the first three millennia BCE. The first descriptions of avalanche-like phenomena on record date from the Roman Empire. Even then, pertinent accounts are few and far between: I have found no more than three authors who have something to say on the matter. None of them has a proper word for 'avalanche', and their comments remain vague and contradictory. The first among them is the Greek geographer Strabo, who writes in the first decades CE. In his description of the Alps, he mentions (*Geography* 4.6.6)

... αἱ κατολισθάνουσαι πλάκες τῶν κρυστάλλων ἄνωθεν ἐξαίσιοι, συνοδίαν ὅλην ἀπολαμβάνειν δυνάμεναι καὶ συνεξωθεῖν εἰς τὰς ὑποπιπτούσας φάραγγας. πολλαὶ γὰρ ἀλλήλαις ἐπίκεινται πλάκες, πάγων ἐπὶ πάγοις γενομένων τῆς χιόνος κρυσταλλωδῶν καὶ τῶν ἐπιπολῆς ἀεὶ ῥᾳδίως ἀπολυομένων ἀπὸ τῶν ἐντὸς πρὶν διαλυθῆναι τελέως ἐν τοῖς ἡλίοις.

... the huge ice slabs sliding down from above, which can seize whole caravans and take them down with themselves into the canyons below. For many slabs lay one upon the other, because one icy stratum upon the other is formed by the snow, and those on top are always prone to break away from those below before melting completely in the sun.

This remarkable passage probably contains the earliest information on record not only on avalanches, but also on the genesis of glacier ice, which is indeed formed from numerous strata of snow accumulating winter by winter. However, the avalanches seem to be ice avalanches, which substantially differ from normal snow avalanches, as they occur when the foremost parts of a glacier snout terminating in steep terrain break loose. Moreover, the two phenomena mentioned are mixed up to form an unworldly conglomerate. Similar confusion and lack of realism reigns in a later section (11.14.4) where people in Armenia are 'swallowed up by the snow' but can be found and freed by others because they carry long rods with which they pierce through to the surface. Whether these persons are buried by avalanches or simply snowed up was probably as unclear to Strabo as it is to us. The two other classical authors, the epic poets Silius Italicus (first century CE) and Claudian (c. 400), perpetuate the idea that avalanches have something to do with slippery ice. In Silius' epic about the Second Punic War, Hannibal tries to unfreeze an ice-bound slope and thereby triggers an avalanche which buries his soldiers (*Punica* 3.518–22). In Claudian's *Getica* (343–48), the Roman general Stilicho forces his way across the Alps in the winter. First, his baggage train, trek oxes and all, founders in the deep snow. Later, 'warm south winds' raise, ice begins to slide and the mountain produces a 'ruina' ('collapse'). In neither case does a clear picture of the actual events emerge.

In late ancient and early medieval times, the expression 'labina' or 'lavina', which is derived from the Latin stem 'lab-' ('to slide') and provides the root of most present words for 'avalanche', is attested for the first time. However, the term originally denotes the act of slipping or slippery ground. Later on, it is sometimes applied to a landslide but for a long time never denotes a snow avalanche.[7] Only at the turn of the early modern era did 'labina' and its vernacular derivations begin to obtain their present meaning.[8]

Nor can a clear understanding of the essence of an avalanche be detected during the Middle Ages, although numerous pilgrims and merchants had to cross the Alps and must have made pertinent experiences. This emerges from the *Gesta abbatum Trudoniensium* ('Deeds of the Abbots of Sint-Truiden'), a Flemish monastery chronicle begun in the twelfth century, which contains a circumstantial account of the fate of a group of pilgrims who, in winter 1128/29, were trapped on their way back from Rome by a dangerous avalanche situation at St. Rhémy, south of the Great St Bernard Pass.[9] The situation was characterised by general fear, which turned into panic when a group of 10 local guides ('marrones') was buried during an attempt to blaze a trail across the snow-covered pass. Everyone thought his last hour had come. In their desperation, the pilgrims confessed to one another because there only was one priest among them. Impressive as all of this is, one learns little about the avalanches by which the whole stir was caused. The description of the phenomenon as 'collapsing snow walls' ('ruebant … nivium aggeres') or as 'a strongly compressed snowball as high as a mountain face' ('rupibus instar montis densissimus nivis globus') leaves the reader without a clear idea of what had happened.

Thus, up to the beginning of the early modern era, not only no correct, but even no wrong notion of an avalanche can be detected in the written sources. In fact, no writer seems to have developed a real notion of the phenomenon in the strict sense of the word. Instead of a consistent concept, whether accurate or inappropriate, one just encounters fragments of one, so to say: slippery ice, drowning in deep snow, snow walls and lumps of snow. These fragments were combined in various ways, but none of them did result in a coherent, meaningful whole.

The 'modern' avalanche

This state of things changed quite abruptly around the middle of the sixteenth century in the Swiss Confederation, when a reasonably clear and in many respects factually correct understanding of the nature of an avalanche emerged. Moreover, isolated remarks were replaced by a veritable avalanche discourse. Much more people had something to say about the topic than before, and they soon began to refer to one another. Their comments formed a tradition which accommodated new information and grew broader and richer over time. However, the basic understanding of the phenomenon did not evolve substantially over the next two centuries or so. The ideas of different authors varied in details, but these were accretions to a solid core which can be found already in the earliest publications. For quite some time, this occupation with avalanches apparently remained a Swiss speciality. Only from the eighteenth century, comparable observations and statements appeared in print in other Alpine countries, and where they did, they were clearly informed by Swiss models.[10]

That a full-fledged avalanche notion should have been created out of nothing overnight is of course highly improbable. One rather must assume that unwritten everyday knowledge that had been widespread among mountaineers for a long time was taken up and put to writing by learned authors who developed an interest in the phenomenon. The intellectual climate in the Confederation was favourable to such a development. At the turn of the modern era, the inhabitants of several European regions began to cultivate a patriotic identity which would turn into modern nationalism at a later stage. Among other things, this trend led to a new interest in one's natural living environment which found literary expression in various types of chorographic writings.[11] The Swiss were particularly susceptible to such proto-nationalist feelings. During the late Middle Ages, they had endured much insult from their neighbours who reviled them with regard to their habitat as rebellious, uneducated and filthy mountaineers.[12] Now they defiantly turned the tables: they accepted the sobriquet 'mountaineers', but wore it with pride. Swiss intellectuals extolled the Alps as a paradisiac place in the true sense of the word[13] and styled them as a kind of national landscape *avant la lettre*[14] – an attitude that marked a turning point in the perception of mountains in general and inaugurated the mountain enthusiasm of modern times.[15] In this context, considerable attention was devoted to the numerous natural and cultural peculiarities of the Alps from glaciers to dairy farming. Once kindled, scholarly curiosity in the high mountains did not even back off from their harsh and dangerous features. One of these features was precisely avalanches.

To concretise what has been said so far, I have tabulated the information about avalanches to be found in ten chorographic texts in the widest sense of the word from and about Switzerland from the mid-sixteenth to the early eighteenth century (Table 6.1). In order to keep the table and this article at a manageable size, I have not aspired to completeness, but since other texts from the era in question contain little additional information, the material selected should be by and large representative.[16] This can all the more be assumed because most of the respective texts were fairly successful and influential, as evidenced by frequent references from the part of later authors to their predecessors.[17] None of the respective works is exclusively dedicated to avalanches,[18] but each has something substantial to say about the topic. The two volumes of Johannes Stumpf's *Gemeiner loblicher Eydgnosschaft beschreibung*, better known as *Schweizerchronik* (Zurich: Froschauer, 1547–48), provide an

Table 6.1 Texts about avalanches composed in Switzerland from the mid-sixteenth to the early eighteenth century

Designation and etymology	Kinds	Causes	Triggers	Course of descent	Damages	Fate of victims	Epiphenomena	Protective devices	Examples
Stumpf, *Schweizerchronik,* vol. 2, fols. 284ᵛ–285ʳ		Warm air and rain in spring	Birds; wind; noise	Small slide grows to great heap gliding downward	Soil; trees; rocks; hamlets, villages		Noise perceptible over long distances, like thunder or earthquake	Avoidance of erecting buildings on the mountain slope	
Tschudi, *Gallia Comata,* pp. 358–59		Spring weather	Gunshots; clamour; singing		Travellers, their luggage and equipment		Quivering; thunder; whole valley shaking	Prohibition of shooting, shouting, singing	(Frequent at Great St Bernard Pass)
D. Chiampell, *Raetiae alpestris … descriptio,* pp. 12–15 Schiess	(Wet avalanche; dust avalanche)	'labina'/'lavina' < Latin *labes, labi;* in German depravated to 'Löwin' [see below]	Birds, small animals; wind; human voice, singing; descent without trigger	Small slide grows to great heap plunging downward	Personal and material damage; rocks; soil; trees; farms	Death through suffocation (also in dust avalanches), except if head sticks out; stones and similar objects; crushed by wet snow	Thunder-like noise	Avoidance of dangerous building places	February 1567: author's nephew and his friend saved; February 1568: three casualties, one man safed from dammed Inn river
Simmler, *De alpibus commentarius,* pp. 222–30 Coolidge	Surface avalanche; ground avalanche	Romansh 'labina' < Latin 'labi'; in German depravated to 'Löuwinen' [see below]	Birds; animals in general; wind; clamour; echo	Rolling snowball grows, then turns into avalanche sliding downward	Rocks; trees; ibexes and other animals; humans; huts	Survival possible, particularly under loose snow avalanches; rescue by locals	Thunder; 'earth-quake'; huge avalanche cones, which remain in place for a long time; dammed streams causing inundations	Avoidance of dangerous building places; protective woods (e.g, over Andermatt); crossing of avalanche lines in the early morning, silently	Accidents around Rheinwald (Grison); 60 Swiss soldiers buried

- D. Chiampell: 'labina'/'lavina' < Latin *labes, labi;* in German depravated to 'Löwin'
- Simmler: Romansh 'labina' < Latin 'labi'; in German depravated to 'Löuwinen'

Source	Terminology	Type	Terrain	Trigger	Movement	Objects affected	Human effects	Noise	Protective / preventive measures	Historical
Rebmann, *Gastmal und Gespräch*, pp. 131–32	'Schneebruch'; technical term: 'Lowin'	Special form: dust avalanche (if there is strong wind)		Wind; birds		Trees; soil; rocks; houses; humans; livestock	Freezing to death; suffocation; rescue by locals possible	Noise	Protective walls; thick house walls; whole winter spent indoors	
Heremita, 'De Helvetiorum moribus epistola', pp. 487–88			Steep terrain	Wind		Farms, hamlets				
Plantin, *Helvetia antiqua et nova*, pp. 34–35	Romansh 'labina' < Latin 'labi'; in German depravated to 'Lauwine'; French 'levantze', 'valantze' < Latin 'vallis'		Thaw in spring; steep, woodless terrain	Minimal trigger	Ball (?) breaks up and flows downward in a broad stream					
Wagner, *Historia naturalis Helvetiae curiosa*, pp. 32–36	Latin 'labina'/ 'labena' < 'labi'; German 'Louwi'/ 'Löüwi'?, i.e., 'lionness'	Surface avalanche; ground avalanche	Steep terrain; sun's warmth	Wind; birds; echo	Slide grows rapidly, slides down towards the valley	Woods; houses; hamlets; rocks; humans; animals	(Salvages after ten hours, one day, three days)		Preventive triggering through cowbells, shouting, gunshots; for crossing avalanche lines, cowbells silenced by stuffing them with hay or similar things	1477: 60 Swiss soldiers buried at the Gotthard Pass; five more accidents 1501–1624

(Continued)

Table 6.1 (Continued)

	Designation and etymology	Kinds	Causes	Triggers	Course of descent	Damages	Fate of victims	Epiphenomena	Protective devices	Examples
Hottinger, *Montium glacialium Helveticorum descriptio,* pp. 49–50	'labina'; 'Louwine'		(Spring; a little wind)	Birds flying by; stronger wind; clamour; echo of voices; other minimal triggers	Small particle of snow begins to move, rolls and grows to a mountain of snow which rushes down towards the valley	Stones; trees; humans				
Scheuchzer, Οὐρεσιφοίτης *Helveticus,* pp. 220–33	Late Latin 'labina'/'labena' < 'labi'; German 'Louwine' and a dozen other forms < 'Löwin', i.e, 'lioness'; French 'lavanze'/'vallantze'; but real origin seems to lie in Romansh 'lavi(g)ne'	Loose snow or dust avalanche; wet snow or ground avalanche	Steep terrain	Rockfall; wind; fresh snow; falling trees; bells, gunshots, shouts, conversations; rain; warmth; chamois; birds		Humans; animals; trees; buildings	Death by suffocation, except if one has some free space to breath	Thundering noise; inundations because of dammed streams; huge avalanche cones lasting through the whole summer	Wedge-shaped ramparts; houses built under rock faces; preventive triggering through gunshots; cowbells silenced; conversations forbidden; churchbells may not toll; avalanche galeries	Accident of 1477 (see Simmler, Wagner); 14 more accidents until 1699

overview of Swiss history with extensive chorographic digressions. Aegidius Tschudi's *Gallia Comata* (composed 1572, printed Konstanz: Waibel, 1758),[19] which focusses on the country's early history, also devotes much space to its geography. Ulrich Campell's *Raetiae alpestris topographica descriptio* (composed 1573) presents a detailed description of the area of the Three Leagues, which more or less corresponds to the present Canton of Grisons.[20] Josias Simmler's *De Alpibus commentarius* (Zurich: Froschauer, 1574)[21] is the first monograph about the Alps in general. Hans Rudolf Rebmann's *Neuw, Lustig, Ernsthaft, Poetisch Gastmal, und Gespräch zweyer Bergen* (Bern: Le Preux, 1606) is a poem of epic proportions, styled as a dialogue between the two mountains Stockhorn and Niesen in the surroundings of Bern, which contains a mixed bag of information of all kinds, but devotes particular attention to the Swiss mountains.[22] Daniel Heremita writes a short description of the Confederation for foreigners in a pertinent collected volume published in the Low Countries,[23] while Jean-Baptiste Plantin's *Helvetia antiqua et nova* (Bern: Sonnleitner, 1656) constitutes the most extensive seventeenth-century account of the country's history and geography. Johann Jacob Wagner focusses exclusively on Swiss nature in his *Historia naturalis Helvetiae curiosa* (Zurich: Lindinner, 1680). Johann Heinrich Hottinger's 'Montium glacialium Helveticorum descriptio' (1706) is a short monograph on glaciers, perhaps the first of its kind, published as a journal article.[24] In his Οὑρεσιφοίτης *Helveticus* (Leiden: van der Aa, 1723), Johann Jacob Scheuchzer informs the reader about nine research excursions to various parts of the Swiss Alps undertaken by him in the first and second decades of the eighteenth century.[25] In the following, I will sketch the image of the avalanche that emerges from these texts by spelling out and briefly discussing the information provided under the various headings of my table.

Terms for 'avalanche' are given in four languages – Latin, German, French and Romansh – and include many phonetic variants. Most authors agree that the various designations are derived from 'labi' ('to slide'). This is worth mentioning not only because they are right, but also because a name's etymology was thought to disclose a thing's real nature at the time, a point to which we will return later.

The authors are aware that not all avalanches are the same. Although some accounts betray a certain confusion in this respect, many make two valid distinctions still in use today and already mentioned at the beginning (n. 3): they distinguish dry snow and wet snow avalanches on the one hand and surface and ground avalanches on the other. As a special form, the dust avalanche is known as well.

The causes leading to a dangerous avalanche situation and the occasions triggering a descent are not always distinguished as neatly as suggested by the table (although an understanding of the difference between these two categories seems to be implicit in some works). Wind in particular, which is actually one of the most important causes of avalanches, because it freights snow and heaps it up at wind-protected places, is regularly mentioned as an occasion. Apart from this, a geographical and a meteorological situation are most often adduced as causing avalanches: steep, woodless terrain and thaw in spring. Simmler appears the most modern in blaming what one would today call a weak bond between old and new snow. Among the occasions, by contrast, an idea stands out which is erroneous but enjoys a certain popularity even today: it is repeatedly stated that avalanches can be triggered acoustically, for instance by clamour, gunshots or even birdsong.[26] As implied in the introduction, the possibility of divine intervention is neither considered nor is an explanation in terms of Aristotelian or early modern physics. In the first case, this may be due to the frequency of avalanches, which occur often enough to appear as part of the regular course of nature, in the last, to the lack of observational, let alone quantifiable data; in the second case, I have no explanation.

The ideas about the actual descent of an avalanche reflect the swiftness of the event and the fact that it can hardly ever be overseen to its whole extent: they are few and their information content is meagre. Basically, it is only agreed that an avalanche begins small but may become huge in the end. Simmler and perhaps Plantin assume it to originate in a snowball, which however breaks up soon – another aspect on which we will pick up later.

The resulting damages, on the other hand, whose emotional impact is of course strong and for whose detailed inspection one could take one's time, are described much clearer and in greater detail. Things and living beings swept away and destroyed include rocks, trees, buildings, cattle and humans. Regarding the fate of the victims, the most important manners of death are known. Chiampell, for instance, gives a list including suffocation, being crushed to death and being battered by rocks, trees and other hard objects. At the same time, authors also know that survival in an avalanche is possible and recount rescue stories of people buried alive and saved after days by knowledgeable locals. Incidentally, the examples given in this context provide good evidence of the evolution of the textual tradition. In the sixteenth century, only a few cases were adduced of which an author had heard by chance. By contrast, Wagner and especially Scheuchzer (both of whom work in the tradition of natural history, a discipline focussed on data gathering) systematically collect everything they find in the written record and supplement it with examples from their own day.

As the most striking epiphenomenon of an avalanche, its roaring noise is often mentioned. This may even have resulted in an alternative etymology, mentioned by Wagner and Scheuchzer, which deduced 'Lawine' ('avalanche') from 'Löwin' ('lionness').[27] Massive avalanche cones proved impressive as well, in part for practical reasons: such cones could cause additional damage, for instance by damming streams and leading to flooding.

Some of the numerous precautions enumerated anticipate measures taken today. A kind of informal spatial planning seems to have been widespread: no buildings were erected in areas menaced by avalanches. Various kinds of protective constructions play an important role as well. Heremita tells of houses with walls so thick that they can withstand an avalanche. Scheuchzer knows wedge-shaped walls supposed to divide and slow down avalanches as well as simple avalanche galleries. Simmler describes a protective forest above the village of Andermatt as 'planted in triangular form' ('triangulari forma consita'), which still looks the same today. Wagner and Scheuchzer even report attempts at the preventive triggering of avalanches, precursors of today's avalanche explosions. The means proposed to this end – cowbells, clamour and gunshots – follow logically from the aforementioned ideas about acoustic triggering. For the same reason, the avoidance of any noise is often mentioned as an important precaution as well.

The snowball avalanche

The story told up to this point sounds fairly straightforward. But as already intimated, the conceptualisation of the avalanche presents a paradox. Alongside the 'modern' tradition sketched so far, there emerges another strand that takes an entirely different direction, imagining avalanches as monstrous snowballs.[28]

As intimated earlier, snowballs are occasionally mentioned in the 'modern' tradition, but only as starting points of avalanches, not in the sense that an avalanche would simply be equated with one. The 'modern' and the snowball tradition also differ in that the latter develops a strong pictorial component early on. While its origins predate the Swiss mainstream discourse and the first examples come from outside Switzerland, it nonetheless finds its most pronounced expression once more in the Confederation of the sixteenth and following centuries. Thus, although the two traditions obviously contradict one another, they partly develop over the same timespan, in the same geographical region, and sometimes in the very same book, as will become clear by some of the following examples.

In the literary record, a snowball avalanche appears as early as the abovementioned *Gesta abbatum Trudoniensium*, where there is talk of 'a strongly compressed snowball as high as a mountain face'. When the smart and curious Dominican Felix Fabri had to cross the Friulian Alps in January 1484 on his return from a pilgrimage to the Holy Land, he found his way blocked by an avalanche, which prompted him to enter the following explanation into his travel diary:

Periculosum valde est hieme per montana transire propter importunum lapsum nivium de montanis, praecipue quando post frigora incipit aura tepescere et glacies et nives resolvi, et nonnumquam contigit, quod globi nivium sic descendentes domos involvunt et cum hominibus evertunt, immo quandoque integras villas cum ecclesia et domibus involvit huiusmodi tempestas et deducit usque ad locum terminationis casus.

It is very dangerous to cross the mountains in winter because of the inconvenient sliding down of snow from the mountains, in particular when the air begins to warm and ice and snow to melt after a frost. Sometimes it has happened that snowballs rolling down this way engulf houses and uproot them along with the inhabitants. Occasionally, such a calamity even engulfes whole villages with their church and houses and takes them down to the place where its fall comes to rest.[29]

A few decades later, the *Theuerdank*, a versified and fictionalised biography of, and perhaps by, Emperor Maximilian I, appeared in Nuremberg (Schönsperger, 1517).[30] The plot of this work is driven by the doings of the treacherous captain Unfallo ('Accidenter') who deploys great ingenuity in finding novel methods to put young Theuerdank, the work's hero and Maximilian's alter ego, to death. Theuerdank therefore has to overcome numerous dangers and adventures, each of which is illustrated by a woodcut. Among other things, Unfallo hires a servant to unleash an avalanche upon Theuerdank, instructing him as follows (adventure 36):

Eylunds hin auf das gepirg lauff/Unnd schaw mit allem fleys darauf/ Wann der Held Tewrdannckh wirt reyten/Unden für an des pergs leyten/So mach von schnee ein pallen/Unnd lass den gmach herab fallen/ Das daraus werd ein leenen gross/Dieselb den Helden zutodt stoss ...

Hurry up, run up the mountain and observe very carefully when the hero Theuerdank will ride below along the mountain's slope. At that moment, make a snowball and let it slowly roll down: it will turn into a great avalanche that will hit and kill the hero.

The servant does as he has been told, and Theuerdank only survives thanks to his speedy horse. The *Theuerdank* image (Figure 6.1), perhaps the first pictorial representation of an avalanche ever, takes its clue from the text, but goes far beyond it: in the woodcut, the avalanche not only originates from a snowball, it is itself composed of a series of snowballs.

A much stronger contrast between text and image can be observed in Stumpf's *Schweizerchronik*. Its text, which states 'that it [the snow] begins to break loose a bit and gradually grows to such a heap that it runs down

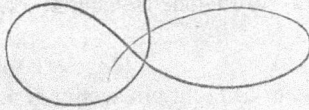

Figure 6.1 Theuerdank narrowly escapes a snowball avalanche. [Credit: Bayerische Staatsbibliothek München, Rar 325a, fol. [l viiᵛ], image 36; urn:nbn: de:bvb:12-bsb00013106-2].

lenden mit vnuerſähenlichen ſchaden mit

Figure 6.2 Another snowball avalanche in Stumpf's *Schweizerchronik* (vol. 2, fol. 285*r*). Zentralbibliothek Zürich, AW 40: 1–2, https://doi.org/10.3 931/e-rara-5076/Public Domain Mark.

into the valley' ('das er anfacht ein wenig rysen/und zestund meeret er sich zuo einem soelichen hauffen/das er gegen tal laufft', fol. 284ᵛ), is flatly contradicted by the woodcut purporting to illustrate it (Figure 6.2). The difference is all the more striking because the *Schweizerchronik*'s images, the work of the renowned artist Heinrich Vogtherr the Elder, generally aim at a true-to-life representation of the phenomena described.

David Herrliberger's *Topographie der Eydgnoßschaft* (Zurich: Ziegler, 1754), a collection of 328 copper engravings with accompanying texts, contains two avalanche depictions. In both of them, the snowball has grown to grotesque dimensions: it is just about to fall on a whole hamlet, which it will inexorably crush (Figure 6.3). In this case, the text is in full accordance with the image, as Herrliberger describes the nature of avalanches as follows:

Sind eigentlich grosse ungeheure Schnee-Schemel, oder in grosse Ballen zusammen gerollete Schneeklumpen, welche von denen höchsten gächstotzigen Bergen mit ungestümen und entsezlichem Krachen und Tossen in die Thäler und Tieffenen herunter stürzen.

In fact, they are big, enormous snow chunks or snow boulders rolled into big balls, which fall from the highest, precipitous mountains down into the valleys and depths with impetuous, terrible crashing and thundering.[31]

To round off this sample, Johann Ludwig Bleuler's painting *Lawinenabgang in Graubünden* (some years before 1850; Figure 6.4) represents a late stage in which the graphic tradition has emancipated itself from the texts and the snowball avalanche has become the subject of an independent tableau. Its date demonstrates the tradition's longevity, which extends to the early nineteenth century.[32]

How to make sense of this state of things? How could a factually untenable notion not only emerge, but also flourish for three centuries in parallel with the development of a basically correct one? Because of the apparently complete lack of contemporary comments on the phenomenon by writers, artists or readers, it is difficult to offer a cogent explanation. One is reduced to educated guesses, for which two starting points present themselves: firstly, the concrete perspective from which avalanches were perceived in early modern times, secondly, the autonomy of different media, which to some degree develop according to their own rules.

Before the arrival of winter sports, avalanches were mostly experienced from below, from the perspective of the dalesman or of the traveller who had to cross a pass. People were confronted with what arrived in the valley, but hardly ever noticed how an avalanche commenced. Moreover, the breaking of an avalanche is a complex event conditioned by numerous factors such as the structure of the snowpack, the amount of friction between snow and ground and between different

Figure 6.3 An avalanche about to crush a hamlet in Herrliberger's *Topographie der Eydgnoßschaft*. Zentralbibliothek Zürich, Res 1364, https://doi.org/10.3931/e-rara-18231/Public Domain Mark.

Figure 6.4 A nineteenth-century representation of the snowball avalanche in a painting by Johann Ludwig Bleuler.

snow layers, tension stress at the top of a slope and compression stress below, to name but a few. All of this lay beyond the physical understanding of the early modern era. Under these circumstances, it presumably was seductive to explain the genesis of an avalanche with an effect familiar to most people who have ever been in contact with snow: the size of a wet snowball rolling down a slope increases rapidly. In fact, avalanches never start this way, but many people in early modern times thought they did. As already mentioned, the author of the *Theuerdank* is joined by Simmler and Plantin in this respect, who speak of balls ('globi') rolling down towards the valley. Both emphasise that rolling soon turns into sliding, the snowball into a true avalanche,[33] and the derivation of the phenomenon's name from 'labi' ('to slide') indicates that this was known to the average mountaineer as well. Unexperienced visitors, by contrast, could indeed hit on the idea that an avalanche as such was made up of one or several snowballs. This notion could even have received a seeming confirmation from the look of some avalanche cones whose rough surfaces appear as composed, if not of true snowballs, at least of numerous lumps of snow.[34]

For a woodcutter, engraver or painter, the idea of the snowball avalanche was probably attractive for a number of reasons. An avalanche descent would have been exceedingly difficult to depict in a truly realistic manner. Provided one was as lucky as to catch sight of an actual descent without being involved in it, there only remained a few seconds to study the event before everything was over. Moreover, a strictly realistic image probably would have failed to convey the fear and awe inspired by the real event, which to a great extent depends on sheer size, sound and the impression of mortal danger. A snowball, by contrast, was easy to draw. It suggested rapid movement and its size could be increased at libitum to suggest the desired degree of menace and fearfulness. In addition, the destructive potential could be visualised by the inclusion of trees, rocks or houses – a possibility used to striking effect by Vogtherr, whose snowball resembles a spiny monster using its tree thorns as weapons to stab travellers and pack animals. Finally, while a real-life avalanche remains on the ground, a snowball could be represented in free fall through the air, hitting its victims from above, to great advantage for an image's emotional impact. Herrliberger and Bleuler exploited this possibility to the maximum.

That the idea of the snowball avalanche proved hard to die in spite of the spread of better knowledge can partly be ascribed to another peculiarity of the graphic medium, namely, to the stubbornness that often characterises pictorial conventions. Until today, corvids are often painted with yellow beaks and feet, although one can check every day that they are entirely black. In early modern times, pertinent examples abounded. Exotic phenomena not easily subjected to a reality check were particularly liable to falling victim to distorting graphic traditions. Camels' heads were preferably depicted like horse heads throughout the sixteenth century, and when the fins of a sperm whale stranded near Antwerp in 1577 were misunderstood as its ears and depicted accordingly, the resulting tradition quickly spread in natural historical publications.[35] The autonomy of the image was enhanced by the techniques of woodcut and copper engraving. Because new illustrations were expensive, old ones were often recycled, even if they did not fit their new context. The author and illustrator usually were two different persons, and the latter will often have been unable to read or understand what the former had written. Then as now, such pictures reached a mass audience, shaping and perpetuating habits of seeing, expectations and convictions on a large scale. Their resistance force against better knowledge was accordingly great.

Aftermath

However, even the greatest perseverance has its limits. From the nineteenth century, more and more people felt attracted to the high regions of the Alps, where they sometimes triggered and often witnessed avalanche descents. Soon, such events were even recorded in photographs and film. Under these conditions, snowball avalanches went out of fashion, at least in serious depictions.[36] However, they live on in comics, cartoons[37] and above all in metaphor, language's repository and recycling factory of antiquated notions. 'To snowball' and 'snowball effect' have remained common idioms until today. In German, the initiation of a momentous development is still expressed by the idiom 'eine Lawine ins Rollen bringen' ('to let an avalanche roll').

The discussions of avalanches in early modern chorographic writings, on the other hand, presented a number of starting points which could be picked up and developed by modern avalanche research and prevention. True, the latter's combination of systematic data gathering in the field and physical interpretation soon went far beyond anything imaginable in earlier centuries. Nonetheless, it presumably is no coincidence that the systematic investigation of avalanches once more commenced in Switzerland, where pioneers like the Grison Johann Wilhelm Coaz (1822–1918)[38] were active and the 'Eidgenössische Schnee- und Lawinenforschungskommission' was founded in 1931.

Notes

1 This article originated in a talk given at the 'Drittes Montafoner Gipfeltreffen: Sterben in den Bergen' (Schruns, October 2016). My audience provided perceptive comments, questions and criticisms. The article's further elaboration benefitted from discussions with Ovanes Akopyan, Florian Baumgartner, William Barton, Dominik Berrens, Johanna Luggin, Irina Tautschnig and Stefan Zathammer. My heartfelt thanks to all of them! – All translations are mine.

2 Having undertaken no systematic research in this respect, the earliest treatise known to me is Johann Wilhelm Coaz, *Die Lauinen der Schweizeralpen* (Bern: Dalp, 1881).

3 An overview of the current state of avalanche research is provided by Jürg Schweizer, Parry Bartelt and Alec van Herwijnen, 'Snow Avalanches', in *Snow and Ice-Related Hazards, Risks, and Disasters*, ed. by Wilfried Haeberli and Colin Whiteman (Amsterdam et al.: Elsevier, 2015), pp. 395–436.

4 A brief summary of this understanding may be useful. What constitutes an avalanche is the sudden loss of stability of a substantial mass of snow, which subsequently slides down a slope. The mass may carry downhill any things or creatures it encounters on its way and bury them at the bottom of the slope, where it comes to a stand, causing death by suffocation, pressure or hard objects (stones, trees, etc.) carried along. The descent of an avalanche usually presupposes an inclination of c. 30°–50° and a snowpack which is unstable with regard to its connection to the ground or internally. Such a snowpack can be the result of a number of factors from heavy snowfall and wind to warmth, sunshine and weak coherence between different layers. The descent may happen by itself or be triggered by an additional strain, for example by a skier. There exist different types of avalanches. Two significant criteria of distinction

concern the kind of snow involved (for instance dry and wet snow avalanches) and the depth to which the avalanche reaches (ground avalanche vs surface avalanche). Two overviews in layman's terms are provided by Eike Roth, *Lawinen. Verstehen – Vermeiden – Praxistipps* (Munich: Bergverlag Rother, 2013), and Rudi Mair and Patrick Nairz, *Avalanche. Recognizing the Decisive Problems and Danger Patterns* (Innsbruck; Vienna: Tyrolia, 2018).

5 See for instance Christian Rohr, 'Leben mit dem "Weißen Tod". Zum Umgang mit Lawinen in Graubünden seit der Frühen Neuzeit', *Bündner Kalender*, 174 (2015), 52–59; Christian Rohr, 'Sterben und Überleben. Lawinenkatastrophen in der Neuzeit', in *Sterben in den Bergen: Realität – Inszenierung – Verarbeitung*, ed. by Michael Kasper, Robert Rollinger and Andreas Rudigier (Vienna; Cologne; Weimar: Böhlau, 2018), pp. 135–60.

6 Raphael Rabusseau, *Les neiges labiles. Une histoire culturelle de l'avalanche au XVIIIe siècle* (Geneva: Presses d'Histoire Suisse, 2007); Peter Utz, *Kultivierung der Katastrophe. Literarische Untergangsszenarien aus der Schweiz* (Munich: Fink, 2013), pp. 115–47.

7 In the meaning 'slippery ground', the term is explained by Isidor of Seville, *Etymologiae* 16.1.4: 'labina, eo quod ambulantibus lapsum infert, dicta per derivationem a labe'; see in general *Thesaurus linguae Latinae*, VII 2, 774.16–33. As 'landslide', the word is first attested in Paulus Diaconus' *Historia Langobardorum*, 3.23 in connection with heavy rainfalls: 'Factae sunt lavinae; possessionum seu villarum hominumque magnus interitus'.

8 The first possible testimony for the meaning 'snow avalanche' is provided by a decree issued by the bishop of Konstanz in 1302, which mentions the 'horridam et execrabilem cladem lowinarum' rendering the way from Morschach to Schwyz (both in Central Switzerland) dangerous and impassable (*Urkunden zur Geschichte der eidgenössischen Bünde*, ed. by J. E. Kopp (Lucerne: Xaver Meyer, 1835), pp. 54–55). However, since the topographical situation lets neither landslides nor avalanches appear quite probable, the passage remains inconclusive. For the vernacular avalanche words in the Alpine area, see Jon Pult, *Die Bezeichnungen für Gletscher und Lawine in den Alpen* (Samedan; St. Moritz: Engadin Press, 1947), pp. 74–119, esp. pp. 105–8 (forms involving metathesis of 'v' and 'l' like Italian 'valanga', Spanish 'avalancha', French and English 'avalanche'), p. 115 (Latin 'lab-' as the common root). Pult also discusses local expressions of different etymological origins (pp. 75–89).

9 Paul de Saint-Trond, *Gesta abbatum Trudoniensium*, 12.6 (*Monumenta Germaniae Historica*, SS 10, pp. 213–448, at p. 307).

10 Tyrol, for instance, defined itself as a 'land in the mountains' ('Land im Gebirge') from the sixteenth century (Martin Korenjak, 'Wie Tirol zum Land im Gebirge wurde. Eine Spurensuche in der Frühen Neuzeit', *Geschichte und Region/Storia e Regione*, 20 (2012), 140–62), but for a long time, nothing substantial was said about avalanches in this context. This may be seen from the short shrift they get in Hippolytus Guarinoni's *Grewel der Verwüstung menschlichen Geschlechts* (Ingolstadt: Angermayr, 1610), an enormous miscellany in which mountain-related topics are otherwise discussed in considerable detail: 'In den Bergen gibts viel Stein und Felsenbrüch oder Lähnen ab, die unversehens von den höchsten Jöchern sich sambt dem Schnee wie auch für sich abledigen ...' (p. 443). Snow avalanches do not appear as a phenomenon in their own right, but only as an epiphenomenon of rockfalls. Among the first more pertinent discussions is Joseph Walcher SJ, *Nachrichten von den Eisbergen im Tyrol* (Frankfurt; Leipzig: [n.s.], 1773), pp. 76–80.

11 On early modern preformations of nationalism in general, see *Diffusion des Humanismus. Studien zur nationalen Geschichtsschreibung europäischer Humanisten*, ed. by Johannes Helmrath, Ulrich Muhlack and Gerrit Walther (Göttingen: Wallstein, 2002), and Caspar Hirschi, *Wettkampf der Nationen: Konstruktionen einer deutschen Ehrgemeinschaft an der Wende vom Mittelalter zur Neuzeit* (Göttingen: Wallstein, 2005). On the importance of chorographical texts and landscape concepts for this movement, Ulrich Muhlack, 'Das Projekt der Germania illustrata. Ein Paradigma der Diffusion des Humanismus?' in *Diffusion des Humanismus*, pp. 142–58, François Walter, *Les figures paysagères de la nation: territoire et paysage en Europe (16e–20e siècle)* (Paris: Editions EHESS, 2004), and Martin Korenjak, 'Deutschland als Landschaft. Konrad Celtis und der Herkynische Wald', in *Würzburger Humanismus*, ed. by Thomas Baier and Jochen Schultheiß (Tübingen: Narr, 2015), pp. 19–35.

12 *In Helvetios – Wider die Kuhschweizer. Fremd- und Feindbilder von den Schweizern in antieidgenössischen Texten aus der Zeit von 1386 bis 1532*, ed. by Claudius Sieber-Lehmann, Thomas Wilhelmi and Christian Bertin (Bern; Stuttgart; Vienna: Haupt, 1998).

13 Martin Korenjak, 'Die Schweizer Alpen als Wasserschloss Europas: Glarean und die *Helvetia mater fluviorum*', *Schweizerische Zeitschrift für Geschichte*, 62 (2012), 390–404.

14 Ulrich Im Hof, *Mythos Schweiz. Identität – Nation – Geschichte 1291–1991* (Zurich: Verlag Neue Zürcher Zeitung, 1992), pp. 106–11, 158–60; *La découverte des Alpes/La scoperta delle Alpi/Die Entdeckung der Alpen*, ed. by Jean-François Bergier and Sandro Guzzi (Basel: Schwabe, 1992) and therein in particular the contribution by Guy P. Marchal, 'La naissance du mythe du Saint-Gothard ou la longue découverte de l' "homo alpinus helveticus" et de l' "Helvetia mater fluviorum" (XVe s. – 1940)', pp. 35–53; Thomas Maissen, 'Die Bedeutung der Alpen für die Schweizergeschichte von Albrecht von Bonstetten (ca. 1442/43–1504/05) bis Johann Jakob Scheuchzer', and Guy P. Marchal, 'Johann Jakob Scheuchzer und der schweizerische "Alpenstaatmythos"', in *Wissenschaft – Berge – Ideologien. Johann Jakob Scheuchzer (1672–1733) und die frühneuzeitliche Naturforschung/Scienza – Montagne – Ideologie. Johann Jakob Scheuchzer (1672–1733) e la ricerca naturalistica in epoca moderna*, ed. by Simona Boscani Leoni (Basel: Schwabe, 2010), pp. 161–78 and 179–94, respectively.

15 Martin Korenjak, 'Why Mountains Matter: Early Modern Roots of a Modern Notion', *Renaissance Quarterly*, 70 (2017), 179–219 (pp. 185–87). The early modern 'discovery of the mountain' in general is described in greater detail by William Barton, *Mountain Aesthetics in Early Modern Latin Literature* (Abingdon; New York: Routledge, 2017).

16 Texts not tabulated include Johann Heinrich Hottinger the Elder, 'Methodus legendi historias Helveticas', in *Dissertationum miscellanearum pentas* (Zurich: Bodmer, 1654), pp. 197–631 (pp. 236–39); Johannes Leonhard, *Brevis descriptio democratici, liberae et a solo Deo dependentis Rhaetiae reipublicae … regiminis* (London: Buckley, 1704), p. 6; David Herrliberger, *Neue und vollständige Topographie der Eydgnoßschaft* (Zurich: Ziegler, 1754), pp. 76–82.

17 For instance, the chronologically last of the authors tabulated by me, Johann Jacob Scheuchzer, cites no less than six out of the nine names preceding him in the table, namely, Stumpf, Tschudi, Simmler, Rebmann, Plantin, and Wagner, alongside works by others.

18 In the hope of unearthing monographic discussions of avalanches, I have browsed, inter alia, bibliographies of early modern dissertations such as

Hanspeter Marti, *Philosophische Dissertationen deutscher Universitäten 1660–1750. Eine Auswahlbibliographie* (Munich: Saur, 1982), but to no avail.

19 Modern reprint: Lindau: Antiqua-Verlag, 1977.

20 Modern editions: Ulrich Campell, *Raetiae alpestris topographica descriptio*, ed. by C. I. Kind (Basel: Schneider, 1884); *Dritter und vierter Anhang zu Ulrich Campells Topographie von Graubünden*, ed. by Traugott Schiess (Chur: Casanova, 1900). The Three Leagues were not formally a part of the Swiss Confederation, but intimately linked to it as a so-called 'Zugewandter Ort'. Economic, cultural, and personal ties were strong.

21 It was printed together with another work, the *Vallesiae descriptio, libri duo*. Modern reprint: Bologna: Libreria Alpina degli Esposti, 1970. Modern edition in William Augustus Brevoort Coolidge, *Josias Simler et les origines de l'alpinisme jusqu'en 1600* (Grenoble: Allier, 1904).

22 See Rosmarie Zeller, 'Die Wunderwelt der Berge. Literarische Form und Wissensvermittlung in Hans Rudolf Rebmanns *Gastmal und Gespräch zweier Berge*', in *Scientiae et artes: die Vermittlung alten und neuen Wissens in Literatur, Kunst und Musik*, ed. by Barbara Mahlmann-Bauer (Wiesbaden: Harrassowitz, 2004), ii, pp. 979–95.

23 'De Helvetiorum, Raetorum, Sedunensium situ, republica, moribus epistola', in *Helvetiorum respublica. Diversorum auctorum, quorum nonnulli nunc primum in lucem prodeunt* (Leiden: Elzevir, 1627), pp. 485–535.

24 *Miscellanea curiosa, Appendix ad Annum IX et X decuriae III Ephemeridum Academiae Caesareo-Leopoldinae Naturae Curiosorum* (1706), 41–75. English translation: Gavin de Beer, 'Johann Heinrich Hottinger's Description of the Ice-Mountains of Switzerland, 1703', *Annals of Science*, 6 (1950), 328–60.

25 See Martin Korenjak, 'Das Gebirge als Erlebnisraum. Johann Jakob Scheuchzers *ΟΥΡΕΣΙΦΟΙΤΗΣ Helveticus sive Itinera per Helvetiae Alpinas regiones*', *Daphnis*, 41 (2012), 203–31.

26 Utz, *Kultivierung der Katastrophe*, pp. 126–30; see esp. p. 128 for the persistence of the idea.

27 Utz, *Kultivierung der Katastrophe*, p. 127.

28 The only discussion of this phenomenon known to me is found in Utz, *Kultivierung der Katastrophe*, pp. 118–22.

29 Felix Fabri, *Evagatorium in Terrae Sanctae, Arabiae et Egypti peregrinationem*, ed. by Konrad Dietrich Hassler (Stuttgart: Societas Litteraria Stuttgardiensis, 1843), iii, p. 445.

30 Modern reprint with an introduction by Stefan Füssel: *Die Abenteuer des Ritters Theuerdank. Kolorierter Nachdruck der Gesamtausgabe Nürnberg 1517* (Köln: Taschen, 2003).

31 See Herrliberger, *Neue und vollständige Topographie der Eydgnoßschaft*, tables 41 and 42 and the accompanying text on pp. 76–82 (pp. 76–77). The text volume and the engravings were sold separately; the latter could be bound into the former (see p. 5), but this was not always done. Both images can be found in Utz, *Kultivierung der Katastrophe*, p. 121 and in Daniela Vaj's and Claude Reichler's database *Viatimages* (https://www2.unil.ch/viatimages/ [accessed 12 October 2020]).

32 The title and date are taken from Rohr, 'Leben mit dem "Weißen Tod"', p. 53. For additional images, see *Viatimages* and Utz, *Kultivierung der Katastrophe*, pp. 118–33.

33 See in particular Coolidge, *Josias Simler*, p. 224: 'is [sc. globus] ubi multis provolutionibus excreverit et iam propter magnitudinem et inaequalitatem volvi amplius non potest, vehementiore tum impetu delabitur …'.

34 The phenomenon is illustrated and explained by Schweizer, Bartelt and van Herwijnen, 'Snow Avalanches', pp. 411–13.

35 Dániel Margócsy, 'The Camel's Head: Representing Unseen Animals in Sixteenth-Century Europe', *Netherlands Yearbook of Art History*, 61 (2011), 63–85. For many more examples, see P. Michel's webpage *Wandernde Bilder* (http://www.enzyklopaedie.ch/dokumente/Bildmigration.html [accessed 12 October 2020]).

36 For some 'realistic' nineteenth century avalanche images, see Utz, *Kultivierung der Katastrophe*, pp. 122–23, 125, 133, and *Viatimages*.

37 For an example from Hergé's *Les aventures de Tintin*, see Utz, *Kultivierung der Katastrophe*, p. 119. Snowball avalanche cartoons are easily found on the internet.

38 Carl Schröter, *Oberforstinspektor Dr. Johann Coaz 1822–1918. Ein Nachruf* (Zurich: Rascher, 1919); Conradin Ragaz, 'Johann Wilhelm Fortunat Coaz', in *Bedeutende Bündner aus fünf Jahrhunderten* (Chur: Calven, 1970), ii, pp. 108–18 (pp. 114–15).

7 Disasters and devotion

Sacred images and religious practices in Spanish America (16th–18th centuries)

Milena Viceconte[1]

In recent years, research into the sociocultural aspects of disasters in Hispano-American territories has shed new light on the important role of images in the reinforcement of appeals to divine invocation in colonial societies. The complex imagery was charged with heterogeneous symbolism deriving from the interaction between the local cosmogony of precolonial deities and the Christian vision promulgated by the Spanish Empire. Paintings, sculptures and prints representing Christ, the Virgin Mary and the saints were the fulcrum of rituals in which they were paraded through city streets and surrounding rural areas afflicted by disasters.[2] As in the Old World, these practices were an effective way of channelling collective fear and mitigating against the social disorder that resulted from catastrophic disruption to normal life. They also served to engage the urban elite, the political authorities, the clergy, the religious orders and a range of other social groups in collective veneration, each targeted according to their social status and ethnicity.

The aim of this essay is to examine the heterogeneous catalogue of imagery used in rites to ward off disasters in America, analysing a range of iconographic and functional aspects. In terms of composition, the most remarkable ones were designed to raise awareness of the value of collective prayer rather than to describe the material consequences of the disaster. We will identify the visual schemata and the narrative formulae of such images, in which European elements were appropriated and elaborated, and simultaneously infused with the symbolism of local spirituality. In our analysis of how these visual devices spread, and the purposes they served, we will address their intrinsic links with the promotion of certain divine agents as protectors against disasters, some of whom remain very popular to this day.[3]

The Crucifix as the main bastion against earthquakes

The geomorphological characteristics of the South American continent predispose it to significant seismic activity. This was compounded by the

DOI: 10.4324/9781003029823-10

numerous volcanoes at high risk of eruption in the area known as *Cinturón de Fuego* ('the Ring of Fire'). In modern times, the continent has seen numerous combined disasters, that is, those where geological events (earthquakes and volcanic eruptions) and hydrometeorological events (floods and tsunamis) occur together.[4] Events of this kind were seen as manifestations of divine power, a view that was not dependent only on the Catholic concepts expounded by the conquistadores. Precolonial cosmogonies also held that there was a close link between natural phenomena and the divine. For example, Mayan cultures saw earthquakes as a tangible sign of the existence of supernatural beings and Andean cultures believed that interacting with volcanoes could establish a spiritual connection with Mother Earth. Moreover, indigenous beliefs maintained that mankind could influence earthquakes and eruptions by calling on local deities through propitiatory rituals and divinatory practices.[5]

With the arrival of the Spanish, the spiritual propaganda promoted by the clergy and the religious orders led to a gradual affirmation of the Judeo-Christian vision. Its moralistic doctrine coexisted with complementary and sometimes antithetical explicative models based on astrology, meteorology and even the practice of magic.[6] Prevalent belief systems that saw earthquakes and other natural phenomena as punishments exacted by God on humanity led to the practice of invoking the powers of sacred items seen as particularly powerful against calamities. Of these, the crucifix emerged as being particularly appropriate for earthquakes, as it allowed a clear analogy with the geological phenomena associated with Christ's death on Calvary such as the strong tremors and other violent natural phenomena described in the Gospel (Matt. 27:45–54).[7]

In the New World, appeals to the power of the crucifix were particularly widespread in the southern Andes. Around the middle of the seventeenth century, a number of earthquakes were associated with the onset of the veneration of three objects depicting Christ on the Cross: two wooden crucifixes and a mural. The earthquakes ceased, which was seen as a clear demonstration of the power of these items. The crucifixes are known as the *Cristo de Mayo* and the *Señor de los Temblores*, and the mural as the *Cristo de los Milagros*.

The wooden crucifixes are derivations of the late Gothic Spanish sculpture in Burgos Cathedral, commonly known as the *Cristo de Burgos* or the *Cristo de San Agustín*. The cross was an object of intense devotion on the Iberian peninsula, given its miraculous and thaumaturgical properties, mainly during droughts.[8] This popularity led to the proliferation of reproductions in the seventeenth and eighteenth centuries, such as devotional prints and paintings. Some of them carry the usual inscriptions — *verdadero retrato* ('authentic reproduction') or *verdadera effigie* ('authentic effigy') —

to certify that they are faithful to the original and therefore endowed with the same miraculous powers.

Iconographic approaches to the realistic representation of Christ on the Cross were particularly in vogue in the New World. It was the Order of St Augustine that was mainly responsible for the spread of devotional images of the evocative Burgos figure, placing faithful reproductions on the altars of its churches and promoting the intensive production of devotional prints, starting with engravings brought over from Spain.[9]

The *Cristo de Mayo* is a polychrome wooden sculpture created around 1613 by the Augustinian criollo Pedro de Figueroa for the church of the Convent of San Agustín in Santiago, Chile.[10] It depicts Jesus as *Christus triumphans*, face turned upwards and eyes open, and includes elements not present in the original sculpture such as a loincloth and hair, which add great expressivity. For this reason, it became known in common parlance as the *Señor de la Agonía*, alluding to the agony Christ experienced before his death. It acquired its status as a miraculous sculpture when Santiago was struck by a powerful earthquake on 13 May 1647. The convent was among the buildings damaged but, according to news reports of the day, the statue escaped any damage other than the crown of thorns slipping from Christ's head to his neck.[11] The statue further demonstrated its powers over the following days, bringing the earthquake to an end. Events of this kind were interpreted as signs of its prodigious capacities and it came to be known as the *Cristo de Mayo*, a reference to the month of May, when the earthquake had struck. The statue quickly became the undisputed focus of religious practices associated with the earthquakes that occurred in the area. In consequence, figures invoked during pre-1647 earthquakes, such as St Saturnin, lost favour.[12] Reproductions began to show the crown of thorns around the neck as an iconographic element differentiating it from other crucifixes. The oldest versions include one in which the Cristo de Mayo is flanked by the Madonna and St John, with Mary Magdalen and a Carmelite member of the faithful, probably the person who commissioned the painting, kneeling at Christ's feet (Figure 7.1).[13] One of the main proponents of this new form of devotion was Gaspar de Villaroel, Bishop of Santiago, who, in 1648, ordered a religious ceremony to be held annually on the thirteenth day of May to venerate the miraculous statue. This ceremony is to this day a much-anticipated event in the religious calendar of Santiago, which demonstrates the enduring widespread devotion to the *Cristo de Mayo*. It has since been elevated to the status of protector of the people of Chile.[14]

The *Señor de los Temblores* in Cusco Cathedral also enjoyed great popularity in the New World, well beyond the Andean region. The crucifix is probably the work of a criollo artist, as it includes a number of clearly indigenous aesthetic elements: materials such as maguey, balsa wood and

Figure 7.1 Anonymous, *Cristo de Mayo Flanked by the Madonna and St John, with Mary Magdalen and a Carmelite Nun* (first half of the eighteenth century), Santiago de Chile, Monastery of Carmen de San José. https://arca.uniandes.edu.co/obras/194

glued canvas, and the dark tonality of Christ's complexion.[15] The figure was traditionally seen as a Catholic appropriation of the ancestral indigenous deity Pachacamac. Its iconographic elements identify it as a form of *Christus patiens*, with the head resting on the right shoulder and the wounds clearly visible on the body. Its status as a bulwark against earthquakes, promoted a few years later than that of the *Cristo de Mayo*, is linked to the Cusco earthquake of 31 March 1650. The miraculous intervention consisted of the sudden cessation of the tremors during a public invocation to the statue.

One of the most well-known visual testimonies to this earthquake is the controversial votive painting commissioned by Alonso Cortés de Monroy and located in the same cathedral. It provides a bird's-eye view of the city of Cusco. The anonymous artist, possibly part of Diego Quispe Tito's circle, used artistic licence to give the city the appearance of one built according to imperial canons. It shows what is happening in the streets of the devastated city centre: hordes of fleeing residents and the faithful extracting sacred objects from churches to carry them in a procession. However, the main focus is on the city's main square, where people are gathered around the statue of Christ on the Cross. The details of this public demonstration of faith, with its focus on the local statue, overshadows what is depicted in the upper-left corner of the painting: the *Virgen de los Remedios* and a nun interceding with the Holy Trinity on behalf of the people of Cusco. This secondary thread is reinforced by the inscription below the scene, which makes no mention of the role of the crucifix in the ending of the earthquake.[16]

Yet, in collective popular memory, it was the figure of Christ that was credited with bringing the seismic shocks to an end. The painting has been the focus of debate as a result of its links to the iconographic conflict between the Bishop of Cuzco, Manuel de Mollinedo y Angulo, a fierce supporter of the Spanish image of Remedios, and the Society of Jesus, decidedly in favour of promoting images that reflected the cultural heritage of the city, once the capital of the Inca empire. The crucifix became the de facto primary devotional figure to invoke when an earthquake struck, generating a significant market in carved and painted reproductions with minimal iconographic variations that preserved its instant identifiability as the protector of the city of Cusco.[17] Some of these are still located where they were originally venerated, while others are the jewel in the crown of eminent private collections of colonial art.[18]

The third image that is associated with earthquakes is the *Cristo de los Milagros*, venerated in Lima in colonial times. The mural depicts Christ on the Cross, with other devotional figures added at a later date. Its role as the focus of devotional practice is linked to the earthquake that hit the viceregal capital on 13 November 1655. The Pachacamilla quarter suffered powerful shocks, and the buildings housing a brotherhood of Angolan slaves were destroyed.[19] The only structure miraculously left standing was the wall on which a few years earlier one of the slaves had created this depiction of the Crucifixion. However, veneration of the image started only years later, when it demonstrated its power once again by healing a tumour and ending the 1687 earthquake.[20] Veneration of the Pachacamilla mural had initially consisted of informal gatherings of local people, but this second miracle led to its formalisation in two ways: two new brotherhoods were established, one run by indigenous people of African-American descent, and one by the Jesuit order, with a shrine built in 1771 specifically for the purpose of venerating the mural.[21] Innumerable reproductions also testify to how widespread veneration of the image became, such as the engravings that now form part of the Oswaldo and Zannie Sandoval collection,[22] and those held in the Biblioteca Nacional de Lima.[23]

The Madonna and disasters in the Andes: From Guápulo to Cayma

The influence of Old World compositional approaches was also fundamental to the development of votive images that associated the Virgin Mary with disasters. Established representations, such as the *Virgen de los Remedios*, the *Virgen de la Merced*, the *Virgen del Rosario* and the *Virgen de la Candelaria*, entered Latin American culture via different routes, augmented with novel symbolic, aesthetic and narrative elements.[24] New legends soon emerged about these images and their identity, such as references to them being found by chance by a local person, to their clear similarities to the physical traits of indigenous people or even to their ability to influence local spirits through mysterious events for the benefit of local communities.

Iconographic research recently undertaken by art historians has reconstructed the range of ways in which Spanish and European approaches to representing the Virgin were assimilated in Latin America, also recognising elements of indigenous spirituality such as graphical devices, the use of colour and specific materials. In Mexico, the historiographical focus has mainly been on the *Virgen de los Remedios* and the *Virgen de Guadalupe*.[25] New light has also been shed on pre-Hispanic virgin deities reincarnated as new forms of the Madonna, including ones linked to the elements of earth and water.[26]

In a well-known contribution to the use of religious images in Latin America, Alfonso Rodríguez de Ceballos elaborated a classification of images based on their function.[27] The approach proposed that narrative or descriptive images, whose function is in essence didactic, should be distinguished from iconic or atemporal ones used in veneration, as well as from those designed to produce an emotional response and were therefore used in public or private devotional practices. However, the classification does not appear to cover images linked to disasters or ones that feature the Madonna, which are hybrid in nature as they perform all three of these functions. This holds for at least the majority of the cases examined, where the Madonna, depicted in the standard devotional form, is set in a specific calamitous situation and creates a more or less explicit relationship with the faithful, who are shown performing acts of devotion in the foreground. To assist the observer, the scene also often contains ornate inserts that provide a brief verbal description or explanation, thus fulfilling the didactic function. The main objective of such compositions is primarily to ensure that the memory of the calamity remains vivid in the local community and at the same time to create an indissoluble link between the disaster and the relevant miraculous action, which fulfils the devotional function.

The clearest examples of this polyfunctional nature of images pertaining to disasters are found in Andean art. Here, more than anywhere else, the disaster appears in scenes associated with local miraculous figures, in particular in images in shrines at some distance from the city and above all where devotional practices were performed by the indigenous community.

In the city of Quito, for example, the *Virgen de la Merced* above the main altar of the cathedral has, since the sixteenth century, been the object of fervent invocation whenever the nearby Pichincha volcano made its presence known through tremors that exposed the residents to danger. For this reason, the statue was also known as the *Virgen del Terremoto* ('Virgin of the Earthquake'). It became the prototype for two other locally venerated figures: the *Virgen del Volcán* ('Virgin of the Volcano') in Recoleta di San Diego and the significantly more popular *Virgen Pelegrina* ('Pilgrim Virgin'), appearing in numerous prints from the seventeenth century and celebrated as a powerful image, in particular against earthquakes. Another figure of the Virgin with Spanish origins, greatly venerated in the province of Quito but less associated with disasters, is the *Virgen de los Desemparados* of Valencia, which is in turn a derivation of the *Madonna della Misericordia*.[28] However, no available sources document the processions of these statues during the earthquake. In contrast, there are some representations of the *Virgen de Guápulo* and the *Virgen de El Quinche*, whose sculptures are located in two shrines outside the city and remain very popular to this day. Both are triangular compositions, drawing inspiration from the *Virgen de Guadalupe* in Extremadura, a Romanesque Black Madonna whose image is found across the Andean region.[29]

The Guápulo statue, recognisable by the ornate floral sceptre held in the right hand, demonstrated its powers on many occasions, including several associated with earthquakes or hydrometeorological disasters. There is a painting that depicts a procession venerating the statue in supplication of the end of a long drought. The work is part of a series created by Miguel de Santiago and his workshop between 1699 and 1706, as part of a major campaign organised by José de Herrera y Cevallos to adorn the shrine.[30] The series depicts a sequence of miraculous interventions and is one of the many attestations of the power of the statue against the force of nature. The works focus on events that involved either the whole community or specific individuals. It is based on the European hagiographical model, which Miguel de Santiago had used in the convent of San Agustín in Quito.[31]

The scene is set in an arid rural landscape with the cordillera and the Pichincha volcano in the background. Rather than depicting the arrival of the rain that brings the disaster to an end, the focus is on the intercessional activities. The Madonna appears twice, once as the statue being carried in procession and also, in the upper-left part of the composition, as an ethereal being enveloped in divine light. The canvas carries an inscription that briefly describes the circumstances of the 1621 miracle, which occurred as part of the special religious activities organised by the local secular authority, the *Real Audiencia de Quito*. This was not the first time that this had happened. Faced with extraordinary events like droughts, earthquakes or epidemics, the image would be brought to the city and carried in solemn procession to allow the invocations to be heard.

The other important votive image in colonial Quito is known as the *Virgen de El Quinche*. The miraculous figure, a wooden sculpture with Renaissance stylistic features, was discovered in the indigenous locality of Oyacachi. At the beginning of the seventeenth century, it was relocated to the nearby city of El Quinche for security reasons and placed in a chapel that would soon be replaced by an imposing shrine. The paintings conserved there include one of special interest, namely, depicting the arrival of the image in Quito (Figure 7.2). The composition reflects that of the painting described above: the miracle has not yet taken place. The anonymous artist, perhaps inspired by the Guápulo series, focuses on the activities of collective prayer, the only true way to produce the rain that the fields so badly need.

In the southern Andean area, the most strongly felt devotion was to the *Virgen de la Candelaria* of Lari Lari (also known as Cayma). Its key iconographic element is the candle held in the right hand. The original is located on Tenerife and is recognised as the protector of the Canary Islands.[32] The best-known faithful reproduction to reach the New World at the beginning of the seventeenth century was that sent to the diocese of Arequipa in what is now southern Peru.[33]

Around 1780, Jacinto Carvajal produced a series of 13 paintings depicting local miracles attributed to the figure for the church of San

Figure 7.2 Anonymous, *Procession of Virgen de El Quinche to Quito during a Drought* (end of the eighteenth century), Quito. Courtesy of the Shrine of El Quinche.

Miguel Arcángel.[34] This interesting series was probably ideated in collaboration with Juan Domingo Zamácola y Jáuregui, the parish priest who commissioned it. Two of the scenes illustrate the dramatic consequences of the eruption of the Huainaputina volcano near the city of Arequipa.[35] In both cases, it is the very nature of the catastrophe that determines its dramatic representation, clearly stretching the artist's inherently modest talent to the limit. The first depicts an act of public prayer that took place on 19 February 1600, when an earthquake and an ash storm followed the eruption, as noted in the medallion placed at the lower left of the image (Figure 7.3).

The composition falsifies the perspective to compress the buildings deformed by the earthquake. Some of the figures are fleeing, others tend to victims, some are on their knees with hands clasped in prayer for divine respite, while the clergy gather round the canopy that shields the statue. The other painting references a procession during what became known in collective memory as *vómito negro* ('black vomit'): the ash storm of 1604 caused by the volcano's volatile nature. Here, too, the execution is pedestrian but effective in rendering the collective panic engendered by the earthquake, perceived as a supernatural event disrupting social order and releasing deep emotions.

Figure 7.3 Jacinto Carvajal and his workshop, *Public Prayer to the Virgen de Cayma during the 1600 Earthquake* (ca. 1780), Cayma. Courtesy of the Church of San Miguel Arcángel.

These examples show how urgent it was to create a memorable identity of supplications to the Madonna in the Andean provinces. This urgency meant that calamities became pretexts to enrich legends about the statue and reinforce its prodigious nature. Further examples of the opportunity disasters provided to make good use of religious images are the mural and the canvas located in the parish churches of Catca and Tiobamba, to the south-east and north-west of Cusco, respectively, whose subject matter is the plague of 1720. The outbreak, one of the most devastating in the history of South America, started in Buenos Aires and quickly spread along the trade routes of Peru, reaching as far as Cusco and Lima.[36]

The Catca work is a large painting commissioned by the priest Tomás Collado and executed by José Benito Calderón between 1806 and 1820, a century after the event. It shows a simplified urban view of Catca, with the catastrophe presented as a social phenomenon affecting the whole population of the city. Particular attention is paid to the Franciscans, Dominicans and Mercedarians providing assistance to the sick.[37] Similarly, the painting in the parish church of Tiobamba, produced in the second half of the eighteenth century by an equally mediocre artist, shows the general turmoil caused by the epidemic, with various groups of agitated figures. The appeal to God is here mediated by the figure of Immaculate Mary at the centre of the scene. The work is displayed together with other canvases that depict miracles attributed to the Madonna in a collection that was first documented by Teresa Gisbert and José Mesa.[38] Unfortunately, it is currently in a precarious state of conservation and in danger of severe fading, and it is hoped that restoration work can be undertaken soon.[39]

Images of disasters in New Spain: Chaos and order

In the capital of the Viceroyalty of New Spain, veneration in the face of disasters focused mainly on the *Virgen de los Remedios*, a practice imported from Spain, and the *Virgen de Guadalupe di Tepeyac*, a local figure venerated in particular by indigenous and criollo people. These images were regularly used in collective supplication, as evidenced by archive material.[40] As early as in the first quarter of the sixteenth century, both were invoked against droughts and agricultural epidemics. Droughts were also particularly frequent in this area in the seventeenth century and throughout the eighteenth century.[41] Over time, the *Virgen de los Remedios* became established as the prime intercessor in the case of hydrometeorological events. The *Virgen de Guadalupe* was seen as having miraculous powers against biological adversities. However, the predilection of the Mexican people for the latter prevailed and she was proclaimed the protector of the capital city of New Spain on 27 April 1737. This prestigious recognition was the result of intervention in an epidemic known as *matlazáhuatl*, a typhoid-like infection that had been ravaging the indigenous population since 1736. In gratitude to Guadalupe, the priest Cayetano Cabrera Quintero published *Escudo de armas de México* in 1746.[42] This contains the famous engravings by Baltasar Troncoso y Sotomayor, from a drawing by José de Ibarra, which depict the apparition of the *Virgen de Guadalupe* to the population afflicted by the disease (Figure 7.4). The composition is elaborated on two planes.

Figure 7.4 José de Ibarra (drawing), Balthasar Troncoso (engraving), *Virgen de Guadalupe Appears to the Population Afflicted by the Plague*, in C. Cabrera y Quintero, *Escudo de armas de México* [...], 1746. © John Carter Brown Library, Box 1894, Brown University, Providence, R.I. 02912. Courtesy of the John Carter Brown Library. (CC BY-SA 4.0)

The background shows a dramatic urban scene. The streets are filled with the dying and the dead, with a semi-naked woman lying next to her lifeless child at the centre of the work. In this human scenario where the epidemic is victorious, José de Ibarra shows his debt to the European figurative tradition, which had offered a range of compositional approaches to scenes of pestilence.[43] The foreground shows the municipal authorities, dressed in their finery and prostrate before the Madonna, who appears in the usual iconographic form, surrounded by an aura of light and accompanied by two cherubim, one of whom holds an inscription in Latin that identifies the figure as a divine shield for the people of Mexico.

The images associated with biological disasters in New Spain include another significant painting. The work, by an unknown artist, is located in the church of San Pedro Apóstol in San Pedro Zacatenco, but it was probably originally located in the Jesuit church of San Gregorio in the capital. The large canvas depicts a procession celebrating the European *Madonna di Loreto*, to whom a chapel inside the San Gregorio seminary in the capital of New Spain was dedicated. The chapel was founded in 1680 and was redeveloped later. It contained a rich collection of sacred artefacts and relics, including a reproduction of the original Italian statue that has since been lost, so we can only assume that it was a faithful reproduction of the figure in the painting.[44]

The scene appears to show a well-documented 1727 ceremony, as noted in the inscription attached to the lower edge of the modern frame that now protects the painting. On 29 October 1727, the *Madonna di Loreto* was carried through the streets of the city in a plea for intercession against an epidemic of measles that had been decimating the population for about a year. Thanks to the detailed representation of the festooned buildings, with people gathered on the balconies, the precise location can be determined, as well as the actual moment, namely, when the statue was carried back to the church of San Gregorio at the end of the novena held in honour of the Madonna at the cathedral.

However, even though this is a relatively faithful reconstruction of an event that was directly related to the disaster, the painting makes no obvious reference to the tragic consequences of the spread of the disease. Instead, there are abundant details of the civic and ecclesiastical personages taking part in the procession, who appear in a very precise order. First, at the head of the procession, come the religious orders (Mercedarians, Augustinians, Franciscans and Dominicans), followed by members of the clergy and the chapter of the cathedral, including Archbishop José de Lanciego y Eguilas and the prelate Antonio Villaseñor y Monroy. Also present are the municipal authorities and those of the *Real Audiencia*, preceded by university scholars, and then a throng consisting of the city's elite.

The hidden message of the painting appears clear: to commemorate the magnificence of the public ceremony in honour of the Madonna and the political repercussions for the civic and religious dignitaries represented rather than the horror experienced by the unfortunate victims of the terrible epidemic. However, ethnically indigenous members of the *Buena Muerte* brotherhood were deliberately omitted from the apparently rigorous representation, even though they were in fact present at the event.[45] In other words, as in the case of the inscription of the *Virgen de Guadalupe* noted above, rather than focusing on the chaos sweeping through the city as a result of the outbreak, the anonymous artist, possibly abiding by the wishes of the unknown person who commissioned the work, chooses to show a celebration of civic religious fervour focused on the Vergine di Loreto: the disaster is thus used as an excuse to exalt political order and the single-minded actions of the highest civic and ecclesiastical authorities.[46]

The intercession of the saints and the veneration of St Emygdius

In this preliminary and partial catalogue of imagery related to disasters, representations of saints cannot be ignored, even if they had only a marginal impact with respect to the significantly more abundant imagery of the Crucifixion and the Virgin Mary. There is documentary evidence of the use of the representations of saints, mainly their statues, in processions and collective prayers in times of disaster. The saints most frequently associated with earthquakes include Nicholas of Tolentino (Mexico), Francis Xavier and Saturnin of Toulouse (Santiago), Martha of Bethany (Arequipa), Francis Solanus (Lima) and the Jesuit Francis Borgia (Bogotá).[47] However, there is currently no evidence of visual sources documenting the circumstances of these practices. This is probably because the choice of which saint to venerate was often determined by drawing lots or by voting rather than as a result of a specific miracle. Moreover, one saint was often substituted for another, simply as a result of being seen as more efficacious on the basis of a range of criteria.[48]

Resorting to the intercession of saints in times of earthquakes, eruptions, floods and epidemics did not therefore necessarily lead to the elaboration of representations of the kind we have been considering for the Virgin. In fact, the few cases discovered do not relate to contemporary miracles but to events that had occurred earlier in the individual's lifetime and were recorded in official documents. One example is the painting in the Augustinian convent in Quito of the miracle performed by St Augustine to counter a locust plague. This is part of a series by Miguel de Santiago and his workshop around the middle of the seventeenth century, based on a series of engravings by Boetius Adamsz Bolswert.[49] There are also the images of St Francis Xavier, invoked against the bubonic plague, of which there are at least two versions, one in the monastery of Carmen de San José in Santiago, Chile, and the other in the Jesuit church in Quito. The composition is inspired by European prototypes disseminated in the press.[50]

Nevertheless, the eighteenth century saw a significant increase in the veneration of Emygdius, the martyred Bishop of Ascoli, specifically in connection with earthquakes. A notable number of paintings are associated with his veneration as an intercessor in calamities of this kind. His first involvement came in early 1703, when powerful shocks struck central Italy.[51] Veneration of the saint subsequently spread to the Iberian peninsula, where it was consolidated around the middle of the century when an earthquake struck Lisbon in 1755. This disaster led to intensive graphical output showing the saint as a force not only against earthquakes but also against tidal waves.[52]

The numerous paintings of St Emygdius found in Latin America appear to follow the same compositional prototype, which has its roots in iconographic approaches elaborated in Spain and disseminated through the press. The full figure is represented in episcopal attire — cape, mitre and crozier — in the act of bestowing a blessing. The saint appears on a cloud below which, at the lower edge of the image, two disasters are depicted: a city devastated by an earthquake on the left, and one or more boats menaced by a storm on the right. The intercessor's gaze is directed either downwards towards the city in ruins, its buildings toppling and its towers cracked, or upwards towards God. In some cases, the figure is accompanied by pairs of cherubim holding the usual iconographic elements, including the palm of martyrdom. A frequent addition is an oration to the saint's special powers against earthquakes, which suggests that the main purpose of images of this kind was to provide visual support for individual or collective veneration.[53]

This iconographic approach was absorbed into local convention largely unmodified, as demonstrated in the larger and smaller paintings that to this day hang on the walls of churches and convents founded in colonial times.[54] Just in the capital of the New Kingdom of Granada we find five examples: in the convents of Santo Domingo and Santa Inés, in the churches of San Francisco and La Tercera, and in the Colegio del Rosario (Figure 7.5). This last one is from the last quarter of the eighteenth century and has been attributed to Joaquín Gutiérrez or his workshop. It carries the usual prayer inscribed in a medallion at the lower right.[55]

An isolated case, but one that is no less interesting, is a painting conserved at the *Recoleta* in Arequipa that depicts a martyred bishop whose veneration is closely associated with a disaster that hit the Kingdom of Naples, namely, the patron saint of the city of Naples, St Januarius. His status as an intercessor in cases of catastrophe is linked in particular to the eruption of Mount Vesuvius in December 1631, which was the first

Figure 7.5 Joaquín Gutiérrez or his workshop, *St Emygdius as Intercessor against Earthquakes* (second half of the eighteenth century), Bogotá, Colegio del Rosario. Courtesy of the Museo de la Universidad del Rosario. Photograph: © Museo de la Universidad del Rosario.

disaster in modern times to spark the production of dedicated artistic and literary works.[56] The iconography of this image is somewhat unusual: the full figure of the saint is portrayed indoors, in episcopal attire and with an ampule of the blood that flowed at his beheading.[57] Two tame lions lie at his feet, in reference to one of the tortures endured by the saint before his martyrdom. The window on the right of the scene reveals a stylised mountain, alluding to Mount Vesuvius in eruption, but the saint does not appear to be interacting with the volcano. We have no way of knowing whether the anonymous artist was aware of the usual compositions depicting St Januarius, in which the saint appears on a cloud in the sky, repelling the lava flowing down Mount Vesuvius with his hands.[58]

Current research sheds little light on why such an unusual painting was produced in the Peruvian city, although some documentary sources indicate that the saint was particularly venerated at the beginning of the seventeenth century, especially when the activity of the nearby Huainaputina volcano caused repeated earthquakes. A relevant corpus of documents identified by Victor Barriga over half a century ago provides information about an important religious commission, which preceded the eruption of Mount Vesuvius in 1631 that confirmed the saint as the protector of Naples. The commissioning clients were the *contador* Sebastián de Mosquera, who financed the construction of a hermitage dedicated to the saint in 1601, the *cabildo* (municipal authority) that provided a wooden statue for the altar, and Antonio de la Raya, Bishop of Cusco, who authorised the veneration of the saint, proclaiming processions and collective prayer whenever the volcano threatened the city.[59] This devotion was short-lived, however. Even though it was quickly supplanted by the veneration of the *Virgen de la Candelaria* and of St Martha, further research might reveal interesting new information about this early veneration of St Januarius, and about any political implications underlying its spread in Andean religious practice.

Conclusion

A number of conclusions can be drawn from this initial overview of religious imagery linked to disasters in Latin America. First, the significant presence of devotional images refers to the spiritual dimension in which calamities were experienced and understood. As in other contexts conditioned by post-Tridentine Christianity, the images were the visual framework within which the interpretations of disasters as forms of divine punishment were articulated. Their presence in chapels, churches and shrines was a way of formalising the event and recording it as a demonstration of the power of prayer in the face of calamity.

The divinities invoked to mediate with God were mainly Christ crucified and the Madonna. The seventeenth century saw, in the Andean region, the proliferation of figurative representations like the *Cristo de Mayo* and *Señor de los Temblores* statues and the *Señor de los Milagros* mural. In contrast, iconographical representations of the Virgin, in various forms across America, were narrative in nature: the miracle took precedence over figurative or aesthetic considerations. The Madonna usually appears enveloped in a supernatural divine aura and assumes the locus of the invocations of the faithful. In addition to these divine figures, certain saints were also venerated in the context of American religiosity. In this regard, the analysis of figurative and literary evidence is bringing out the most popular ones, in particular St Emygdius, usually invoked against earthquakes. From the eighteenth century onwards, votive images of this saint, with clear roots in European iconography, were promoted across Latin America.

Beyond the formal and iconographic aspects, the figurative sources considered in this contribution are also significant for their socio-historical value, namely, as tools for the promotion and renewal of devotions in times of calamities. As we have seen, the need to build a narrative of legends was often designed to demonstrate a direct relationship between the images and the local communities, and in particular the indigenous community. In this sense, disasters were seen as a valid pretext for disseminating narratives in which the miracles performed by such new images were presented as a direct consequence of collective invocation. Thus, these considerations open the door for further research into the presence and meaning of religious images in the American context, which would indeed improve our knowledge of the different interpretations and representations of disasters in the early modern period.

Notes

1 This work was supported by the DisComPoSE project (*Disasters, Communication and Politics in Southwestern Europe*), which has received funding from the European Research Council (ERC) under the European Union's Horizon 2020 research and innovation programme (grant agreement No. 759829).

2 Virginia García Acosta, 'Divinidad y desastres. Interpretaciones, manifestaciones y respuestas', *Revista de Historia Moderna. Anales de la Universidad de Alicante*, 35 (2017), 46–82; María Eugenia Petit-Breuilh Sepúlveda, 'Religiosidad y rituales hispanos en América ante los desastres (siglos XVI-XVII): las procesiones', *Revista de Historia Moderna. Anales de la Universidad de Alicante*, 35 (2017), 83–115.

3 It is beyond the scope of this paper to consider the use of religious images in the more generic context of private devotion. We will therefore not address votive paintings used in individual invocations, such as preventing death or healing illness. Such items undoubtedly played a key role in popular Hispano-American imagery, and their use in private veneration makes them complementary to those designed specifically to meet the needs of collective rituals.

4 *Historia y desastres en América Latina*, ed. by Virginia García Acosta, 3 vols. (Bogotà; Lima; Mexico City: La Red – Ciesas, 1996–1997, 2008).

5 María Eugenia Petit-Breuilh Sepúlveda, *Naturaleza y desastres en Hispanoamérica: la visión de los indígenas* (Madrid: Sílex, 2006).

6 Gerrit Jasper Schenk, '*Disastro*, Catastrophe, and Divine Judgment: Words, Concepts and Images for 'Natural' Threats to Social Order in the Middle Ages and Renaissance', in *Disaster, death and the emotions in the shadow of the apocalypse, 1400–1700*, ed. by Jennifer Spinks and Charles Zika (London: Palgrave Macmillan, 2016), pp. 45–68.

7 The cross was also the symbol that was best suited to the monolithic pre-Hispanic sculptures used in the veneration of local idols. See Federico González Frías, *El simbolismo precolombino: cosmovisión de las culturas arcaicas* (Buenos Aires: Kier, 2003).

8 In Andalusia, a region repeatedly struck by hydrometeorological disasters, there are two paintings depicting *pro-pluvia* rogations to the *Cristo de Burgos*. See William A. Christian, 'Images as Beings in Early Modern Spain', in *Sacred Spain. Art and Belief in the Spanish World*, ed. by Ronda Kasl (New Haven; London: Yale University Press, 2009), pp. 75–100 (p. 90), and Almudena Ros de Barbero, 'Procesión de rogativa por falta de lluvias del Santo Crucifijo de San Agustín, ante el Ayuntamiento de Sevilla', in *De Zurbarán a Picasso. Artistas andaluces en la Colección Abelló*, ed. by Benito Navarrete Prieto and Almudena Ros de Barbero (Seville: Instituto de la Cultura y las Artes de Sevilla, 2015), p. 38, cat. 13.

9 Antonio Iturbe Sáiz, 'El Cristo de Burgos o de San Agustín en España, América y Filipinas', in *Los crucificados, religiosidad, cofradías y arte. Actas del Simposium*, ed. by Francisco Javier Campos y Fernández de Sevilla (San Lorenzo de El Escorial: Real Centro Universitario Escorial-María Cristina, 2010), pp. 683–714. For Ibero-American variants of the *Cristo de Burgos*, see also Hector H. Schenone, *Iconografía del arte colonial: Jesucristo* (Buenos Aires: Fundación Tarea, 1998), pp. 304–6.

10 Guillermo Carrasco Notario, 'La primera escultura chilena. El Señor de Mayo en el arte y la literatura nacionales', in *Iconografía agustiniana*, ed. by Rafael Lazcano (Rome: Institutum Historicum Augustinianum, 2001), pp. 285–88.

11 The damage to the statue was seen as proof of its divine power, as it had survived the earthquake. See Lucila Iglesias, 'Imágenes castigadas y redentoras. Estrategias narrativas para la consolidación de devociones vinculadas con terremotos en la Sudamérica colonial', in *Paisaje y naturaleza. IX Encuentro Internacional sobre Barroco*, ed. by Paola Maurizio and María Cecilia Avegno (La Paz: Fundación Unión Latina, 2019), pp. 303–8.

12 Alfredo Palacios Roa 'Antecedentes históricos de la "abogacía telúrica" desarrollada en Chile entre los siglos XVI y XIX', *Historia crítica*, 54 (2014), 171–93.

13 Luis Mebold K., *Catálogo de pintura colonial en Chile. Obras en monasterios de religiosas de antigua fundación* (Santiago: Ediciones Universidad Católica de Chile, 1987), p. 175.

14 Mauricio Onetto Pavez, *Temblores de tierra en el jardín del Edén. Desastre, memoria e identidad. Chile, siglos XVI–XVIII* (Santiago: Dirección de Bibliotecas, Archivos y Museos – DIBAM, Centro de Investigaciones Diego Barros Arana, 2017). See also the illustrations reproduced in Alfredo Palacios Roa, *Historia ilustrada de los megaterremotos ocurridos en Chile entre 1647 y 1906* (Valparaíso: Ediciones Universitarias de Valparaíso, 2016), pp. 46, 50–51, figs. 5–7.

15 Hector H. Schenone, *Iconografía del arte colonial: Jesucristo* (Buenos Aires: Fundación Tarea, 1998), pp. 323–27.

16 Maya Stanfield-Mazzi, *Object and Apparition: Envisioning the Christian Divine in the Colonial Andes* (Tucson: The University of Arizona Press, 2013), pp. 97–116; Thomas B. F. Cummins, 'Argumentos milagrosos: pintura y política cultural tras el terremoto de 1650', in *Pintura cuzqueña*, ed. by Ricardo Kusunoki and Luis Eduardo Wuffarden (Lima: Asociación Museo de Arte de Lima, 2016), pp. 73–91; Patrick Thomas Hajovsky, 'Shifting Panoramas: Contested Visions of Cuzco's 1650 Earthquake', *The Art Bulletin*, 100, 4 (2018), 34–61.

17 For the spread of images of the *Señor de los Temblores*, see Alfonso Rodríguez G. de Ceballos, '"Trampantojos a lo divino': iconos pintados de Cristo y de la Virgen a partir de imágenes de culto en América meridional', in *Actas del III Congreso Internacional de Barroco Iberoamericano: Territorio, Arte, Espacio y Sociedad*, ed. by José Manuel Almansa et al. (Seville: Universidad Pablo de Olavide, 2001), pp. 24–33.

18 Luis Eduardo Wuffarden, 'El Cristo de los Temblores', in *Los siglos de oro en los virreinatos de América, 1550–1700* (Madrid: Sociedad Estatal para la Conmemoración de los Centenarios de Felipe II y Carlos V, 1999), pp. 354–56, cat. 109; Thomas B. F. Cummins, 'Señor de los Temblores', in *La colección Petrus y Verónica Fernandini. El arte de la pintura en los Andes*, ed. by Ricardo Kusunoki (Lima: Museo de Arte de Lima, 2015), pp. 126–31; Luis Eduardo Wuffarden, 'Señor de los Temblores', in *Pintura virreinal en los Andes. Colección Celso Pastor de la Torre*, ed. by Alberto Servat (Lima: Instituto Cultural Peruano Norteamericano, 2017), pp. 34–35; Isabel Cruz de Amenábar (ed.), *Terremotos. Cristos milagrosos en el arte virreinal surandino* (Santiago: Fundación Joaquín Gandarillas Infante, 2019), pp. 56–69. See also Ramón Mujica Pinilla, *La imagen Transgredida. Estudios de iconografía peruana y sus políticas de representación simbólica* (Lima: Fondo editorial del Congreso del Perú, 2016), pp. 526–27, 532.

19 Rubén Vargas Ugarte, *Historia del Santo Cristo de los Milagros* (Lima: H. Vega Centeno, 1957); Raúl Banchero Castellano, *Lima y el mural de Pachacamilla. Historia del Señor de los Milagros de Nazarenas, del monasterio y de la hermandad* (Lima: Aldo Raúl Arias Montesinos, 1972).

20 Flavia Tudini, 'Narrating the 1687 Lima Earthquake: Institutions and Devotions in the Face of Catastrophe', in *Heroes in Dark Times. Saints and Officials Tackling Disaster (16th–17th Centuries)*, ed. by Milena Viceconte, Gennaro Schiano, Domenico Cecere (Rome: Viella, 2023), pp. 259–85.

21 Julia Costilla, '"Guarda y custodia' en la Ciudad de los Reyes: la construcción colectiva del culto al Señor de los Milagros (Lima, siglos XVII y XVIII)', *Fronteras de la Historia*, 20, 2 (2015), 152–79.

22 Luis Eduardo Wuffarden, 'El mural y su estela iconográfica', in *El Señor de los Milagros. Historia, Devoción e Identidad,* ed. by Ramón Mujica Pinilla et al. (Lima: Banco de Crédito del Perú, 2016), pp. 171–91 (p. 178); Luis Eduardo Wuffarden, 'Cristo de los Milagros', in *Arte colonial. Colección Museo de Arte de Lima*, ed. by Ricardo Kusunoki and Luis Eduardo Wuffarden (Lima: El Museo de Arte de Lima, 2016), pp. 168–69.

23 Ricardo Estabridis Cárdenas, *El grabado en Lima virreinal. Documento histórico y artístico (siglo XVI al XIX)* (Lima: Fondo Editorial Universidad Nacional Mayor de San Marcos, 2002), pp. 126–28, 151, 153.

24　Francisco Montes González, 'Vírgenes viajeras, altares de papel. Traslaciones pictóricas de advocaciones peninsulares en el arte virreinal', in *Arte y patrimonio en España y América*, ed. by María de los Ángeles Fernández Valle et al. (Seville: Universidad Pablo de Olavide, 2014), pp. 89–117. For more on the role of the Andalusian route, see *Religiosidad andaluza en América. Repertorio iconográfico*, ed. by Rafael López Guzmán and Francisco Montes González (Granada: Editorial Universidad de Granada, 2017).

25　Francisco Miranda Godínez, *Dos cultos fundantes: Los Remedios y Guadalupe (1521–1649). Historia documental* (Zamora: El Colegio de Michoacán, 2001). On Guadalupe, see in particular Jeanette Favrot Peterson, *Visualizing Guadalupe from Black Madonna to Queen of the Americas* (Austin: University of Texas Press, 2014). Teresa Gisbert has addressed the Andean representations of the Madonna in several works. One of the most recent is Teresa Gisbert, *Arte poder e identidad* (La Paz: Gisbert, 2016), pp. 111–57. For devotion to the Madonna in the New Kingdom of Granada, see Olga Isabel Acosta Luna, *Milagrosas imágenes marianas en el Nuevo Reino de Granada* (Madrid; Frankfurt: Iberoamericana; Vervuert, 2011).

26　María Eugenia Petit-Breuilh Sepúlveda, 'Diosas, vírgenes y chamanes femeninos en el mundo indígena hispanoamericano durante el Antiguo Régimen (siglos XVI-XVIII)', in *Vírgenes, reinas y santas. Modelos de mujer en el mundo hispano*, ed. by David González Cruz (Huelva: Universidad de Huelva, 2007), pp. 215–34.

27　Alfonso Rodríguez G. de Ceballos, 'Usos y funciones de la imagen religiosa en los virreinatos americanos', in *Los siglos de oro*, pp. 89–106.

28　Ángel Justo Estebaranz, 'Advocaciones marianas españolas en el arte de la Real Audiencia de Quito', *Atrio. Revista de Historia del Arte*, 20 (2014), 24–39.

29　Ronda Kasl, '*Milagros por la Similitud*: Our Lady of Guadalupe in the Colonial Andes', *Hispanic Research Journal*, 16, 5 (2015), 456–70. See also Françoise Crémoux, 'Las imágenes de devoción y sus usos. El culto a la Virgen de Guadalupe (1500–1750)', in *La Imagen religiosa en la monarquía hispánica: usos y espacios*, ed. by María Cruz de Carlos Varona et al. (Madrid: Casa de Velázquez, 2008), pp. 61–82.

30　Ángel Justo Estebaranz, 'La representación del milagro en el Quito barroco: La serie de pinturas de la Virgen de Guápulo', in *Exvotos y religiosidad popular en Ecuador Siglos XVII–XX* (Quito: Casa de la Cultura Ecuatoriana, 2017), pp. 49–61; Adriana Pacheco Bustillos, 'La imagen de Nuestra Señora de Guadalupe en el Santuario de Guápulo, pueblo de indios en la Real Audiencia de Quito', in *La Virgen de Guadalupe de Extremadura en América del Sur. Arte e iconografía*, ed. by Rafael Jesús López Guzmán and Pilar Mogollón Cano-Cortés (Cáceres: Fundación Academia Europea e Iberoamericana de Yuste, 2019), pp. 87–118.

31　Nancy Morán de Guerra and Alfonso Ortiz Crespo, 'Procesión durante la sequía', in *Los siglos de oro*, pp. 338–39, cat. 99.

32　Pablo Francisco Amador Marrero and Carlos Rodríguez Morales, 'Aportaciones a la iconografía de la Candelaria isleña en América', in *Imagen y reliquia. Nuevos estudios sobre la antigua escultura de la Candelaria*, ed. by Carlos Rodríguez Morales (San Cristóbal de La Laguna: Ayuntamiento de San Cristóbal de La Laguna, 2018), pp. 129–43.

33　Alejandro Málaga Núñez-Zeballos, 'La Virgen Candelaria en el obispado de Arequipa: origen y milagros', in *Incas e indios cristianos: élites indígenas e identidades cristianas en los Andes coloniales*, ed. by Jean-Jacques Decoster

(Cusco: Centro de Estudios Regionales Andinos Bartolomé de las Casas, 2002), pp. 251–58; Alejandro Málaga Núñez-Zeballos, 'El enojo de los dioses. Terremotos y erupciones en Arequipa del siglo XVI', in *El hombre y los andes. Homenaje a Franklin Pease*, ed. by Rafael Varón and Javier Flores Espinoza, 3 vols. (Lima: PUCP, 2002), vol. II, pp. 905–13.

34 Suzanne Stratton-Pruitt, 'Our Lady of Cayma', in *The Virgin, Saints and Angels. South American Painting 1600–1825 from the Thoma Collection*, ed. by Suzanne Stratton-Pruitt (Milan: Skira, 2006), pp. 152–54.

35 María Eugenia Petit-Breuilh Sepúlveda, 'Miedo y respuesta social en Arequipa: la erupción de 1600 del volcán Huaynaputina (Perú)', *Obradoiro de Historia Moderna*, 25 (2016), 67–94.

36 Gabriela Ramos, *El cuerpo en palabras. Estudios sobre religión, salud y humanidad en los Andes coloniales* (Lima: Instituto Francés de Estudios Andinos, 2020).

37 The mural is the subject of a recent photographic project undertaken by the Center for the Study of Material & Visual Cultures of Religion, <https://mavcor.yale.edu/material-objects/giga-project/ccatcca-plague-mural-zoom> [accessed 31 July 2023].

38 José de Mesa and Teresa Gisbert, *Historia de la pintura cuzqueña*, 2 vols. (Lima: Fundación Augusto N. Wiese, 1982), vol. 1, p. 191; vol. 2, fig. 266.

39 Richard L. Kagan, *Imágenes urbanas del mundo hispánico 1493–1780* (Madrid: Ediciones El Viso e Iberdrola, 1998), p. 216.

40 *Desastres agrícolas en México. Catálogo histórico. Tomo 1. Épocas prehispánica y colonial (958–1822)*, ed. by Virginia García Acosta et al. (Mexico City: Centro de Investigaciones y Estudios Superiores en Antropología Social – Fondo de Cultura Económica, 2003); Francisco de Solano, *Las Voces de la Ciudad. México a través de sus impresos (1539–1821)* (Madrid: CSIC, 1994), pp. 211–14.

41 Adrián García Torres, 'La religiosidad popular frente a las sequías en la ciudad de México (1700–1760)', *Temas americanistas*, 38 (2017), 32–56.

42 Cayetano Cabrera Quintero, *Escudo de Armas de México. Celestial protección de esta nobilísima ciudad, de la Nueva España y de casi todo el nuevo mundo* [...] (Mexico City: Joseph Bernardo de Hogal, 1746).

43 The direct reference for this scene might be the engraving by Aegidius Sadeler based on a drawing by Maerten de Vos *David sacrifices to God* from the *Stories of David* (1580-1596). See on this topic Paula Mues Orts, 'Estampas y modelos. Copia, proceso y originalidad en el arte hispanoamericano y español en el siglo XVIII', *Librosdelacorte.es*, 5 (2017), 96–118. Further European elaborations of this theme stem from Neapolitan art following the 1656 plague: Rose Marie San Juan, 'Contaminating Bodies: Print and the 1656 Plague in Naples', in *New Approaches to Naples c.1500 – c.1800: The Power of Place*, ed. by Melissa Calaresu and Helen Hills (Farnham: Ashgate, 2013), pp. 63–78; Ida Mauro, *Spazio urbano e rappresentazione del potere. Le cerimonie della città di Napoli dopo la rivolta di Masaniello (1648–1672)* (Naples: FedOA, 2020), pp. 276–90.

44 Luisa Elena Alcalá, Patricia Díaz Cayeros, Gabriela Sánchez Reyes, 'Solemne procesión de la imagen de Nuestra Señora de Loreto: la epidemia de sarampión en 1727', *Encrucijada*, 1 (2009), 23–51; Luisa Elena Alcalá, Patricia Díaz Cayeros, Gabriela Sánchez Reyes, 'On the Path to Good Health: Representing Urban Ritual in Mexico City during the Epidemic of 1727', *Miradas*, 4 (2018), 51–72. On the devotion to the Madonna of Loreto in the Mexican context, see the recent book by Luisa Elena Alcalá, *Arte y localización de un culto global. La Virgen de Loreto en México* (Madrid: Abada Editores, 2022).

45 Luisa Elena Alcalá, 'The Image of the Devout Indian: The Codification of a Colonial Idea', in *Contested Visions in the Spanish Colonial World*, ed. by Ilona Katzew (Los Angeles; New Haven: Los Angeles County Museum of Art; Yale University Press, 2011), pp. 227–49 (pp. 238–39).

46 Tomás Pérez Vejo, 'La ciudad borbónica: expresiones artísticas de una submetrópoli imperial', in *La Ciudad de México en el arte. Travesía de ocho siglos*, ed. by José María Espinasa and Alejandro Salafranca Vázquez (Mexico City: Secretaría de Cultura, Museo de la Ciudad de México, 2017), pp. 77–93 (p. 83).

47 For more on other saints associated with earthquake disasters, see Monica Azzolini, 'St Filippo Neri as Patron Saint of Earthquakes', *Quaderni Storici*, 52, 3 (2017), 727–50, as well as her essay in this volume.

48 Rogelio Altez, 'Historias de milagros y temblores: fe y eficacia simbólica en Hispanoamérica, siglos XVI-XVIII', *Revista de Historia Moderna*, 35 (2017), 178–213.

49 The Bolswert engravings were for *Iconographia Magni Patris Aurelli Agustini* (1628). See Ángel Justo Estebaranz, *Miguel de Santiago en San Agustín de Quito. La serie de pinturas sobre la vida del santo* (Quito: Fonsal, 2008).

50 *De Augsburgo a Quito. Fuentes grabadas del arte jesuita quiteño del siglo XVIII*, ed. by Almerindo Ojeda and Alfonso Ortiz (Quito: Fundación Iglesia de la Compañía de Jesús, 2015), cat. 15–16. See also Alfonso Rodríguez G. de Ceballos, 'Las pinturas de la vida de San Francisco Javier del Convento de la Merced de Quito: fuentes gráficas y literarias', *Anales del Museo de América*, 15 (2007), 89–101.

51 Viviana Castelli and Romano Camassi, 'A che santo votarsi. L'influsso dei grandi terremoti del 1703 sulla cultura popolare', in *Settecento abruzzese: eventi sismici, mutamenti economico-sociali e ricerca storiografica*, ed. by Raffaele Colapietra et al. (L'Aquila: Colacchi, 2007), pp. 107–30.

52 See, for example, the engraving in Miguel Ruiz de Saavedra, *Nueva descripción de la admirable vida, hechos, sagrado culto, y gloriosos milagros del esclarecido martyr de Jesus-Christo, San Emygdio* [...] (Madrid: Gabriel Ramírez, 1756).

53 See Milena Viceconte, 'La iconografía novohispana de San Emigdio, abogado contra los temblores', *Revista de Arte Ibero Nierika*, 24 (2023), 60–86.

54 A representative corpus of colonial images of St Emygdius can be found in the following database: *ARCA® – Cultura visual de las américas*, ed. by Jaime H. Borja Gómez, <https://arca.uniandes.edu.co/> [accessed 31 July 2023].

55 *Tesoros del Colegio mayor de Nuestra Señora del Rosario. 350 años*, ed. by Benjamín Villegas (Bogotá: Villegas, 2003), pp. 114–15.

56 *Napoli e il gigante. Il Vesuvio tra immagine scrittura e memoria*, ed. by Rosa Casapullo and Lorenza Gianfrancesco (Soveria Mannelli: Rubbettino, 2014); *Disaster Narratives in Early Modern Naples. Politics, Communication and Culture*, ed. by Domenico Cecere et al. (Rome: Viella, 2018); Vera Fionie Koppenleitner, *Katastrophenbilder: der Vesuvausbruch 1631 in den Bildkünsten der Frühen Neuzeit* (Berlin; Munich: Deutscher Kunstverlag, 2018).

57 The painting is reproduced in Luis Enrique Tord, *Arequipa artística y monumental* (Lima: Banco del Sur del Perú, 1987), p. 89.

58 *San Gennaro tra fede arte e mito*, ed. by Pierluigi Leone De Castris (Pozzuoli: Elio de Rosa, 1997).

59 Víctor M. Barriga, *Los terremotos en Arequipa 1582-1868. Documentos de los Archivos de Arequipa y de Sevilla* (Arequipa: La Colmena, 1951), pp. 145–47, 152–55, 169–76, 227–28.

8 Straightening the Arno

Artistic representations of water management in Medici Ducal and Grand Ducal Florence

Felicia M. Else

Early modern Florence could boast of its art and architectural beauty, but when it came to water, it could resemble a hazard zone. Markers on its streets going back to the fourteenth century reminded inhabitants of how dangerously high floodwaters could reach.[1] The sixteenth century was especially bad, with devastating floods in 1547, 1557 and 1589. As scholars of disaster studies have argued, phenomena like flooding can shed light on a society's 'knowledge, cultural expectations and vulnerability' and the development of 'strategies of resilience to cope with environmental threats'.[2] Suzanne Butters observes that for the Medici Dukes and Grand Dukes of Tuscany, 'floods offered rulers an opportunity to demonstrate charity and technical mastery'.[3] One such demonstration of charity appears in an anonymous pamphlet on the 1557 flood which describes a concerned and empathetic Duke Cosimo I de' Medici, who, along with his wife and children, visited homes ruined by the flood, breaking down in tears.[4] Similarly, Grand Duke Ferdinando I de' Medici rose to the occasion after the flood of 1589, praised by Scipione Ammirato as having braved a perilous and filthy passage back to the flooded city to bring support to the impoverished, a sight that brought tears to his eyes.[5]

The demonstration of technical mastery is also featured in the reigns of Cosimo and his successors, Francesco and Ferdinand. Giorgio Spini's landmark study showed the challenges of water management throughout the Medici territories, including flooding from rivers and infestations from standing water in swamps. Cosimo's restructuring of the Capitani di Parte Guelfa in 1549 and the appointment of Ufficiali dei Fiumi produced copious documents on water-related inspections, proposals and conflicts, from grandiose bridges to clogged drains.[6] As Butters puts it, the Medici were reliant on 'efficient administration by magistracies in charge of public spaces and thoroughfares, including waterways'.[7] Gerrit Schenk points out the development of an 'increasingly institutionalised management of

DOI: 10.4324/9781003029823-11

natural hazards' where flood damage repair and water regulation became more effective as a centralised and professionalised working body, with 'permanent staff with clearly defined tasks – from bookkeeping to on-site engineering works' as opposed to the 'improvised, ad hoc committees' of the preceding two centuries.[8]

Thus, it was for good reason that Cosimo and his heirs attended to water regulation and that their efforts were celebrated by court artists and humanists. In the visual arts, however, how to portray such feats had no clear iconographic traditions. The arduous nature of such labour, be it repairing embankments, digging ditches, moving earthworks, dredging swamplands or clearing waste from wells and cesspits, was matched only by the dreadful conditions specific to water management, workers toiling in wet, muddy, foul-smelling circumstances. The geography of Tuscany challenged even the best engineers, due to the tempestuous nature of the Arno and the insalubrious conditions of marshlands and stagnant waters.[9] Such feats did not lend themselves to the idealised, Michelangelesque style favoured by the Florentines, but court artists and iconographers rose to such challenges. Scholars have shown how water control underscored bold, elaborate programmes behind Medici fountains and villas, like the hydrographic scheme planned for Castello.[10] This study will explore visual works that more directly address the regulation of rivers and canals and the reclamation of swamplands, many of which, though less prominent, are important for understanding how Florence responded to natural disasters and hazards. The selected works will focus on Florentine sites of political significance, from Giorgio Vasari's frescoes in the Palazzo Vecchio carried out under Cosimo to the Sala di Bona in the Palazzo Pitti commissioned by Ferdinand I in the early seventeenth century. I will explore how these tasks were recast with rich visual and iconographic content and placed in some remarkable and rarefied courtly settings and occasions. To create these images of princely power and virtue, court humanists and artists drew creatively on ancient mythology and Roman Imperial history in ways meant to impress learned and informed audiences. Technical mastery over nature and the resulting prosperity and well-being of Ducal and Grand Ducal subjects were expressed with bold hydrographic figures like river gods as well as with markers of contemporary engineering and bureaucracy, like reports, maps and plans. In art and reality, contemporaries viewed the regulation of waterways and the drainage of swamps as correctives to the hazards of floodwater, allowing for the fertility of the land and the promotion of agriculture and commerce. As will be shown, the Medici quite literally took the bull by the horns.

Cosimo I de'Medici: Managing waterways like Hercules, Augustus and Apollo

The Palazzo Vecchio, the Republican city hall turned Ducal Palace, served as the site for various depictions of managed waterways, imagery tied to the increasingly bold assertions of territorial power under Cosimo's reign. In 1555, Vasari, under the guidance of humanist advisors Vincenzo Borghini and Cosimo Bartoli, embarked on a grand cycle of painted rooms employing a vast repertoire of ancient, historical and contemporary subjects. Paintings in the Sala di Cosimo I highlight the important stages of the Duke's reign and show him actively engaged with members of his court and personifications of his territories. As Karla Langedijk has discussed, some scenes show Cosimo in the guise of an architect, bearing instruments of measure and maps. One tondo shows the Duke with a compass and T-square surrounded by artists and engineers, and in another, he holds a plan of fortifications and points to a cartographic view of an important coastal territory, Portoferraio on Elba (Figure 8.1).[11]

Figure 8.1 Giorgio Vasari, *Cosimo visits the fortifications on the island of Elba, between allegories of Arezzo and Pisa*, 1556 (Sala di Cosimo I, Palazzo Vecchio, Florence, Italy). Photo Credit: © Alinari Archives/Raffaello Bencini/Art Resource, NY.

202 *Felicia M. Else*

In scenes flanking each tondo, accurate topographical views serve as backdrops for personifications of Tuscan cities and geographic features, including river or mountain gods, showing the importance of hydrography to Ducal territory. The Duke, dressed like an ancient Roman emperor, interacts supportively but firmly with each, including two examples that reference water and drainage schemes. To the right of the Elba tondo, Pisa appears as a beautiful young woman looking up at the Duke, who points his left finger forward and places his right hand against a cornucopia. Vasari describes the benefits of Pisa which, 'His Excellency has made flourish by draining and cultivating the swamps of that city, which previously provided only stench and pestilence ... she accepts the law from the Duke'.[12] In the representation of Prato to the right of the *Election of Cosimo* tondo, the Duke extends a piece of white paper at an attentive youth and an aged bearded figure (Figure 8.2).

Vasari described the scene as 'Prato, whom His Excellency orders to amend the Bisenzio River, which is represented below holding a horn of

Figure 8.2 Giorgio Vasari, *Allegory of the City of Prato and View of Prato*, fresco, 1556 (Hall of Cosimo I, Palazzo Vecchio, Florence, Italy). Photo Credit: © Alinari Archives/Art Resource, NY.

plenty'.[13] The paper is a fascinating though ambiguous detail, as its small oblong dimensions show illegible black lines of writing and an illustration of an outlined triangular form, perhaps the map of a structure, area or proposed plan. Maps showing plans of fortifications and even of canalisation schemes featured in other representations, like the nearby tondo of Cosimo at Elba as well as portraits by Bronzino, including one of Luca Martini, the *provveditore* at Pisa responsible for water regulation.[14] Vasari's motif could represent a folded report or response, perhaps relating to the Ufficiali dei Fiumi. The choice of the Bisenzio may resonate with the flood of 1547, as Giovan Battista Adriani's account mentions the 'great damage made by the Bisenzio through the entire valley'.[15] In the Pisa and Prato scenes, Vasari used terms like 'law' and 'order' to characterise the Duke's actions, perhaps invoking the crucial, organisational process behind water management. The Prato scene combines mythological references, like river gods and cornucopias, with the ultimate tool of administrators past and present, paperwork.

In terms of sheer numbers, however, few things can beat the river god for its distinctiveness and ubiquity in the Palazzo Vecchio and beyond. The river god type, an aged, muscular bearded figure in a reclining pose, developed in the early sixteenth century based on ancient statuary in Rome. It could easily be manipulated to represent a variety of geographic locations and served as signifiers of fertility and territorial power.[16] The water deities in the Sala di Cosimo are subsidiary figures, truncated from the waist down. In the Salone dei Cinquecento, they take on fuller form. The redecoration of this important hall of state began in 1563 and was completed in time for the Entrata of Johanna of Austria, Francesco de' Medici's bride-to-be, in 1565. Vasari and Borghini worked intensely on the complex narratives and allegorical references required from recreating Florence's ancient foundations to its campaigns against Pisa and Siena.[17] Cosimo requested alterations to their programme to include greater geographical and hydrographical content. Sixteen corner panels intended to show the banners of the city's quarters were replaced by Tuscan regions to better represent 'all our territory'. They include river and mountain deities, standards, emblems, mythological figures, references to local industries and agricultural goods and topographical renderings of major cities in the backgrounds.[18] As Randolph Starn and Loren Partridge observe, these panels served as an 'allegorical inventory of the duchy's natural resources', part of a broader triumphal display of Cosimo's 'trophies' of conquest.[19] As can be seen in the panels of *Certaldo*, *Chianti*, *Volterra* and *San Gimignano/Colle Val d'Elsa* (Figure 8.3), river gods are not only present but visually take centre stage, dominating their compositions with Hellenistic muscularity and Mannerist reworkings of figures by Michelangelo.

The whole ceiling presents 18 river gods and in some cases a river's mountain source, along with overflowing water vases and cornucopias.

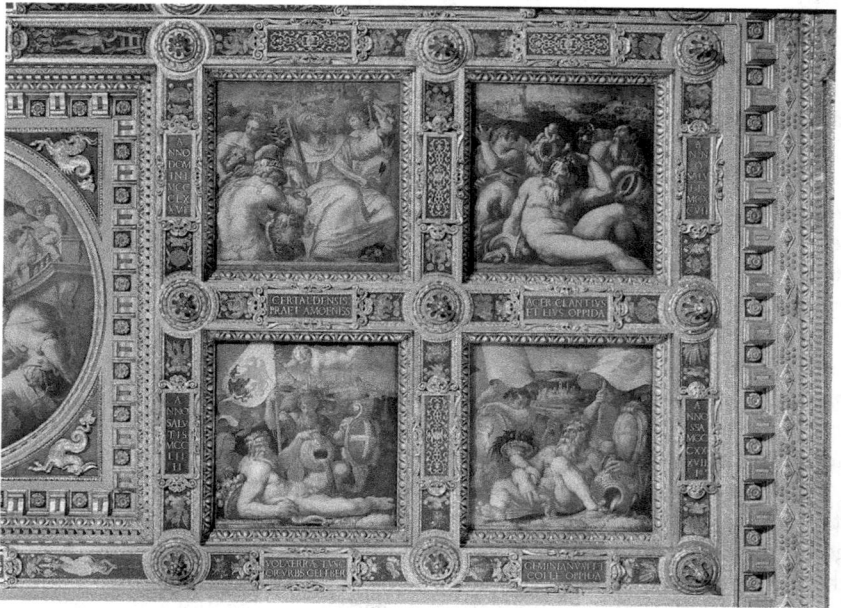

Figure 8.3 Giorgio Vasari, Ceiling panels of *Certaldo, Chianti, Volterra and San Gimignano/Colle Val d'Elsa*, 1563–1565 (Salone dei Cinquecento, Palazzo Vecchio, Florence, Italy). Photo Credit: Scala/Art Resource, NY.

Starn and Partridge point out that the number of water deities comes second only to putti among the *all'antica* personifications in the Salone: 'twenty-seven water deities, twenty-three goddesses, thirteen gods, ten zodiacal signs, four mountain ranges, and one hundred and twenty-five putti. Probably no pagan temple had conjured up a stronger numen'.[20] Though relegated to smaller corner panels, the river gods themselves are as large, if not larger, than figures in the more central panels.

Elsewhere for the 1565 Entrata, Borghini developed a new visual representation for water management, placing it among a special selection of Medici Ducal military and civic achievements. For the courtyard of the Palazzo Vecchio, Borghini devised *all'antica* medallions for the lunettes above views of Austrian cities. As Rick Scorza has analysed, Borghini undertook rigorous research, probing through extensive textual sources, incorporating compositions from ancient coins, even listing the Duke's achievements alongside references from Suetonius' life of Augustus, to whom Cosimo was compared. The subjects included the fortifications of Portoferraio, the Knightly Order of Santo Stefano, the conquest of Siena, the draining of the Pisan swamps, the channelling of the Arno and various public works, such as the Uffizi, the Pitti Palace and the Neptune Fountain. Borghini's hard work and clever idea paid off. Contemporaries like Giovanni Cini lauded the 'glorious deeds of the magnanimous Duke ... so similar to those of the first Octavianus Augustus that it would be difficult to find any greater resemblance'.[21] The painted ovals have since faded, but the images were struck in 1567 as medal reverses by Pier Paolo Galeotti.[22] The scale of the numerous achievements and the adoption of the *all'antica* medal form to celebrate a princely ruler broke new ground in its time, dubbed by Philip Attwood as 'the most important sixteenth-century Florentine contribution to the development of the medal' and by Scorza as unprecedented since antiquity.[23]

Often overlooked in the cycle, the image for the channelling of the Arno perfectly encapsulates Borghini's creative skill at mining ancient sources and recasting material to reflect contemporary aspirations and conventions (Figure 8.4).[24]

The profile of a raging bull thrusts its horns menacingly forward, its head angled downward and slightly towards the viewer creating a striking silhouette of both horns. The forceful power of the creature is evident, but it may not be immediately clear how a landed animal relates to an aquatic subject. The inscription, also carefully contrived by Borghini, reads *IMM-INVTVS CREVIT*, a somewhat contradictory phrase translated by Attwood as 'having been mutilated it prospered'.[25] In fact, Borghini's image has eluded even prominent scholars today who have misidentified it as a reference to Tuscan fortifications.[26]

Borghini strove to follow rules laid out by humanists like Paolo Giovio, who recommended that subjects for *imprese* not be immediately clear but also not require 'the wisdom of a Sibyl to unravel'. However, the image in question is a *rovesco* (medal reverse), 'an explicit form of public glorification', so its meaning should not be ambiguous.[27] The confusion is understandable because, as Cesare Johnson points out, the image of the bull was taken from an ancient coin of Augustus, probably known to Borghini from Enea Vico's compendium on numismatics of the ancient emperors.[28] In antiquity, this coin did not likely relate to managing riverways but to Augustus' strength, determination and divinely sanctioned rule, associations that could be applied to fortifications.[29] However, the devil is in the details, and if one looks closely, there are two horizontal indentations at each horn's base, meant to be read as cuts. This alteration references the myth of Achelous. As recounted in Ovid's *Metamorphoses*, Achelous was a river god who fought with Hercules for the hand of Deianira. When he changed into a bull, Hercules defeated him by breaking off one of his horns which he then gave to the Naiads, who created the cornucopia by filling it with fruits of the harvest.[30] Cini explains that the bull represents 'the straightening of the River Arno in many places, carried

Figure 8.4 Pietro Paolo Galeotti, *Medal showing bull with broken horns*, ca. 1567. British Museum, London, UK, G3, TuscM.217. ©The Trustees of the British Museum. All rights reserved.

out with such advantage by the Duke'.[31] As Scorza points out, the raging bull stands as a metaphor for 'a devasting flood' and therefore Hercules' cutting off of the horn is an analogy for the cutting or channelling of the Arno to help manage the river and create fertile lands. Scorza's translation of the inscription, 'having been mutilated, the river prospered', clarifies the analogy more.[32] Ferretti furthermore points out the influence of Vincenzo Cartari's mythographic texts where the ferocity of rivers like the Po and Tiber were likened to bulls and whose riverbanks were twisted like horns.[33] By this time, Florence had experienced yet another terrible flood in 1557, taking out Bartolomeo Ammannati's Santa Trinita bridge, still unrepaired by the 1565 Entrata, surely an eyesore that reinforced the urgency and relevance of this image.[34]

In this relatively simple composition, Borghini cleverly summoned the achievements of two powerful ancient figures, Augustus and Hercules. The association with Augustus tied into a myriad of other representations likening the Duke to this Imperial predecessor, from their shared star sign of Capricorn to architectural and engineering feats, which included water provision.[35] Suetonius wrote that Augustus 'as a precaution against floods, cleared the Tiber channel which had been choked with an accumulation of rubbish', a suitable analogy for the Duke and the Arno.[36] The ties to Hercules are no less impressive and just as wide-ranging, from grand fountain statuary to leonine details on portrait armour.[37] Both Augustus and Hercules were considered by court humanists to be involved in the foundations of Florence and Tuscany. Historians argued that Augustus, then Octavian, had founded the city, a scene representing in the Salone. A legend circulated by members of the Florentine Academy asserted the primacy of Tuscany via its Etruscan roots, founded by the 'Egyptian Hercules' who cut channels to drain all the marshes in the Arno Valley. The very name of the Arno relates to the word for lion, referencing the demi-god's attribute.[38]

Another relevant work has been overlooked because it may have been misidentified in the sixteenth century. From 1556 to 1557, Vasari and Marco da Faenza painted a room dedicated to Hercules in the Palazzo Vecchio. The subject, as Vasari explained in the *Ragionamenti*, likens Hercules' carrying out of his labours to 'great princes ... [who] labour every hour that they govern to combat the vices of envy, avarice, lasciviousness and the like'.[39] One panel shows a somewhat gruesome scene, as Hercules sits astride a bull whose head flows with blood from the stump of his left horn which the demi-god has torn off and handed to a woman, transformed into a cornucopia.

This scene clearly shows the story of Hercules and Achelous, providing another precedent for Borghini's image. The anguished look on the bull's face even echoes the Ovidian passage where Achelous is reminded of his battle with Hercules and 'with a groan and a hand raised, feebly, towards his forehead'.[40] However, Vasari described the scene as 'when Hercules captured the bull which the victorious Theseus destroyed in Crete', and this identification has informed subsequent scholarship.[41] Vasari likely got his bull narratives mixed up, as Hercules' defeat of the Cretan bull did not involve the breaking of its horn or the cornucopia. Vasari goes on to observe that Cosimo, like Hercules, 'masters and breaks the powerful horns of the proud bull, filling the empty, dry horn full with virtuous fruits'.[42]

Borghini's image was a bold new take on the arduous task of water management, a prime example of his creative process. As Ferretti puts it, Hercules gives 'a heroic base to such hydraulic works', a perfect showcase for Borghini's ability to transform classical forms to express contemporary needs.[43] This representation was distributed through various visual and textual expressions. As part of the 1565 Entrata, it would have been visible to court nobles who attended as well as others who could have read about it in descriptions published by Cini and Domenico Mellini. Its reproduction on medals placed it in a dynamic and intimate context. As discussed by Attwood, medals served as gifts, 'sent from one Prince to another' or as rewards for an individual's service. Artists distributed and displayed them as evidence of their skill, and antiquarians and learned humanists circulated such items, whether for self-promotion, friendship or as part of the growing interest in portrait collections. Their small size encouraged a closer, more personal manner of looking and was often stored in boxes and cabinets in more private studies or bedrooms. Medals could complement collections of ancient coins, whose format and illustrious pedigree they sought to emulate. Cosimo had hundreds of ancient coins in his possession, and Vasari noted that the Duke's study was laid out so he could keep his medals well organised and visible in cedarwood caskets. Medals were buried in tombs, cast into building foundations and featured among wonders of art and nature in cabinets of curiosities.[44] A study of the distribution specific to the Galeotti medals has yet to be done, but given its scope and quality, laden with the dense references to antiquity for which Borghini and the Florentines were known, it surely would have served any collector or courtier well. One can imagine learned nobles peering at Borghini's bull to see the tiny indentations on its horns, sharing their knowledge of Augustus, Hercules and water management under the Medici.

Borghini's *rovesci* were so effective that they were reproduced in two more different materials and displayed in other important architectural sites. In 1571, Raffaello Fortini carved these images on eight marble blocks, each measuring around 1.9 feet or 57.5 cm. These remarkable but little-understood works are visually striking, the images rendered in relief on white Carrara marble contrasting with a variegated green marble backdrop (Figure 8.5).

Each block comprises two reliefs, the bull paired with a Grand Ducal crown and sceptre, a later subject celebrating Cosimo's acquisition of this honour in 1569. Marco Chiarini has identified these blocks as supports for seats ('sedili') placed above three coloured marble steps in front of windows in the former Salone dei Forestieri (Foreign Prince's Apartment), now the White Room, in the Palazzo Pitti.[45] This series, grander in scale and coming not long after Cosimo's acquisition of the Grand Ducal title, reinforces one of many links between the two major Grand Ducal residences in Florence, the Palazzo Pitti and the Palazzo Vecchio, whose courtyard paintings might have still been visible.[46] The bull has been rendered in greater detail than Galeotti's medal, with a more defined groundline of rocks and plants.

Figure 8.5 Raffaello Fortini, *Sedile marmoreo con l'impresa del toro*, 1571 (Palazzo Pitti, Florence, Italy).

Curiously, Fortini has not maintained the cuts in the horns, perhaps translating them to a more decorative pattern of curved lines.

These images were also featured in the funeral of Grand Duke Cosimo in 1574, rendered in large grisaille ovals along the nave of San Lorenzo.[47] Like the 1565 Entrata, published texts described the decorations, explaining that the bull with a broken horn represented the Arno whose redirected course rendered it more navigable and lands more fertile.[48] For the oval representing the draining of swamps, which in the 1565 Entrata and Galeotti medal comprised just an inscription, a new image was conceived, drawing on a different mythological tale. It showed Python, the serpent slain by Apollo, 'pierced by several arrows' which alluded to 'the very difficult draining of the swamps in the Maremma of Pisa and of Siena', making their lands 'very fertile and habitable all year round'.[49] Apollo was yet another classical deity used to evoke the wisdom and benefits brought by Cosimo, a figure represented by Domenico Poggini in medals and a statue.[50] Ferretti points out writings by humanists like Benedetto Varchi, who described Python as coming forth from the slime following a flood sent by Jove, its death caused by the rays of the sun drying away its waters.[51]

Francesco I and Ferdinando I de' Medici: Images celebrating luxurious materials, social hierarchies and a vibrant port city

For Cosimo's heir, Granduke Francesco, a different image of water management, was developed, also part of a celebratory series of achievements. Dating to 1585–1587, Cesare Targone produced small gilt reliefs based on models by Giambologna and Antonio Susini representing Francesco's succession as Grand Duke and major building projects, such as fortifications at Portoferraio, Livorno and Belvedere; the façade of the Florentine Cathedral and the gardens and fountains at Pratolino.[52] One relief is dedicated to the regulation of riverways, showing the Grand Duke on horseback, engaging a group of architects and engineers to his left and taking hold of a large sheet, presumably a plan or map (Figure 8.6).

To his right, a figure with a crossbow walks alongside Francesco's horse, looking up attentively at the Grand Duke. The image is more literal and pictorial than Borghini's, recalling compositions mentioned earlier by Bronzino and Vasari where rulers or engineers are shown with documents and plans, instruments that reference technical expertise and administrative effectiveness. Maps appear in the gilt reliefs of Portoferraio and Belvedere bearing outlines of fortifications. The image on the sheet in the river management scene is less defined, showing two long parallel lines. While simple, it efficiently conveys the general orientation of straightening a river course. In fact, various actual water-related proposals and plans feature such long horizontal linear forms.[53]

Figure 8.6 Cesare Targone after Giambologna and Antonio Susini, *Regulation of the Arno*, 1585–1587 (Museo degli Argenti, Palazzo Pitti, Florence, Italy).

The two corners of the lunette show pairs of river gods, perhaps suggesting various rivers or related branches. There may even be a visual dialogue between the two river gods in the foreground. The one on the left is shown from the front, leaning on an urn pouring water towards the central scene. The river god on the right is shown from the back resting on an urn that is closed, perhaps evoking the need to close off some water courses to better control others. This scene presents the princely ruler elevated on horseback and placed directly on site in the countryside, delegating the dirty work to others. The tradition of showing rulers on horseback hails from ancient sources, used in many paintings by Vasari in the Palazzo Vecchio rooms, from battle scenes to triumphant entries. A more direct precursor may lie in an ephemeral painting of the draining of the Pisan swamps by Giovanni Battista Naldini for the 1565 Entrata, placed on a wall of the Salone dei Cinquecento.[54] Alessandro Cecchi has identified a preparatory drawing in Lille, showing Cosimo on horseback giving orders to men around him, including one attendant, presumably Luca Martini, who holds a large sheet or plan (also with simplified parallel lines).[55] The gilt relief presents the princely ruler in a similarly elevated manner gazing downwards at his courtly retinue, though in a smaller scale and with fewer figures, and it is not entirely clear whether Francesco is handing the plan to his engineers or vice-versa.[56]

Another detail on this gilt relief suggests a link between water, Medici power and the practice of hunting. Detlef Heikamp observed how these scenes, while reminiscent of the Vasarian style under Cosimo, are less grand and heroic, showing genre elements alongside the prince's actions. In the river management scene, he identifies the figure to Francesco's right as a hunter, a bundle of dead ducks hanging from his crossbow.[57] Indeed, he is a striking figure, set apart from the other more courtly servants of the Grand Duke, bent down as he strides forward, his crossbow creating a bold silhouette. Donning a plain hat and carrying his weapon and spoils, he resembles other visual depictions of rustic labourers and hunters, such as Giambologna's statuette of a *Bird Catcher*.[58] Compositions by another Medici court artist, Giovanni Stradano, show similarly capped hunters of birds also armed with crossbows, part of a series dedicated to the hunt that was extraordinarily popular.[59] Such genre figures were an established type, even represented in large-scale marble sculptures in princely villas.[60] Among the mythological deities at Pratolino, a massive stone laundress carries out her washing, a *contadino* empties a wine cask and an over-life-size peasant cuts reeds in an artificial pond accompanied by a giant salamander. Such rustic characters had their origins in ancient traditions like the pastorale, and privileged visitors to Francesco's villa offered learned interpretations of them, some elevated, some jocular but all reinforcing their place in an established social hierarchy.[61] Francesco de' Vieri observed that 'all these works stand for corrupt or depraved people', describing countrymen as 'ill-mannered in their behaviour' and standing for 'our uncultivated nature'.[62] In the context of the water management gilt relief, the Grand Duke's interventions provide fertile lands for hunting, an idealised image of beneficent ruler and grateful subject.

But, as Butters rightly warns, just because lower social classes were depicted among the visual imagery produced for Francesco does not mean his actual subjects enjoyed benign or even reasonable treatment. In fact, harsh conditions of the workers at Pratolino, many of whom were *comandati*, commandeered peasant labour, were made all the more miserable by its involvement with water. The creation of an artificial lake drew special outcry from chronicler Bastiano Arditi who noted 'those commandeered countrymen were in that lake working in water and mud up to their knees all day' and that their hardships and personal loss were such that 'one heard of nothing else in the whole countryside except poverty and death from starvation'.[63] The dire conditions resulted in truancy problems, fines and physical punishments and an incident of vandalism in 1582, when statues' noses were cut off and fountain pipes damaged.[64] The reality here could not be more at odds with the idealised representation of water and prosperity. The references to hunting in the gilt relief can be viewed as a blunt expression of princely power and privilege. The maintenance of rivers, roadways, forests and marshes under the aegis of 'public works' helped provide fertile hunting and fishing grounds often restricted to private use by the Medici.[65] Pratolino was known for its extensive hunting grounds, and D. R. Edward Wright examines how such activity fit within different areas of the villa, pointing out that the fountain featuring the stone laundress would have been adjacent to where the hunting party was gathered for breakfast and thereby convenient for servants to wash the dishes. Legislation on hunting and fishing walled off reserved areas 'to the Prince for his pleasure and recreation', and a renewed *bando* of 1581 prohibited specifically the hunting of partridge and pheasant.[66]

The rarefied nature of the gilt relief becomes even clearer when we consider its materials and intended context. Measuring about 6 inches (15.5 cm) long, the brilliant gilding of the delicate figures contrasts against slick, dark green jasper. Such a precious object was made to adorn a magnificent *stipo*, or cabinet, set in one of the most splendid spaces in Grand Ducal Florence, the Tribuna of the Uffizi, whose lush red walls and scintillating octagonal dome covered in mother-of-pearl shells evoke wonder from visitors even today (Figure 8.7).

Built by Bernardo Buontalenti for Francesco, the Tribuna served as a showcase for the Medici's most valuable treasures. Documents reveal that the gilt reliefs were planned for a 'Studiolo Nuovo', also commissioned by Francesco from Buontalenti, a cabinet composed of ebony that took the form of a 'tempietto' placed in the centre of the room. The gilt reliefs were to decorate the front of small drawers ('cassette') which held precious objects, like ancient coins and medals, the kind of setting one imagines for Galeotti's medals. Francesco's untimely death in 1587 has left questions about the incomplete state of the gilt series as well as whether it was ever mounted on this cabinet, which was moved out of the Tribuna by 1629. His successor, Ferdinand, commissioned yet another ebony *stipo* ('Studiolo Grande') also from Buontalenti for the Tribuna, bearing a rectangular architectural façade and located in a niche across from the entrance. Eighteenth-century documents place the four gilt lunette scenes, which include the river management subject, in four archways on the cabinet's façade.[67] Both cabinets have been lost to us, but contemporary descriptions marvel at their colours and rich materials, including alabaster, pearls, lapis lazuli, agate, sapphire, emeralds and rubies.[68] Clad of gold and jasper, the river management scene must have been at home here for its materials and workmanship as well as its glorification of the Medici. Although Ferdinand disliked his brother, the Tribuna was, as Heikamp puts it, an 'allegory of Medici power', and water-related initiatives were a legacy of its dynasty.[69] Just like Cosimo before him, Francesco's regulation of riverways featured in his funeral in 1587 in an image that showed the Granduke 'along a riverbank discussing with the architect and

Figure 8.7 Bernardo Buontalenti, *View of the Tribuna*, 1581–1583 (Galleria degli Uffizi, Florence, Italy). Photo Credit: Scala/Art Resource, NY.

engineer how to constrain its impetuous course and hold it within its boundaries' making the river 'more navigable and rendering fruitful the uselessness of the terrain', comparable to great men like Caesar who corrected the Tiber and other rivers.[70]

The reign of Ferdinand provides a final set of examples that build on the importance of water management in key occasions and locations of Medici power and privilege. For Ferdinand's entry into Pisa in 1588 celebrating his installation as Grand Duke, the Ufficio de' Fossi recreated the marshy landscape of Lake Bientina, featuring a flowing stream, a bridge, a swamp with marsh flora, a watermill, a floodgate and live animals, including some frogs whose presence incurred disdain from the Ferrarese ambassador.[71] Ferdinand intentionally set himself apart from Francesco in his dedication to public works and his generous nature – where Francesco was despised for private indulgences and enforced labour, Ferdinand was celebrated for providing good-paying jobs and looking after the general welfare.[72] Ferdinand's reign used artworks to pay tribute to his father's legacy, when, as Langedijk observes, 'the veneration of Cosimo reached its peak'.[73] One such commission involves an important cycle of paintings by Bernardino Poccetti in the Sala di Bona of the Palazzo Pitti, where images glorify Cosimo, Medici military might and control over waterways.

Poccetti's paintings for the Sala di Bona are considered one of the most important wall decorations of Seicento Florence, carried out from 1607 to 1609 during the last years of Ferdinand's reign and in time for the wedding of his son Cosimo II. Outfitted with military scenes and *all'antica* Michelangelesque imagery, this bombastic statement of Medici power was intended to echo Vasari's Salone dei Cinquecento in the Palazzo Vecchio, yet another link between the two Grand Ducal palaces.[74] The subjects emphasise maritime and coastal feats, including Ferdinand's victories over the Ottomans and Barbary corsairs at Preveza and Annaba (Bona), 'the two greatest land assaults' in the history of the Knightly Order of Santo Stefano.[75] On the ceiling, Cosimo appears in the boldest triumphant manner yet, as though in an apotheosis, his heroically nude form enthroned among the clouds. Flanked by allegorical female figures associated with wisdom and prudence, Cosimo holds compasses and a T-square, attributes of the ruler as architect.[76] One wall shows an extraordinary representation of the thriving Medici port city of Livorno (Figure 8.8).

The viewpoint from the sea creates a striking expanse of blue water, filling half the composition, complemented by an aggrandised rendering of its magnificent harbour, complete with ships, lighthouses and towers. Poccetti has painted a highly detailed plan of the city, its fortifications and, as Stefania Vasetti points out, the precise 'topography of the surrounding territory, showing the canalisation made to connect Livorno to Pisa and to the Arno, favouring thus commerce and agriculture'.[77] In the immediate

Figure 8.8 Bernardino Poccetti, *Plan of Livorno with allegorical figures of the Tyrrhenian Sea and the Arno River*, 1607–1609 (Sala di Bona, Palazzo Pitti, Florence, Italy). Photo Credit: Scala/Art Resource, NY.

picture plane below sit two massive water deities, the Arno River and the Tyrrhenian Sea with a lush cornucopia between them. Vasetti further notes the extensive praise for Ferdinand's managing of waterways which made uncultivated fields fertile and Tuscany well-provided for, even as famine struck the rest of Italy.[78] This room served an important social and political function, located in the quarters where illustrious guests were housed, perhaps as a waiting room where visitors could view the great deeds of Cosimo and Ferdinand.[79] I would further point out that the Sala di Bona linked directly to the Salone dei Forestieri, a large hall where banquets and entertainments took place and where Fortini's marble reliefs of Cosimo's deeds, including the Bull representing the channelling of the Arno (as on Figure 8.5), would have been displayed.[80] Just as Cosimo's achievements were reinforced in the Palazzo Vecchio courtyard and Salone, here in the other Grand Ducal residence of the Palazzo Pitti, too, adjacent ceremonial spaces present Ferdinand's deeds, the *sedili* perhaps offering a tired noble or court attendant a seat and view out the window.

When it comes to disasters and hazards of water, the Medici Grand Dukes had their work cut out for them, facing a time marked by cold, floods and famine. They owed much to their engineers, architects and labourers but also to their administrators, humanists and artists. In the dynamic and ever-evolving arena of the visual arts, the threat of disaster could be seen as an opportunity for expanding the representations of princely power. Transforming the arduous and often unsightly work of water management into idealised images worthy of the Medici required extensive research and imaginative reworkings combining ancient sources with contemporary practices and customs. Thanks to their efforts, water management held a prominent place among the achievements of these rulers with its own lofty iconography celebrating water control and prosperity, placed among important public and private settings.

The legacies of water control expressed through the visual arts were just one of the many ways Early modern societies reacted to the hazards of their environment and the fear of natural disasters. This chapter has built on the extensive research on Medici patronage of the arts for the promotion of their dynastic rule and the special role water and engineering played in that dynamic. This study expands on this knowledge by looking specifically at lesser-known imagery of managing rivers, a subject of key importance in how the Medici responded to one of the period's greatest challenges, flooding. Such a subject also did not have a clear iconographic tradition and presented special visual challenges to its artists and iconographers. In unpacking the strategies taken to develop these images, this chapter explores motifs and sources often overlooked or even misunderstood. Furthermore, this study shows the importance this subject had in its time by detailing the spaces, events and social dynamics that originally framed these images. The selected examples presented here also add to our understanding of artistic and thematic links between the Grand Ducal residences of the Palazzo Vecchio and the Palazzo Pitti. While the images discussed here may not have the clarity and punch of a colossal Neptune or evoke the grandeur of projects like the Uffizi, what they represented was an initiative that was equally if not more so impactful for a greater portion of the citizenry, as the rivers of Tuscany reached far beyond Florence.

Notes

1 For the plaques on the floods of 1333, 1547 and 1557, see Piero Bargellini and Ennio Guarnieri, *Le strade di Firenze*, 4 vols (Florence: Bonechi, 1977–1978), vol. 1 (A–F), pp. 211–12; vol. 2 (G–O), pp. 320–24; and vol. 3 (P–S), pp. 213 and 320–31.

2 Domenico Cecere, Chiara De Caprio, Lorenza Gianfrancesco, and Pasquale Palmieri, 'Disaster Narratives and Texts. A Meeting Ground for Different Cultural Domains', in *Disaster Narratives in Early Modern Naples. Politics, Communication and Culture*, ed. by Domenico Cecere et al., trans. by Enrica Maria Ferrara (Rome: Viella, 2018), pp. 9 and 22.

3 Suzanne B. Butters, 'Princely Waters: An Elemental Look at the Medici Dukes', in *La civiltà delle acque tra medioevo e rinascimento*, 2 vols, ed. by Arturo Calzona and Daniela Lamberini (Florence: Olschki, 2010), vol. 1, p. 400.

4 This pamphlet at the British Library, describes Cosimo 'crying like a baby' ('piangere come un bambino'). See *Il Miserabiliss. et inestimabil. danno fatto dal fiume Arno e la Città di Firenze* (n.p., 1557), British Library, London (8775 c 21), no pagination.

5 Giuseppe Aiazzi, *Narrazioni istoriche dell più considerevoli inondazioni dell'Arno e notizie scientifiche sul medesimo* (Verona: L'Arco dei Gavi, 1966), pp. 21–29 [reprint of Florence: Tipografia Piatti, 1845] and Butters, 'Princely Waters', pp. 400–1.

6 See studies in *Architettura e politica da Cosimo I e Ferdinando I*, ed. by Giorgio Spini (Florence: Olshki, 1976) such as Giorgio Spini, 'Introduzione generale', pp. 9–77; Anna Cerchiai and Coletta Quiriconi, 'Relazione e rapporti all'Ufficio dei Capitani di Parte Guelfa. Parte I: Principato di Francesco I dei Medici', pp. 187–257; and Anna Maria Gallerani and Benedetta Guidi, 'Relazioni e rapporti all' Ufficio dei Capitani di Parte Guelfa. Parte II: Principato di Ferdinando I', pp. 261–329. Subsequent studies include Emanuela Ferretti, '"Imminutus crevit". Il problema della regimazione idraulica dai documenti degli Ufficiali dei Fiumi di Firenze', in *La città e il fiume (secoli XIII–XIX)*, ed. by Carlo Travaglini (Rome: École Française de Rome, 2008), pp. 105–28; Butters, 'Princely Waters', pp. 389–90; and Felicia M. Else, *The Politics of Water in the Art and Festivals of Medici Florence: From Neptune Fountain to Naumachia* (Abingdon; New York: Routledge, 2019), pp. 141–42.

7 Butters, 'Princely Waters', p. 389.

8 Gerrit J. Schenk, 'Managing Natural Hazards: Environment, Society, and Politics in Tuscany and the Upper Rhine Valley in the Renaissance (ca. 1270–1570)', in *Historical Disasters in Context: Science, Religion, and Politics*, ed. by Andrea Janku, Gerrit J. Schenk, and Franz Mauelshagen (Abingdon; New York: Routledge, 2012), pp. 33–8.

9 Franco Borsi likens Cinquecento Tuscany to 'a kind of leopard skin', riddled with spots of unhealthy stagnant waters. See Suzanne B. Butters, 'Pressed Labor and Pratolino: Social Imagery and Social Reality at a Medici Garden', in *Villas and Gardens in Early Modern Italy and France*, ed. by Mirka Beneš and Dianne Harris (Cambridge: Cambridge University Press, 2001), pp. 61–62 and 69–75; eadem, 'Princely Waters', pp. 389–96; Franco Borsi, *L'architettura del principe* (Florence: Giunti Martello, 1980), pp. 125–57 and studies cited in note 6.

10 Claudia Lazzaro, *The Italian Renaissance Garden* (New Haven; London: Yale University Press, 1990), pp. 167–87. Further bibliography can be found in Else, *Politics of Water*, pp. 4–5.

11 Karla Langedijk's insightful discussion draws on ancient, medieval and Renaissance influences. See Karla Langedijk, *The Portraits of the Medici 15th–18th Centuries*, 2 vols (Florence: Studio per Edizioni Scelte, 1981–1987), vol. 1, pp. 139–65. On the Sala di Cosimo, see Ettore Allegri and Alessandro Cecchi, *Palazzo Vecchio e i Medici. Guida storica* (Florence: Studio per Edizioni Scelte, 1980), pp. 143–53 and Ryan Gregg, *Panorama, Power, and History: Vasari and Stradano's City Views in the Palazzo Vecchio* (Ph.D. thesis, Baltimore: Johns Hopkins University, 2008), pp. 138–40 and 349–50.

12 Translation from Vasari's *Ragionamenti* by Jerry Lee Draper. See Jerry Lee Draper, *Vasari's Decoration in the Palazzo Vecchio: The* Ragionamenti *Translated with an Introduction and Notes* (Ph.D. thesis, Chapel Hill: University of North Carolina at Chapel Hill, 1973), p. 363 and Allegri and Cecchi, *Palazzo Vecchio e i Medici*, pp. 147–48.

13 Draper, *Vasari's Decoration in the Palazzo Vecchio*, p. 366; Allegri and Cecchi, *Palazzo Vecchio e i Medici*, pp. 146–48; and Felicia M. Else, 'Vasari, the River God and the Expression of Territorial Power under Duke Cosimo I de' Medici', *Explorations in Renaissance Culture*, 39 (2013), 79–80.

14 The portrayal of water management in Pisa is its own rich subject, especially the influence of Luca Martini, a humanist, art patron and engineer. See Jonathan Nelson, 'Creative Patronage: Luca Martini and the Renaissance Portrait', *Mitteilungen des Kunsthistorischen Institutes in Florenz*, 39 (1995), 282–303.

15 Author's translation from extract of Book Six of *Istoria de' suoi tempi di Gio. Batista Adriani* in Aiazzi, *Narrazioni istoriche*, p. 12.

16 River gods ornamented fountains and villas, like the Medici Villa at Castello. On its origins and dissemination, see Ruth Rubinstein, 'The Statue of the River God Tigris or Arno', in *Il cortile delle statue. Der Statuenhof des Belvedere im Vatikan*, ed. by Matthias Winner, Bernard Andreae, and Carlo Pietrangeli (Mainz: Verlag Philipp von Zabern, 1998), pp. 275–85; Claudia Lazzaro, 'River Gods: Personifying Nature in Sixteenth-Century Italy', *Renaissance Studies*, 25, 1 (2011), 70–94; and Else, 'Vasari, the River God', 71–83.

17 For an overview of the decoration of the Salone, see Allegri and Cecchi, *Palazzo Vecchio e i Medici*, pp. 231–67 and Randolph Starn and Loren Partridge, *Arts of Power. Three Halls of State in Italy, 1300–1600* (Berkeley: University of California Press, 1992), pp. 175–84, 186–93, 203–12 and 294–304.

18 Robert Williams, 'The Sala Grande in the Palazzo Vecchio and the Precedence Controversy between Florence and Ferrara', in *Vasari's Florence: Artists and Literati at the Medicean Court*, ed. by Philip Jacks (New York: Cambridge University Press, 1998), pp. 169–79; Starn and Partridge, *Arts of Power*, pp. 189–91; and Ugo Muccini, *The Salone dei Cinquecento* (Florence: Le Lettere, 1990), pp. 82–4.

19 Starn and Partridge, *Arts of Power*, pp. 187–88.

20 Starn and Partridge, *Arts of Power*, p. 177 and Else, 'Vasari, the River God', 81–2.

21 Giovanni Battista Cini, 'Description of the Festive Preparations for the Nuptials of the Prince Don Francesco of Tuscany (Florence, 1565)', in Giorgio Vasari, *Lives of the Painters, Sculptors and Architects*, 2 vols, trans. by Gaston du C. de Vere (New York: Knopf, 1996), vol. 2, p. 958.

22 Cini, 'Description of the Festive Preparations', pp. 958–61; Domenico Mellini, *Descrizione dell'entrata della serenissima Reina Giovanna d'Austria* (Florence: Giunti, 1566), pp. 118–21; Cesare Johnson, 'Cosimo I de' Medici e la sua "storia metallica" nelle medaglie di Pietro Paolo Galeotti', *Medaglia*, 6 (1976), 15–46; Allegri and Cecchi, *Palazzo Vecchio e i Medici*, pp. 275–85; Langedijk, *The Portraits of the Medici*, vol. 1, pp. 139–43 and 486–94; Rick Scorza, '*Imprese* and Medals: *Invenzioni all'antica* by Vincenzo Borghini', *The Medal*, 13 (1988), 22–32; and Philip Attwood, 'Catalog Entry no. 98', in *The Medici, Michelangelo, and the Art of Late Renaissance Florence*, ed. by Cristina Acidini Luchinat et. al. (New Haven; London: Yale University Press, 2002), pp. 235–41.

23 Attwood and Scorza point out the cycle's influence on subsequent rulers, including Louis XIV. See Philip Attwood, *Italian Medals c. 1530–1600 in British Public Collections*, 3 vols (London: The British Museum, 2003), vol. 1, p. 22; Scorza, '*Imprese* and Medals', 27; and Rick Scorza, 'Ricerca storica e invenzione: la collaborazione di Borghini con Cosimo I e Francesco I, I suoi rapporti con gli artisti, gli apparati effimeri', in *Vincenzo Borghini: Filologia e invenzione nella Firenze di Cosimo I*, ed. by Gino Belloni and Riccardo Drusi (Florence: Olschki, 2002), p. 77.

24 Some compositions used images from earlier medals by Domenico Poggini; some, like the drainage of the Pisan Swamps, bear an inscription and no image; others displayed an image of the monument or structure, like the *Neptune Fountain*, whose celebration of the city's aqueducts has been discussed by Ferretti, Henk Th. Van Veen and Else. See note 22 and Henk Th. Van Veen, *Cosimo I de' Medici and His Self-Representation in Florentine Art and Culture* (Cambridge: Cambridge University Press, 2006), pp. 107–10; Emanuela Ferretti, *Acquedotti e fontane del Rinascimento in Toscana. Acqua, architettura e città al tempo di Cosimo I dei Medici* (Florence: Olschki, 2016), pp. 152–56 and Else, *Politics of Water*, pp. 82–3.

25 Attwood, *Italian Medals*, vol. 1, p. 355.

26 Cesare Johnson's misidentification likely informed subsequent scholars. See Johnson, 'Cosimo I de' Medici', 19 and 36–37; Detlef Heikamp, 'Catalog Entry no. 688.1', in *Palazzo Vecchio: committenza e collezionismo medicei 1537–1610*, ed. by Paola Barocchi (Florence: Centro Di, 1980), pp. 340–41; and Giuseppe Toderi and Fiorenza Vannel, *Le medaglie italiane del XVI secolo*, 3 vols (Florence: Polistampa, 2000), vol. 2, p. 525.

27 Scorza's studies of Borghini show him to be a stickler for detail, and he rightly concluded that Borghini 'over-estimated the sophistication of his audience'. See Rick Scorza, 'Vincenzo Borghini and *Invenzione*: The Florentine *Apparato* of 1565', *Journal of the Warburg and Courtauld Institutes*, 44 (1981), 59 and 63–64; and Scorza, '*Imprese* and Medals', 21–22.

28 Johnson, 'Cosimo I de' Medici', 19 and 36. For the Vico, see Scorza, '*Imprese* and Medals', 25 and Ferretti, *Acquedotti e fontane*, pp. 157–59.

29 Edward Sydenham, *Historical References on Coins of the Roman Empire from Augusuts to Gallienus* (London: Spink & Son, 1917), p. 26 and Paul Zanker,

The Power of Images in the Age of Augustus (Ann Arbor: The University of Michigan Press, 1988), pp. 225–26.

30 Ovid, *Metamorphoses*, trans. by Rolfe Humphries (Bloomington: Indiana University Press, 1983), IX, 78–89, p. 211.

31 Cini, 'Description of the Festive Preparations', p. 959.

32 Scorza, '*Imprese* and Medals', 25.

33 Scorza notes the 1535 entry of Charles V into Naples where the Bagrada River was represented with broken horns to symbolise its submission. Butters discusses a 1589 text by Giovanni Botero which praised the canalisations of the Ticino and Adda rivers by the lords of Milan, invoking Hercules and Achelous. See Scorza, 'Ricerca storica', p. 85, n. 83; Butters, 'Princely Waters', p. 395; and Ferretti, *Acquedotti e fontane*, pp. 157–60.

34 Borghini covered up the sight of the 'ponte rovinato' with an elaborate arch of marine deities, a water and wine fountain and references to the Duke's maritime empire. See Felicia M. Else, 'Fountains of Wine and Water and the Refashioning of Urban Space in the 1565 *Entrata* to Florence', in *Architectures of Festival in Early Modern Europe: Fashioning and Re-fashioning Urban and Courtly Space*, ed. by J. R. Mulryne et al. (Abingdon; New York: Routledge, 2017), pp. 80–3.

35 Kurt Forster, 'Metaphors of Rule: Political Ideology and History in the Portraits of Cosimo I de' Medici', *Mitteilungen des Kunsthistorisches Institutes in Florenz*, 15 (1971), 85–89 and 98–99; Paul Richelson, *Studies in the Personal Imagery of Cosimo I de' Medici, Duke of Florence* (New York; London: Garland, 1978), pp. 25–78; Janet Cox-Rearick, *Dynasty and Destiny in Medici Art. Pontormo, Leo X, and the Two Cosimos* (Princeton: Princeton University Press, 1984), pp. 257–58 and 276–83; and Scorza, '*Imprese* and Medals', 23–7.

36 Suetonius, *The Twelve Caesars*, trans. by Robert Graves (London: Penguin, 1989), II, 30, p. 70.

37 Forster, 'Metaphors of Rule', 78–82; Richelson, *Studies in the Personal Imagery*, pp. 79–106; and Cox-Rearick, *Dynasty and Destiny*, pp. 253–54.

38 Nicolai Rubinstein, 'Vasari's Painting of *The Foundation of Florence* in the Palazzo Vecchio', in *Essays in the History of Architecture Presented to Rudolf Wittkower*, ed. by Douglas Fraser, Howard Hibbard, and Milton J. Lewine (London: Phaidon, 1967), pp. 64–73; Forster, 'Metaphors of Rule', 82; and Ferretti, *Acquedotti e fontane*, p. 160.

39 Quotation from Draper, *Vasari's Decoration in the Palazzo Vecchio*, p. 192. On the Sala di Ercole, see Allegri and Cecchi, *Palazzo Vecchio e i Medici*, pp. 97–101 and Draper, *Vasari's Decoration in the Palazzo Vecchio*, pp. 187–98. This image can be seen on Google Art Project at https://commons.wikimedia.org/wiki/File:Marco_Marchetti_from_Faenza_-_Hercules_kills_the_bull_from_Crete_-_Google_Art_Project.jpg.

40 Ovid, *Metamorphoses*, VIII, 884, p. 208.

41 Draper points out that Vasari was mistaken in setting Theseus' killing of the bull in Crete as it was in Marathon. See Draper, *Vasari's Decoration in the Palazzo Vecchio*, pp. 189 and 441, n. 4, and Allegri and Cecchi, *Palazzo Vecchio e i Medici*, p. 98.

42 Draper, *Vasari's Decoration in the Palazzo Vecchio*, p. 196.

43 Scorza similarly points out how Borghini 'combines an authentic classical motif with a contemporary metaphor'. See Ferretti, *Acquedotti e fontane*, p. 160 and Scorza, '*Imprese* and Medals', 26.

44 Attwood, *Italian Medals*, vol. 1, pp. 53–5.

45 Marco Chiarini's discovery is discussed in Eike D. Schmidt, 'Catalog Entry no. 36', in *Palazzo Pitti. La reggia rivelata*, ed. by Gabriele Capecchi et al. (Florence: Giunti, 2003), pp. 512–13. For the attribution to Raffaello Fortini and the date of 1571, see Amedeo Belluzzi, 'Gli interventi di Bartolomeo Ammannati a Palazzo Pitti', *Opus Incertum*, 1, 1 (2006), 66 and 72 and Emanuela Ferretti, 'Cosimo I, la magnificenza dell'acqua e la celebrazione del potere: la nuova capitale dello Stato territoriale fra architettura, città e infra-sturtture', *Annali di Storia di Firenze*, 9 (2014), 17–18.

46 The two residences are linked by Vasari's Corridor above and by an aqueduct that brings water from the Pitti to the city centre and Palazzo Vecchio. See Lazzaro, *The Italian Renaissance Garden*, p. 191 and Ferretti, *Acquedotti e fontane*, pp. 54–55 and 77–90.

47 Eve Borsook, 'Art and Politics at the Medici Court I: The Funeral of Cosimo I de' Medici', *Mitteilungen des Kunsthistorischen Institutes in Florenz*, 12 (1965), 44–45 and Forster, Forster, 'Metaphors of Rule', 80–1.

48 *Descritione della pompa funerale fatta nelle essequie des Ser.mo Sig. Cosimo de' Medici Gran Duca di Toscana* (Florence: Giunti, 1574), n.p [p. 31].

49 *Descrittione della pompa*, n.p [p. 27].

50 Ulrich Middeldorf and Friedrich Richelson, 'Forgotten Sculpture by Domenico Poggini', *Burlington Magazine*, 53 (1928), 11–17; Richelson, *Studies in the Personal Imagery*, pp. 37–40; Langedijk, *The Portraits of the Medici*, vol. 1, pp. 93–95 and 492; Attwood, *Italian Medals*, vol. 1, pp. 335, 344, and vol. 2, plate 176; and Toderi and Vannel, *Le medaglie italiane del XVI secolo*, vol. 2, pp. 490–91 and vol. 3, tavola 303.

51 Ferretti, *Acquedotti e fontane*, pp. 160–62.

52 There were at least eight different scenes, two in a rectangular format and the others a lunette. The Museo degli Argenti has seven wax models and seven gilt reliefs for these subjects. There is no wax model for Portoferraio and there is no gilt relief for the subject of Cosimo naming Francesco his successor. See Langedijk, *The Portraits of the Medici*, vol. 1, pp. 141–46 and vol. 2, pp. 901–4; Detlef Heikamp, 'Catalog Entry no. 12: Sette bassorilievi con gli Atti di Francesco I de' Medici', in *Splendori di pietre dure. L'Arte di corte nella Firenze dei Granduchi*, ed. by Annamaria Giusti (Florence: Giunti, 1988), pp. 96–101; Amelio Fara, 'Catalog Entries nos. 33–39', in *Magnificenza alla corte dei Medici. Arte a Firenze alla fine del Cinquecento*, ed. by Cristina Acidini Luchinat and Maria Sframeli (Milan: Electa, 1997), pp. 74–77; and Barbara Bertelli, 'Catalog Entries nos. 33–35: Le imprese di Francesco I, 1585–1587', in *Giambologna, gli dei, gli eroi*, ed. by Beatrice Paolozzi Strozzi and Dimitrios Zikos (Florence; Milan: Giunti, 2006), pp. 226–31.

53 For examples, see a plan by Girolamo di Pace da Prato in Ferretti, *Acquedotti e fontane*, tavola IIIb and a canal embankment proposal by Ammannati in Felicia M. Else, 'Bartolomeo Ammannati: Moving Stones, Managing Waterways and Building an Empire for Duke Cosimo I de' Medici', *The Sixteenth Century Journal*, 42, 2 (2011), 421, figure 7.

54 Mellini, Mellini, *Descrizione dell'entrata*, p. 134; Piero Ginori Conti, *L'apparato per le nozze di Francesco de' Medici e di Giovanna d'Austria* (Florence: Olschki, 1936), p. 60; Allegri and Cecchi, *Palazzo Vecchio e i Medici*, p. 257; and Starn and Partridge, *Arts of Power*, p. 301. On a subse-quent unrealised proposal by Borghini to include this subject among a 'pro-getto per i basamenti' of the Salone, see Scorza, 'Ricerca storica', pp. 93–7.

55 Alessandro Cecchi, 'Disegni inediti o poco noti di Giorgio Vasari', in *Kunst des Cinquecento in der Toskana*, ed. by Monika Cämmerer (Munich: Bruckmann, 1992), pp. 245–47.

56 Some scholars even reference the gilt relief as the draining of the Pisan swamps. In the wax version, one can make out the presence of a handkerchief between the Grand Duke's hand and the plan. See Anna Maria Massinelli, 'Magnificenze Medicee: gli stipi della Tribuna', *Antologia di belle arti*, 35–38 (1990), 122–23 and Bertelli, 'Catalog Entries nos. 33–35: Le imprese di Francesco I, 1585–1587', pp. 226–27, figure 3.

57 Heikamp, *Splendori di pietre dure*, p. 98. He also notes examples in the other gilt reliefs such as a dog and a sculptor at work on a Medici stemma.

58 Charles Avery, *Giambologna: The Complete Sculpture* (London: Phaidon, 1993), pp. 46 and 266, and Alan P. Darr, 'Catalog Entry no. 76', in *The Medici, Michelangelo, and the Art of Late Renaissance Florence*, pp. 212–13.

59 These include hunts of warblers, blackbirds and partridges. Two scenes feature similar bundles of dead birds. See Alessandra Baroni Vannucci, *Jan Van Der Straet detto Giovanni Stradano, flandrus pictor et inventor* (Milan; Rome: Jandi Sapi, 1997), pp. 254–55 and 371–84. On the scope and popularity of the series, see Manfred Sellink, 'Johannes Stradanus and Philips Galle. A Noteworthy Collaboration between Antwerp and Florence' and Marjolein Leesburg, 'Catalog Essay: *Venationes Ferarum, Avium, Piscium*, nos. 32–49', in *Stradanus 1523–1605. Court Artist of the Medici*, ed. by Alessandra Baroni and Manfred Sellink (Turnhout: Brepols, 2012), pp. 111 and 245–58.

60 Detlef Heikamp, '"Villani" di marmo in giardino', in *Il giardino d'Europa. Pratolino come modello nella cultura europea*, ed. by Alessandro Vezzosi (Milan: Mazzotta, 1986), p. 63.

61 Lazzaro discusses how these figures echo the flow and hierarchy of water use. The peasant and salamander one is especially relevant, as it relates to the unpleasant nature of marshes. Known now only in a preparatory drawing, the peasant is shown with a similar cap. See Claudia Lazzaro, 'From the Rain to the Wash Water in the Medici Garden at Pratolino', in *Renaissance Studies in Honor of Craig Hugh Smyth*, 2 vols, ed. by Andrew Morrogh et al. (Florence: Giunti Barbèra, 1985), vol. 2, pp. 317–26; Lazzaro, *Italian Renaissance Garden*, pp. 136 and 150–52; and Butters, 'Pressed Labor and Pratolino', pp. 67–70 and 78–87.

62 Excerpts taken from Suzanne Butters' translation of Francesco de' Vieri: Butters, 'Pressed Labor and Pratolino', pp. 78–9.

63 Excerpts from Butters' translation of Bastiano Arditi: Butters, 'Pressed Labor and Pratolino', pp. 70–75. See also eadem, 'The Medici Dukes, *Comandati* and Pratolino: Forced Labour in Renaissance Florence', in *Communes and Despots in Medieval and Renaissance Italy*, ed. by John E. Law and Bernadette Paton (Farnham; Burlington, VT: Ashgate, 2010), pp. 252–71.

64 Butters, 'Pressed Labor and Pratolino', p. 87.

65 Spini discusses the Medici acquisition of swamps and woodlands as a way for the Grand Dukes to 'unload their passion for the hunt' ('sfogare la passione per la caccia'). Butters discusses the designation of Medici holdings like Pratolino as 'public' and therefore open to commandeered labour although their actual use was private, 'belonging to members of the dynasty, surrounded by hunting parks that were essentially off-limits to all but them'. See Spini, 'Introduzione generale', pp. 34–38 and Butters, 'The Medici Dukes, *Comandati* and Pratolino', pp. 255–59 and 270–71.

224 *Felicia M. Else*

66 The legislation cited dates to 1568 with the 1581 renewal allowing for the hunting of songbirds in privately owned blinds. See D. R. Edward Wright, 'Some Medici Gardens of the Florentine Renaissance: An Essay in Post-Aesthetic Interpretation' in *The Italian Garden: Art, Design and Culture*, ed. by John Dixon Hunt (Cambridge: Cambridge University Press, 1996), pp. 50–9.

67 Detlef Heikamp, 'La Tribuna degli Uffizi come era nel Cinquecento', *Antichità viva*, 3, 3 (1964), 11–30; Massinelli, 'Magnificenze Medicee', 111–34; Heikamp, *Splendori di pietre dure*, pp. 96–98; Detlef Heikamp, 'Le sovrane bellezze della Tribuna', in *Magnificenza alla corte*, pp. 329–45; and Bertelli, 'Catalog Entries nos. 33–35: Le imprese di Francesco I, 1585–1587' pp. 226–31.

68 Drawn from excerpts by Giovanni Bocchi and Giovanni Cinelli in Heikamp, 'La Tribuna degli Uffizi', 13 and 16.

69 Amelio Fara actually challenges whether some scenes portray Francesco. See Fara, Fara, 'Catalog Entries nos. 33–39', pp. 74–75; Massinelli, 'Magnificenze Medicee', p. 125; and Heikamp, 'La Tribuna degli Uffizi', 29.

70 Author's translation of Giovanni Battista Strozzi's excerpt in Heikamp, *Splendori di pietre dure*, p. 100.

71 Butters, 'Princely Waters', p. 409; Maria Ines Aliverti, 'Water Policy and Water Festivals: The Case of Pisa Under Ferdinando de' Medici (1588–1609)', in *Waterborne Pageants and Festivities in the Renaissance. Essays in Honour of J. R. Mulryne*, ed. by Margaret Shewring (Farnham; Burglinton, VT: Ashgate, 2013), pp. 119–20 and 128–37; and Else, *Politics of Water*, pp. 150–52.

72 Eric Cochrane, *Florence in the Forgotten Centuries: 1527–1800* (Chicago; London: University of Chicago Press, 1973), pp. 129–30 and Butters, 'Pressed Labor and Pratolino', pp. 75–77.

73 Langedijk, *The Portraits of the Medici*, vol. 1, pp. 161–67.

74 Stefania Vasetti, 'I fasti granducali della Sala di Bona: sintesi politica e culturale del principato di Ferdinando', in *Palazzo Pitti. La reggia rivelata*, pp. 229–39 and Nadia Bastogi, 'La Sala di Bona', in *Fasto di Corte: la decorazione murale nelle residenze dei Medici e dei Lorena. Vol. 1: Da Ferdinando I alle Reggenti (1587–1628)*, ed. by Elisa Acanfora et al. (Florence: Edifir, 2005), pp. 87–97.

75 Katherine Poole, 'The Medici, Maritime Empire, and the Enduring Legacy of the Cavalieri di Santo Stefano', in *Push Me, Pull You: Physical and Spatial Interactions in Late Medieval and Renaissance Art*, ed. by Sarah Blick and Laura D. Gelfand (Leiden; Boston: Brill, 2011), pp. 160–63.

76 For Langedijk's excellent discussion of these motifs and their appearance in Cesare Ripa's *Iconologia*, see Langedijk, *The Portraits of the Medici*, vol. 1, pp. 161–66.

77 Vasetti points out Poccetti's use of differing perspective views, combining an expansive view of the harbour with a bird's eye view of the city. See Vasetti, 'I fasti granducali', pp. 236–37.

78 Vasetti, 'I fasti granducali', pp. 236–37 and Benedetto Buonmattei, *Orazione di Benedetto Buommatti fatta in morte del Sereniss. Don Ferdinando Medici Gran Duca Terzo di Toscana* (Florence: Antonio Caneo, 1609), fol. 9r-9v.

79 Vasetti, 'I fasti granducali', p. 231.

80 This room is now the Sala Bianca, covered with white stucco in the eighteenth century. On the function of the Salone dei Forestiere, see Leon Satkwoski, 'The Palazzo Pitti: Planning and Use in the Grand-Ducal Era', *Journal of the Society of Architectural Historians*, 42 (1983), 341.

9 Responses to a recurrent disaster

Flood writings in Rome, 1476–1598

Pamela O. Long

The Tiber is a 406-kilometre-long river that begins in the Apennine mountains east of Florence and empties into the Tyrrhenian Sea at Ostia, 20 kilometres west of Rome. It flows through Rome and flooded the city on a regular basis, often bringing disastrous consequences.[1] In the early modern period, both major floods and smaller floods occurred. Responses to flooding depended not only on the physical event, but also on social, cultural and political factors.[2] In this paper I investigate specific responses to the major floods between 1476 and 1598, and the ways in which these responses changed through this 122-year period.[3]

In a recent essay on disasters, Judy F. Irwin and Jenny Leigh Smith emphasised that all disasters 'should be understood as stemming from a mixture of environmental, technological, and human factors'. Similarly, Gerrit Jasper Schenck has noted that natural disasters are no longer understood as purely physical events, but as social/cultural constructs as well, which can be studied in the context of both urban and environmental history.[4]

This essay reflects these interdisciplinary orientations. In the long term, building a city on a flood plain, particularly on the flood plain of the Tiber, in itself suggests the intermixture of the human, the natural and the technological. In late fifteenth- and sixteenth-century Rome, a growing urban population clustered near the banks of the river, increasing the potential for disaster during floods. Flooding was not just one event but a combination of interrelated events – high water through the city; the collapse of houses; drownings of humans and of animals; the putrefaction of the dead bodies of animals leading to widespread disease; the destruction of the river mills that were used to grind the city's grain;[5] and the ruination of vineyards, olive groves and vegetable gardens that undermined the urban food supply, leading to widespread hunger.

Responses to Tiber flood disasters were varied and complex. Irwin and Smith suggest that premodern disasters were blamed on 'inauspicious planetary alignments, angry gods, or sinful behaviors' in contrast to the 'systematic observation-based knowledge that has largely left behind

DOI: 10.4324/9781003029823-12

astrology, religion and superstition'. This is an inaccurate description of the earlier period.[6] Responses did sometimes include a discussion of divine signs and human sins. But they could also omit such topics, and increasingly, in the period focused upon here, included discussion of the natural causes of flooding, historical accounts of floods in Rome and elsewhere, detailed empirical descriptions of particular floods, and suggestions and plans for practical solutions.

In the major flood of 8 January 1476, the water rose 17.32 masl (metres above sea level).[7] Stefano Infessura (ca. 1435–1500), a law professor, secretary of the Roman Senate and diarist, wrote that on that day the water became so high that one could not walk to St. Peter's, and thus the papal audience (given by the pope Sixtus IV, r. 1471–1484) was held at Santa Maria Maggiore, located safely on the Esquiline Hill.[8] Another diarist, Jacopo Gherardi (1434–1516) from Volterra, provided more details in his brief report on the flood. Gherardi, who had become a priest at age 24 and eventually a bishop, enjoyed a distinguished career as secretary to cardinals and popes, becoming apostolic secretary to Sixtus IV in 1479. He reported that it had rained day and night for 12 days before the flood and that the Via Sistina (present-day Borgo Sant'Angelo) flooded. More frightening than the flood itself, he emphasised, was the certainty of the pestilence that invariably followed Roman floods. For fear of this, the pope withdrew from Rome, first going to Vetrulla, then to Orte, then to Narnia, and finally to Foligno, 'with both great expense and great inconvenience to all'. Gherardi also noted that an inscribed stone (no longer extant) indicating the height of the flood could be found in the house of his employer, Cardinal Jacopo Ammannati Piccolomini (1422–1479) in the Borgo.[9]

Cardinal Ammannati Piccolomini himself reported on the flood in a letter he wrote to a friend in Siena shortly after the event. The letter vividly describes the rise of the Tiber to the height of its channel on 6 January; on 7 January it overflowed the riverbed; and on 8 January it flooded the city unimpeded. It transcended the walls of the cardinal's garden and flowed into the atrium of his house. The residents were forced to withdraw to the upper stories. Ammannati Piccolomini reports that a huge number of snakes (carried from caves in the sea he explains) had been stirred up and swept into Prati, the open area north of the Vatican. Nothing could be carried into or out of the city through the 'Gate of Hadrian' (*Porta Adriani*) with the waves rising up into it. (This refers to Arco di Portogallo on the Via del Corso, an arch demolished in the seventeenth century. It was situated midway down the main street into the city from the north.)[10] The report notes that the papers of emissaries were floating in whirlpools. Roads to the papacy and the pope were blocked. In all parts of the city, low-level storerooms were destroyed. In 'the building of St. Paul' (i.e., St. Paul's Outside the Walls), the water climbed to the first step of the high altar. Many wood structures were destroyed. Trees were torn up everywhere, small buildings and furniture ended up in the river and the poles in the vineyards were gone. The cardinal added that for the past 60 years, no one recalled higher snow (in the mountains), longer rains, or a greater flood. The return of the times of Noah's flood was feared. The cardinal himself had been forced to leave his house and stay with relatives in another part of town.[11]

Another letter writer, Enrico de Ampringen, was the *scriptor litterarum* or secretary of the penitentiary tribunal of the Roman Curia (the tribunal in charge of issues having to do with forgiveness of sins) and also had been a chaplain at a monastery near Basel (in present-day Switzerland). He described the 1476 Tiber River flood in a letter to Johannes, the Bishop of Basel. The river rose so high, Enrico reported, that no one could go into Saint Peter's or out of the gates of the city except on the boats of the Ponte Sant'Angelo. It was rumoured that the pope might leave the city (which, as we know from Gherardi's diary, he did) and that the pope also mandated that his familiars could not enter the city without a special license.[12]

Finally, a letter of 8 January 1476, to the Duke of Milan from a secretary, Johannes Marchus, reported from Rome that it had been raining for 28 days, making it impossible to grind grain, causing a shortage of bread everywhere, even among cardinals. The Tiber rose over its banks and many animals drowned. The Cardinal of Pavia (this is Cardinal Ammannati Piccolomini discussed above) could not leave his house without a boat. Cardinals could not go to the Vatican Palace for the consistory. The pope (Sixtus IV) was in the Castel Sant'Angelo for hours looking at the flood. No living person had seen such a flood. There were live snakes in the Tiber passing under the Ponte Sant'Angelo, which the pope had come to see. The letter writer noted that he had not seen these snakes but over a thousand people had. There were four of them, half a braccio in size, green and with wings.[13] Leaving aside this apparition of dragon-like snakes, these reports were primarily factual descriptions of the flood by elite men, usually close to the papal court. Much of their reporting concerned the effect of the flood on the functioning of the court.

Almost 20 years later, in the flood of 4 December 1495, the water rose to 16.88 masl.[14] The reports concerning this new flood contain notable detail on how it affected ordinary people, as well as elite courts. From this time, reports from a variety of sources become increasingly detailed and are often offered as eyewitness accounts. There is an increased interest in 'natural particulars' – that is, in precise observations of the way particular floods happened, as well as details about what happened to individuals during those floods.[15]

A notable account of the flood of 1495 flood is found in a letter by an unknown writer in the retinue of the Venetian ambassador of Rome

(Girolamo Zorzi, 1430–1507) addressed to Maximilian I, emperor of the Holy Roman Empire. Writing on the day of the flood, the author reported that the river swelled more than it had ever done in the memory of men and did inestimable damage to every kind of person. It occurred on the day of the cardinalate consistory when a great many cardinals were attending the meeting at the Castel Sant'Angelo. The water rose into the street in an instant so that people could barely save themselves out of their houses; it rose continuously until vespers of the following day. People were caught in their beds, and many died. Many others lost goods and property. Those who tried to escape through doors failed, and in the end, they tried to save their lives by climbing onto their roofs or onto other high places. All night, cries for help could be heard. Because many waited (presumably to save their property) many perished.[16]

The only way to get around the city was in small boats 'as one does in our lagoons', added this Venetian writer. In Rome such boats were scarce. It was necessary to build barges, and with those, supply necessities to those in need. If you had not witnessed this, the writer asserts, you would think it a fairy tale. There was nothing to satisfy thirst. The wells were destroyed. The people of Trastevere (a *rione* or district on the right bank of the river) feared losing their bridges. Many buildings fell, many died, many lost property and the pavements of the churches were all lost. Almost all of the animals of the region died – some in the countryside, others in their dens. Shepherds abandoned their animals and climbed into trees to save themselves, tying themselves so they did not fall from exhaustion. There they died from cold and hunger. Others perched in trees that fell and were carried by the water, half dead, into the city. The barges of the city, along with people and goods on them, were destroyed. A newborn baby was found being carried down the river and was saved – just as God had saved Moses and Romulus. The pope (Alexander VI, r. 1492–1503) ordered solemn processions to implore the clemency of God.[17]

Another familiar of the Venetian embassy in Rome, a *coadiutor* or assistant whose name is unknown, addressed a letter four days later (on 8 December) to an unnamed high-ranking Venetian ('Your Magnificence'). The letter provided further vivid details of the flood. It first described the weather conditions leading up to the event. On 25 November, very cold weather began. After some snow, it rained so hard that water seemed to come down in buckets, continuing for two days. Then the weather cleared, but the Tiber began to rise, covering many places in the city. It rose until it almost covered the entire Ponte Sisto (the bridge built by Sixtus IV between 1473 and 1475). It rained with such fury that it seemed that the world was ruined, pulling down mills, bridges and houses. The writer and his companions went out in a small boat and saw such piteous things that 'for that day we did not wish to see more and returned home'.[18]

The writer describes heroic actions. The servants with huge effort brought wine from the cellar and then closed the cellar off. It remained sealed for nine hours before the water broke the barrier and flowed into the house. The courtyard was full of water and those in the cellar (presumably having returned to save more goods) came close to drowning. Neighbours fled with what they could carry, 'lamenting and crying over their things left in the water'. Rome was like Sodom and Gomorrah. Some said it was the will of God. People went about in barges, rafts and boats 'as one does through Venice'. They carried food and water to those who needed it. On Monday, the water was no longer on the streets, but the courtyards and streets remained full of dead animals. Rome would not recover for 25 years, the correspondent lamented.[19]

This same flood of 1495 was reported in a very different way by Giuliano Dati (ca. 1445–1523). Dati was a priest from Florence who is first recorded in Rome in 1485 as a penitentiary (a priest giving confessions and administering the sacrament of penance) at the Church of San Giovanni Laterano. He pursued a successful clerical career in Rome, culminating in 1518, when Leo X named him bishop of San Leone in Calabria. Dati was also a prolific author and translator. His writings included lives of saints, histories, accounts of the stations of the cross in Rome, a tract on the buildings of Rome, lives of the popes, a metrical paraphrase of a letter of Christopher Columbus, two volumes on 'songs of the Indies' and a poem written in octaves concerning the flood of 1495. This and his other writings were aimed towards a popular readership and were a way, as Paola Farengo has observed, for him to express his pastoral commitments.[20]

Dati interpreted the flood as a sign of divine intervention – a catastrophe that also announced future catastrophic events. He provides a litany of such events from ancient times to his own, taken from the Bible and from a variety of chronicles and histories. He emphasised that each of these past disasters was preceded by ominous signs. In his own age, he pointed to the comet that appeared in Puglia on 20 January 1485, a divine sign of misfortunes to come, such as the death of Pope Innocent III in 1492, the death of King Ferdinand of Naples in 1494, the invasion of Charles VIII of France in the same year, the Turkish conquests in the eastern Mediterranean and the arrival of pestilence.[21] His noting of celestial signs undoubtedly is a reference to the astrological view that celestial bodies influence events on Earth, a frame of reference almost universally accepted in this era, despite occasional dissenting voices.[22]

The first half of Dati's poem provided, from his own point of view, a moral and historical context for the flood. In contrast, he devoted the last half of his 109-verse poem to a detailed description of the flood itself, to which he was an eyewitness. Dati's account is primarily topographical – focused on the districts (*rioni*) of Rome, on the buildings within those districts, and on other structures. He begins (always using rhymed octaves) with a list of the gates of Rome that were flooded. He continues, itemising numerous flooded chur-ches, and then proceeds by districts, starting with the Borgo (the area near the Vatican), and continuing with the Campo Marzio, Ponte, Parione and other *rioni* at the centre of the city. He mentioned the flooding of sewers, damage to the Tor di Nona, the prison across the river from the Castel Sant'Angelo and the hospital of Santo Spirito. Dati itemises the many palaces of nobles, merchants and cardinals that were damaged. His emphasis on damaged churches includes some of their altars and statues. He refers to many in-dividuals, but primarily in the context of their ruined palaces. Dati mentions only one incident of individuals being rescued – three companions from Urbino were in the area of Prati. They climbed to the roof of a house, and as the house itself seemed to be crumbling, shouted at a boat. Although their shouts could not be heard because of the roar of the river, they waved their hats. The men in the boat saved them just before the structure collapsed, and brought them, 'each of them trembling like leaves' to St. Peter's.[23]

More catastrophic than any flood discussed so far was the flood of 8 October 1530, during which the water reached 18.95 masl[24] and which arrived as the traumatised city was just beginning to recover from the terrible Sack of 1527, in which troops of the imperial army of Charles V, led by Charles III, Duke of Bourbon (1490–1527) had ransacked the city and murdered, tortured, castrated and raped its inhabitants for ten long months.[25]

Reports of the 1530 flood include a letter of 13 October 1530 from Giovanni Battista Sanga (1496–1532), a humanistically trained literary figure and secretary to cardinals and popes. Sanga addressed the letter to Alessandro de' Medici (1510–1537), who would rule Florence from 1531 until his death. He reported that the Tiber rose so much that it flowed all through Rome, in some places as much as eight times higher than it had reached at the time of pope Alexender VI (i.e., the flood of 1496). Sanga wrote that 'we all were besieged in our houses' and that 'to a city afflicted and consumed as this one, it seemed like another Sack'. New wine was all lost, as was almost all of the old. So much grain was destroyed that the price quadrupled so that no one could think of living through the year without grain from Sicily. Fodder, hay and firewood were almost all carried away as was 'an infinite number of possessions'. The rise of the river was so sudden that none of these things could be saved. The river had also carried away many animals as well as people who resided in low-lying houses.[26]

Sanga described the physical flood in detail. The water left the riverbed at 7 o'clock the previous Friday, it rose all day Saturday, and all of Sunday, making it impossible to travel through Rome except by boat. The flood 'left streets and houses so deformed that it is frightening to go around Rome'. Many houses were ruined; others were propped up because the water had undermined the foundations. The Ripa (port), where boats came to Trastevere, was carried away. The houses on Via Giulia were destroyed with few signs that they ever would be restored. What happened to Messer Eusebio, once secretary to Cardinal San Giorgio, 'gave the whole city a very great terror'. He was in his house with perhaps 30 other people when the river lifted the terrain from under the house and the structure collapsed. All people and animals within were killed.[27]

An anonymous author provided further vivid detail in a tract printed in Bologna in November 1530. On Thursday, 6 October, it began to rain: 'it seemed as if all the cataracts of the sky had opened'. It rained for two days and two nights. Most of the 'aqueducts' (meaning undoubtedly sewers and drains) broke and became one with the river. The river itself was unable to exit into the sea because of the wind and waves, and 'in a moment the dry land in that place transformed into a great sea'. On Friday the 7th, at 9 in the evening, the water began to pour through the city. Underground cellars, where wine and firewood were stored, filled up with water. The streets began to flood. Some, who were asleep in their beds, drowned. Others fled to rooms above or on roofs.[28]

The anonymous author reports that 'people on foot and on horses searched for a place to save themselves from the horrible fury of the water'. He describes the terrifying separation of families and friends. 'Fathers did not wait for sons, sons did not care for fathers, brothers did not [look after] brothers,, nor did friends think to save anyone but themselves'. The author describes the piteous 'laments, shrieks, shouts, tears, clapping of palms, pulling of clothes, face scratching, breast beating that filled the air'. He continues the list of lost people – the waves carried children away. Mothers could not save them. Fathers, mothers, husbands, brothers, sisters, wives and friends – none of them knew how to save themselves or the others. [29]

The flood of 1530 also occasioned the first Latin treatise on Tiber River flooding. It was written by Luis Gómez (ca. 1482–1542), a Spanish prelate who served as the auditor of the Rota in the papal court (a judge in one of the papal tribunals called the 'Sacra Romana Rota'). An expert in civil and canon law, Gómez wrote lengthy tracts and commentaries on legal matters, such as on the decisions of the Rota and on decretals, or papal decrees.[30] His treatise on the floods of the Tiber River, distinctly different from his other writings, clearly was engendered by his experience of the flood itself.[31]

Gómez begins his treatise with a discussion of the Tiber River in antiquity, based on ancient sources. Notably, he discusses not only the ancient floods themselves, but various solutions to the problem of Tiber flooding that ancient Roman emperors had proposed and/or carried out. This included the large-scale diversion of streams and rivers, proposed by the emperor Tiberius (r. 14–37 AD) and rejected by the Roman Senate; the emperor Trajan's desire to dig an enormous canal so that the swollen waters of the Tiber would have an outlet; and the emperor Aurelian's construction of retaining walls on the river banks. Gómez asks whether the floods were greater in antiquity, providing reasons for pro and con sides of the question.[32]

He concludes that floods in his own time were worse for several reasons. The ancient maintenance of the riverbed through cleaning had been forgotten, and now, the Tiber was full of debris, making it flow sluggishly. In addition, unlike in ancient times, many houses and other structures had been built in the low-lying areas near the river banks. Gómez ends this section with a note that he would be considered an unreliable judge, if he considered 'the most recent and very harsh flood of the Tiber, in which I suffered serious damage', to be less severe than ancient floods which he had never witnessed. He further suggests it would ease the pain of the present suffering if he provided a history of all floods that happened in Rome 'so that we may be consoled by the adversities of past times and bear more easily the present troubles as if we shared the same fate'.[33]

Gómez provides this history in the second part of his treatise in which he describes the major Tiber River floods from antiquity to the 1530 catastrophe, beginning with the 'first flood' in 415 BCE (as reported by Livy) and ending with the 'twenty-third' flood, that of 1530. In his accounts, Gómez remarks that 'the Tiber never overflowed without portending something remarkably bad'.[34] As he arrives at the flood within living memory (i.e., the 1495 flood), he provides greater detail and emphasises the moral implications of the flood. Using a medical analogy, he notes that Rome at the time of the 1495 flood possessed an excess of wealth and abundance of all things. He suggests that the flood was like a phlebotomy (referring to the practice of bloodletting, a standard medical procedure at the time in which what was thought to be too much blood was removed from a sick person). Abundant things, Gómez remonstrates, take people away from themselves and from understanding themselves and lead them 'to neglect and disregard the Savior of all things'. The flood removed the abundance of things and forced people to come to their senses and 'to recognize what was true and just'.[35]

Arriving at the 1530 flood, Gómez captures its full horror. It occurred 'on that unspeakable day' of Saturday, 8 October. On that day, the Tiber moved away from its riverbed 'to the great amazement and injury of all … Great fear caused by the groan of collapsing buildings wore out gentle souls'. Moreover, 'scars . . . were still fresh' from the memories of past misfortunes (here referring to the Sack of Rome). Gómez notes the debates that had ensued about the causes of the flood – it was sent by God for the vindication of sins, or it was caused by an eclipse of the sun and the moon, or perhaps by (other) natural events.[36]

But pure fright eclipsed arguments about causes. The water rose rapidly. People saw the houses of their neighbours become engulfed in water with no way of escape. 'Fear had so invaded the improvident and wretched hearts of mortals that people seemed out of their minds and speechless'. Gómez likens the flooding Tiber to the head of an army of rivers attacking the city. After unsuccessfully trying to destroy the Ponte Sant'Angelo, the troops turned to siege the Via Giulia, which (Gómez adds) was a street that had once been part of the riverbed. The attack shook every building on the street. Many collapsed, including the house of Eusebio (mentioned above) with its tragic deaths.[37]

Vividly portraying the terror of the people of Rome, Gómez suggested that they suffered more trauma and terror than they had under the 'siege of enemy barbarians' (i.e., the Sack). He describes the deaths, the lack of wine and drinkable water, the lack of firewood, the lack of grain and the cold, hunger and thirst. Parents were separated from their children, spouses from spouses and friends from friends. Even when the Tiber receded, the squalor and filth, the destroyed buildings and the certainty of pestilence brought on by filth meant that the misery of the people remained.[38]

The third section of the treatise, entitled 'The Consequences of Floods', includes a further discussion of causes as well as possible preventive remedies. The destruction of dwellings and the death of humans and animals could be avoided if 'the power and nature of rivers' were investigated. Riverine power could then be weakened by engineering measures, such as trenches for runoff during floods, and the construction of embankments and levees. Another consequence of floods, hunger, could be avoided by maintaining a surplus supply of grain and making contingency arrangements for grain with nearby territories. Gómez devotes most of the section to a third consequence, pestilence, that comes from foul vapours emitted after sewers and other rank caverns are excavated by the floodwaters. Here he refers to the miasmatic theory of disease, commonly held at the time, which viewed smells as disease-causing entities. He cites examples from medical texts, including Galen. He also points to cities that do not have adequate sewers (making bad vapours a constant experience) and ends with the idea that it is impossible to clean all the sewers and caverns. Thus, such things are placed in the hands of God almighty.[39] Although Gómez did not offer any of his own solutions to the problem of Tiber River flooding, it is significant that he at least mentioned, as we have seen, concrete solutions by noting the flood prevention projects of ancient Roman emperors.

For almost thirty years, thereafter, Rome was spared a major flood, a respite that ended with the massive inundation of 15 September 1557 at 18 masls. Numerous reports of this catastrophe are extant, detailing the drownings of humans and animals, the collapse of houses, the destruction of the city's mills and consequent hunger, the mud, filth and putrefaction, and the spread of disease.[40]

In addition, the responses reveal significant developments in thought and cultural orientation. First, many writers exhibited an intense interest in the causes of the floods (meaning Aristotelian causes), because they considered knowledge of causes crucial to finding permanent solutions. Second, they carefully discussed particulars, including the weather and other conditions before the flood, and details of the flood itself. Finally, there is extensive discussion of possible solutions to Tiber River flooding: concrete solutions involving engineering projects and other practical remedies that might finally tame the unruly river. These interests can be shown in the flood writings of two men from very different walks of life: the physician and naturalist, Andrea Bacci (1524–1600), and the architect/engineer, Antonio Trevisi († ca. 1567).

Andrea Bacci was by far the most prolific author on Tiber flooding in the second half of the sixteenth century. He was a professor of medicinal simples (herbal and botanical medicine) at the Roman university, La Sapienza, and a physician to Felice Peretti before and after Peretti became Pope Sixtus V (r. 1585–1590). Bacci wrote numerous tracts on medicine, natural philosophy and his central topic – springs, baths and rivers.[41] He published a tract on the Tiber River in 1558, a year after the flood. An enlarged, revised edition appeared in 1576. Then, after the terrible flood of 1598, he published his final revision.[42]

Bacci dedicated his 1558 tract on the Tiber River to Alfonso Carafa, nephew to the pope (Paul IV Carafa, r. 1555–1559), and divided it into two parts. The first part focuses on the nature of water (one of the four elements, the other three being earth, air and fire), and the differences between waters in diverse locations. He described the Tiber River, the history of its uses, particularly by the ancients, and the characteristics of its (very potable, in his opinion) water. In the second part, Bacci discusses various opinions concerning the causes of flooding, emphasising that flooding in different rivers (such as the Nile and the Tiber) has different causes. He then turns to the flood of 1557, describing the weather in the months before, including the heavy, dark and foggy air that infected almost everyone with cough and fever. He describes the rise of the water on 15 September and blames the terrible damage to the city on the failure in his own times to care for the river in the way that the ancients had cared for it.[43]

An intricate empirically based historical study follows, based on ancient texts and on inscriptions. Bacci investigates remedies that were suggested in antiquity, and the remedies that the ancients actually carried out. Suggestions included dividing the Tiber into branches and diverting its tributaries so that they did not add to its volume. Remedies, which Bacci urged on his contemporaries, included cleaning the bottom of the river to eliminate the huge amount of debris and filth; creating a magistrate charged with caring for the ports and the riverbed; straightening some of the bends of the river; fortifying its banks; and keeping the sewers and drains of the city clean. Bacci also suggested that the arches of the bridges might impede the water. Finally, if houses near the river were built with stronger foundations, they would be less likely to collapse during a flood.[44]

In contrast to Bacci, the architect/engineer Antonio Trevisi, who was from Lecce in southern Italy, had practical experience in building walls, fortifications and palaces. Trevisi travelled to Rome in 1559, probably called to work on flood control. His tract on Tiber River flooding, published in 1560, was dedicated to Federico Borromeo (1535–1562), nephew of the pope, Pius IV (r. 1560–1565). In it, he describes the Aristotelian cosmos and the four elements, including water, and the various 'kinds' of water including springs, rivers, seas and rainwater. He discusses the causes of flooding, detailed the damage made by the 1557 flood and reports his own measurements of the heights reached by the water in different locations in the city (probably evident from mud lines). In his tract, Trevisi mentions his 'design' or 'model' of a flood control plan, but he does not make clear exactly what it is.[45]

In another publication, his 1560 edition of the famous woodcut map of Rome originally created by Leonardo Bufalini in 1551, Trevisi attaches letters to the bottom edge of some of the maps. Four letters from two different extant maps are known. They are addressed to the conservators of Rome (the three magistrates who headed the civic government), to Cardinal Carlo Borromeo (1538-1584, the pope's nephew), to 'Readers' and to 'Virtuous Architects'. In each letter, he urges the reader to adopt his plan for flood prevention, which was to dig a large diversionary canal for water runoff that would begin to the north of the city at Ponte Molle (i.e., Milvio) and end in a valley south of the Vatican in times of flooding.[46]

Unlike most of the earlier writings about flood control discussed here, Bacci and Trevisi wrote in a context in which the popes and also the conservators of Rome were discussing and even attempting to carry out concrete solutions to the problem of flooding. Bacci served on a number of cardinalate committees that considered various solutions. Trevisi presented a concrete plan and was undoubtedly attempting to obtain a contract to carry it out. Although Trevisi's flood plan was not adopted, Pope Pius IV did have large trenches dug around the Castel Sant'Angelo which served as a run-off channel during floods.[47]

These trenches may have been partially effective but did not permanently solve the problem. On 24 December 1598, the worst flood ever recorded in Rome submerged the city at 19.56 masl.[48] A flood of writings also appeared, describing the event in detail, providing histories of Tiber flooding and discussing possible solutions. An obscure poet and writer,

Giacomo Castiglione, dedicated his treatise on the Tiber River floods to Cardinal Pietro Aldobrandini (1571–1621), nephew of Pope Clement VIII (r. 1592–1605). Castiglione reports ancient references to floods , discusses the writings of his near contemporaries on the topic and carefully records the inscriptions of many of the flood markers around the city. He notes that the pope had ordered a cardinalate congregation (an official committee headed by cardinals) to find a remedy for the Tiber River floods. He seems to have written his treatise as an informative account of the topic for the elucidation of that committee.[49]

He describes the river and weather conditions in Rome in the six months prior to the flood. He vividly details the catastrophe. 'People were in their houses astonished, seeing such a horrible spectacle'. They did not know what result it would have (for themselves), and they saw their goods and belongings destroyed, and horses dying in their stalls. Castiglione laments 'those poor people and families gathered under the ruin of some small houses which, not being able to resist so much force, fell'. Others were caught in their houses when water rushed through the windows and over the roofs. Hunger came to Rome as mills for grinding grain were destroyed along with food supplies.[50]

Castiglione outlines the causes of floods according to diverse authors – snow, wind or the sand of the sea, celestial influences or the seasons, or rains. He provides a catalogue of 30 floods, clearly relying on the prior writings of Gómez and Bacci. Finally, he discusses in detail the remedies that had been carried out or suggested by the ancients and also by his contemporaries. These include appointing a curator of the river, making levees (*argini*), constructing walls at the banks of the river, creating variously placed run-off canals, damming and diverting Tiber tributaries, and cleaning and deepening the riverbed. He discusses the role of Tiber River bridges in worsening floods (by the impediment of their arches). He itemises possible remedies for the diseases and famines brought by the floods, such as burning the bodies of dead animals, cleaning the streets and cellars and storing a supply of grain in high places. Finally, he appends an account of the rescue of many of the people who lived around the Castel Sant'Angelo and the beneficence of the Aldobrandini and others, who saved lives and helped the hungry and homeless.[51]

A second substantial treatise engendered by the 1598 flood was Andrea Bacci's. As mentioned above, Bacci had published an expanded version of his 1558 treatise in 1576, and a third even further revised and expanded edition in 1599. In the 1576 edition, he had summarised the remedies of the ancients and included accounts of the known difficulties for each. He recommended the more doable measures: fortify the banks and plant trees along them for stabilisation and beauty; straighten the course of the river within the city; enlarge and clean the riverbed; and clean the sewers regularly and expand the moats around the Castel Sant'Angelo, which Pius IV had constructed (to provide an outlet during flooding).[52]

Bacci's 1599 version added an extended summary/translation in Italian of Gómez's Latin treatise of 1531. It also brought the history of Roman floods up to date (i.e., up to 1598), and includes a long description of that flood. Bacci completed his new edition as part of a group of men (including many with practical, skilled backgrounds) brought together by the pope to solve the problem of flooding. He focuses on solutions, detailing the pros and cons of three kinds. The first involves diversions of tributaries of the Tiber. The second comprises defences at Rome itself, including run-off ditches and canals. The third centres on the care of the river within Rome: straightening the course, cleaning and lowering the riverbed, modifying the bridges by remodelling them with fewer piers, and regulating the disposal of sewage into the river.[53]

Another member of the commission was the architect/engineer, Giovanni Fontana (1540–1614), who was the older brother of the famous Domenico (renowned for moving the Vatican obelisk). Giovanni was architect of the pope Clement VIII (r. 1592–1605). He was well-known for hydraulic engineering projects and had directed or worked on the construction of several aqueducts. He believed that solutions to flooding in Rome could only be found if knowledge was acquired of the rivers and streams that were tributaries of the Tiber, including knowledge of the volume of water they contained. Along with some colleagues and assistants, he measured all the rivers in the area north of Rome, recording the depth and width of the water both in normal times and during floods.[54] He wrote a small tract that carefully recorded these measurements and suggested that Tiber River flooding could only be controlled by building two additional canals, each the size of the Tiber riverbed.[55]

Giacomo della Porta (1532–1602) also served on the commission and prepared a memo offering detailed advice. Della Porta, from northern Italy, was the son of a sculptor and trained as a stucco worker. In Rome, he became involved in and eventually headed numerous hydrological and architectural projects, including the repair of the Acqua Vergine. He eventually served as the architect of the Roman People (i.e., the Capitoline Council), as architect to the noble Farnese family, and as architect of St. Peter's.[56]

A highly skilled practitioner who had become by this time a trusted architect, Della Porta's advice to prevent Tiber River flooding was based on his close knowledge of actual floods. He describes the floods as beginning at Ponte Molle (Ponte Milvio) north of the city and describes how the water rises at Ponte St. Angelo. He suggests opening the closed arches that were part of the ancient structure of the Ponte St. Angelo and doubling the size of the ditches around the Castel Sant'Angelo. Between this bridge and the Ponte Sisto, the riverbed should be widened and the excavated earth used to reinforce the embankments. The Ponte Sisto should be given an additional arch. The many mills on the Tiber, Della Porta explains, impede the flow of water, and when they break away from their chains in floods, they crash into and damage the down-river bridges. All of them between the Ponte Sant'Angelo and the Ponte Santa Maria (that is, through the centre of Rome) should be removed along with their associated pilings and rebuilt farther downriver. Further, the winding river should be straightened at four places between Ostia and Rome, allowing for a swifter flow of the water. And finally, larger embankments should be constructed along the river through the centre of the city.[57]

The measures proposed by these and other individuals after the flood of 1598 were not carried out at that time. Indeed, the tradition of floods and proposals of plans to prevent those floods continued unabated until 1880, when engineers constructed the high embankment walls along the river that remain to this day. These walls not only prevented (and still prevent) flooding in the city centre, but also separated Rome from its river.[58]

Between 1476 and 1498, responses to the intermittent catastrophe brought on by floods occurred within a changing urban context. During these years, Rome experienced a rapidly rising population along with a great building boom of palaces and churches. In addition, several of the ancient aqueducts were reconstructed or built anew. With the consequent huge increase in available freshwater, numerous public fountains were built. Streets were straightened and widened, and every effort was made (often without success) to keep the streets clean. Ancient Egyptian obelisks, carried to Rome from Egypt by the ancient Roman emperors, were transported from their ancient locations to newly Christianised sites. At the same time, Rome increasingly became a site for the antiquarian studies of scholars and of artisanally trained men, sometimes working together, seeking understanding of the ancient past. The imperial past of the ancient Roman emperors became an ideal to be achieved in the form of the much-desired magnificence of a Christianised, papal Rome.[59] These developments undoubtedly made the destruction of the Tiber River floods particularly galling. To give just one case in point, the inundation of Via Giulia in the 1530 flood involved the devastation of a wide, stately street along the river that had been built by Pope Julius II (r. 1503–1513).[60]

This study has focused on Rome, but the flood events in that city and responses to them can also be seen as part of a wider Mediterranean and European phenomenon. For one thing, this region (and beyond) in this period was experiencing what is called the 'Little Ice Age', a cooling of the climate that exacerbated flooding in many places. The floods in Rome of 1557 and 1598 were similar to catastrophic floods in Florence and Palermo at the same time, for example. In addition, the Roman writings that have been discussed here reflected a wider phenomenon in that individuals from all walks of life took pen to paper to write about the floods and often to suggest solutions. Roman authors discussed in this paper include jurists, poets, physicians and architect/engineers. The variety of their backgrounds is often typical of responses to public problems and catastrophes in this period, a time of expanding participation of practitioners in the culture of writing, and a time in which 'trading zones' developed – that is, arenas in which substantive communication between university-trained and workshop trained individuals occurred.[61]

In Rome itself, responses to Tiber River flooding changed over this 122-year period. I suggest that those changes also reflect developments in the wider culture of sixteenth-century Europe. Responses to the 1476 flood for the most part took the form of brief news reports in letters and diaries, and mostly concerned the ways in which the flood affected elite Romans and the papal court. Often divine signs and divine retribution are mentioned. In successive floods, increasingly, details of natural phenomena, such as climatic conditions in the days and months before the flood, are discussed, as well as detailed descriptions of the flood itself, usually based on observation. They also include specific accounts of the ways in which elite individuals, ordinary people and specific buildings were affected.

By the later sixteenth century, comments about divine retribution have mostly disappeared. Historical studies of prior floods (which first became prominent in Luis Gómez's Latin treatise of 1531), appear, as do discussions of ancient remedies. Concrete proposals, some quite detailed, for what should be done in the present, become prominent. This is accompanied by papal attempts to find solutions to the perennial problem, including proposals for large-scale (and very costly) engineering undertakings, such as building large run-off canals, widening the river and building large-scale embankments. Other more limited (but by no means trivial) measures were also proposed: moving the mills downriver (or eliminating them altogether), cleaning the bottom of the river of debris and providing for an official in charge of the river.

This is to say that responses to Tiber River flooding reflect empirical values and practical orientations, including the importance of individual observation of the floods themselves and the conditions that preceded them. They began to involve the participation not only of elite purveyors of news, but both university-educated scholars and physicians, and artisanally trained practitioners. No one solved the problem of the Tiber River flooding in the late sixteenth century. However, flood writings, taken as a whole between 1476 and 1598, can be seen as part of a profoundly important cultural development of the second half of the sixteenth century that increasingly valued direct observation, the understanding of natural particulars, the value of skilled practitioners and the devising of practical solutions for ending the intermittent catastrophe of Tiber River flooding.

Notes

1 Several works provide a list of all known Tiber River floods. I have used Gregory S. Aldrete, *Floods of the Tiber in Ancient Rome* (Baltimore: Johns Hopkins University Press, 2007), pp. 241–46; and Pio Bersani and Mauro Bencivenga, *Le piene del Tevere a Roma dal V secolo a C. all'anno 2000* (Rome: Servizio Idrografico e Mareografico Nazionale, 2010).

2 See Andrea Janku, Gerrit J. Schenk, and Franz Mauelshagen, 'Introduction', in *Historical Disasters In Context: Science, Religion, and Politics*, ed. by Andrea Janku, Gerrit J. Schenk, and Franz Mauelshagen (London; New York: Routledge, 2012), pp. 1–14, who stress that 'disasters have to be understood as major forces shaping historical processes and therefore need to be studied not as isolated events but in their historical context' (p. 2). See also Gerrit Jasper Schenk, 'Dis-astri: modelli interpretativi delle calamità naturali dal Medioevo al Rinascimento', in *Le calamità ambientali nel tardo Medioevo europeo: realtà, percezioni, reazioni*, ed. by Michael Matheus et al. (Florence: Firenze University Press, 2010), pp. 23–75, a wide-ranging discussion of the use and meaning of the word 'disaster' and models of interpretation used to understand them.

3 See especially Vittorio Di Martino, Roswitha Di Martino, and Massimo Belati, *Huc Tiber Ascendit: le memorie delle inondazioni del Tevere a Roma* (Rome: Arbor Sapientiae, 2017), pp. 7–89; Cesare D'Onofrio, *Il Tevere: l'Isola tiberina, le inondazioni, i molini, i porti, le rive, i muraglioni, i ponti di Roma* (Rome: Romana Società Editrice, 1980), pp. 301–30; Anna Esposito, 'I "diluvi" del Tevere tra '400 e '500', *Rivista Storica del Lazio*, 20 (2002), 17–26; Anna Esposito, 'Le inondazioni del Tevere tra tardo Medioevo e prima età moderna: leggende, racconti, testimonianze', *and*, Silvia Enzi, 'Le inondazioni del Tevere a Roma tra il XVI e XVIII secolo nelle fonti bibliotecarie del tempo', both in *Mélanges de l'École française de Rome, Italie et Méditerranée*, 118 (2006), 7–12 and 13–20, respectively; Pietro Frosini, *Il Tevere: Le inondazioni di Roma e i provvedimenti presi dal governo italiano per evitarle* (Rome: Accademia Nazionale dei Lincei, 1977), pp. 129–233; Pamela O. Long, *Engineering the Eternal City: Infrastructure, Topography, and the Culture of Knowledge in Late Sixteenth-Century Rome* (Chicago; London: University of Chicago Press, 2018), pp. 19–41; and Maria Margarita Segarra Lagunes, *Il Tevere e Roma: storia di una simbiosi* (Rome: Gangemi Editore, 2004), pp. 69–133.

4 Judy F. Irwin and Jenny Leigh Smith, 'Introduction: On Disaster', in 'Focus: Disasters, Science, and History', *Isis*, 111 (2020), 98–103 (98–99); and Gerrit Jasper Schenk, 'Historical Disaster Research: State of Research, Concepts, Methods, and Case Studies', *Historical Social Research*, 32 (2007), 9–31.

5 For the Tiber River mills, see especially Maria Margarita Segarra Lagunes, 'Le attività produttive del Tevere nelle dinamiche di trasformazione urbana: i mulini fluviali', *Mélanges de l'École française de Rome, Italie et Méditerranée*, 118 (2006), 45–52.

6 Irwin and Smith, 'Introduction: On Disaster', 99–100. For a comparative study of three early modern disasters focused on the close relationships of confessionalism, and political and religious responses to catastrophes, see Elaine Fulton, 'Acts of God: The Confessionalization of Disaster in Reformation Europe', in *Historical Disasters in Context*, pp. 54–74.

7 For the 1476 flood, see especially Anna Esposito, 'Roma e i suoi "diluvi"', in *Giuliano Dati, Del diluvio de Roma del MCCCCXCV a dì IIII de decembre,*

ed. by Anna Esposito and Paola Farenga (Rome: Roma nel Rinascimento, 2011), pp. 5–26 (pp. 13–15); Esposito, 'Le inondazioni del Tevere', pp. 7–8; Di Martino, Di Martino, and Belati, *Huc Tiber ascendit*, pp. 48–50; and Frosini, *Il Tevere*, pp. 153–54. For the height of the flood, see Aldrete, *Floods of the Tiber*, p. 244.

8 Stefano Infessura, *Diario della città di Roma*, ed. by Oreste Tommasini (Rome: Forzani E. C. Tipografi del Senato, 1890), p. 80.

9 'Il diario Romano di Jacopo Gherardi', ed. by Enrico Carusi, in *Raccolta degli Storici Italiani del cinquecento al millecinquecento*, ed. by Lodovico Antonio Muratori (Città di Castello: Casa Editrice S. Lapi, 1904), p. 31: "maximo cum sumptu, tum incommode omnium." For Gherardi, see Stefano Calonaci, 'Gherardi, Jacopo', *Dizionario Biografico degli Italiani,* 100 vols. (Rome: Istituto della Enciclopedia Italiana, 1960–2020) (hereafter *DBI*), 53 (2000), 573–76; and Egmont Lee, *Sixtus IV and Men of Letters* (Rome: Edizioni di Storia e Letteratura, 1978), pp. 74–80. The towns in the pope's itinerary are all in central Italy not far from Rome. For the flood marker, see Di Martino, Di Martino and Belati, *Huc Tiber Ascendit*, pp. 48–50.

10 The phrase is 'Porta Adriani fluctibus in eam delatis, inferri quicquam efferrique non potuit'. The Arco di Portogallo was sometimes referred to as the Archus Hadriani, and it is to this that Ammannati Piccolomini must be referring. Although I know of no evidence that there was an actual gate or door on the arch, it was a necessary passageway into the city on Via Lata (now Via del Corso). See Lawrence Richardson, Jr., *A New Topographical Dictionary of Ancient Rome* (Baltimore: Johns Hopkins University Press, 1992), p. 21. I thank Ann Olga Koloski-Ostrow and Rabun Taylor for clarifying this reference.

11 For the letter, which was addressed to Goro Loli-Piccolomini, and the attached poem, see Iacopo Ammannati Piccolomini, *Lettere (1444–1479)*, 3 vols, ed. by Paolo Cherubini (Rome: Ministero per I Beni Culturali e Ambientali and Ufficio Centrale per I Beni Archivistici, 1997), I, pp. 154–56 and III, pp. 2010–14 (Letter 842). See Giuseppe Calamari, *Il confidente di Pio II: Cardinale Iacopo Ammannati-Piccolomini (1422–1479)*, 2 vols (Rome: Augustea, 1932), II, 347–53 for the flood; and Edith Páztor, 'Ammannati (poi Ammannati Piccolomini)', *DBI*, 2 (1960), 802–3. See also Esposito, 'Roma e I suoi "diluvi"', pp. 13–15; and Di Martini, Di Martini, and Belati, *Huc Tiber Ascendet*, pp. 48–49. For St Paul's Outside the Walls, see Nicola M. Camerlenghi, *St. Paul's Outside the Walls: A Roman Basilica, from Antiquity to the Modern Era* (Cambridge: Cambridge University Press, 2018). For the broad context of the theme of Noah's flood in this period, see especially Lydia Barnett, *After the Flood: Imagining the Global Environment in Early Modern Europe* (Baltimore: Johns Hopkins University Press, 2019).

12 The letter is reproduced in 'Johannis Knebel capellani ecclesie Basiliensis Diarium', ed. by Wilhelm Vischer and Heinrich Boos, *Basler Chroniken*, 2 (1880), 408–9; and see Esposito, 'Roma e i suoi "diluvi"', pp. 13–4.

13 The letter is published in *Bollettino Storico della Svizzera Italiano*, 6 (1884), 107. Johannes Marchus about whom little is known was the secretary to the Orator of Venice. A braccio in Rome at this time equalled about 26 inches or 66 centimetres.

14 Aldrete, *Floods of the Tiber*, p. 244, for the masl. See also Di Martino, Di Martino, and Belati, *Huc Tiber Ascendit*, pp. 50–54; Esposito, 'Roma e i suoi "diluvi"', pp. 15–19; Esposito, 'Le inondazioni del Tevere', pp. 8–11; and Frosini, *Il Tevere*, pp. 155–60.

15 Anthony Grafton and Nancy Siraisi, 'Introduction', in *Natural Particulars: Nature and the Disciplines in Renaissance Europe*, ed. by Anthony Grafton and Nancy Siraisi (Cambridge, MA: MIT Press, 1999), pp. 1–21.

16 *Annali Veneti Dall'Anno 1457 al 1500 ordinati e abbreviati dal Senatore Francesco Longo*, ed. by Domenico Malipiero (Florence: Gio. Pietro Vieusseax, 1848), pp. 409–10.

17 *Annali Veneti Dall'Anno 1457 al 1500*, p. 411: "come si fa nelle nostre lagune."

18 *Annali Veneti Dall'Anno 1457 al 1500*, pp. 411–15 (p. 412): "che per quel giorno non volemo veder altro, et tornamo a casa."

19 *Annali Veneti Dall'Anno 1457 al 1500*, p. 413: "lamentandosi et pangendo della loro roba lasciata nell' aqua'; 'come si fa per Venetia'.

20 G. Curcio and Paola Farenga, 'Dati, Giuliano', *DBI*, 33 (1987), 32–5; Dati, *Del diluvio de Roma*; and Paola Farenga, 'Dati e il suo 'Diluvio', in Dati, *Del diluvio de Roma*, pp. 27–44 (p. 31: "si espresse l'impegno pastorale").

21 Dati, *Del diluvio de Roma*, pp. 71–94 (verses I–XLVII).

22 For an introduction to the large literature on astrology in this period, see especially Monica Azzolini, *The Duke and the Stars: Astrology and Politics in Renaissance Milan* (Cambridge, MA and London: Harvard University Press, 2013), esp. pp. 1–64; and H. Darrel Rutkin, 'Astrology,' in *The Cambridge History of Science*, Vol. 3: *Early Modern Science*, ed. by Katharine Park and Lorraine Daston (Cambridge: Cambridge University Press. 2006), pp. 541–561. The most important dissenting voice was Pico della Mirandola (1463–1494), whose *Disputationes adversus astrologiam divinatricem* [Disputations against Divinatory Astrology] was published in 1496, a year after Dati's poem on the Tiber River flood.

23 Dati, *Del diluvio de Roma*, pp. 94–125 (verses XLVIII–CIX; XCVIII, C for the rescue: "ciascun di lor tremando come foglia").

24 Aldrete, *Floods of the Tiber*, p. 244, for the masl. See also Di Martino, Di Martino, and Belati, *Huc Tiber Ascendit*, pp. 62–65; Esposito, 'Roma e i suoi "diluvi"', pp. 22–26; Esposito, 'Le inondazioni del Tevere', pp. 11–2; and Frosini, *Il Tevere*, pp. 160–65.

25 See especially André Chastel, *The Sack of Rome, 1527*, trans. by Beth Archer (Princeton: Princeton University Press, 1983); Kenneth Gouwens, *Remembering the Renaissance: Humanist Narratives of the Sack of Rome* (Leiden; Boston: Brill, 1998); Judith Hook, *The Sack of Rome, 1527* (London: Macmillan, 1972); Massimo Miglio et al., *Il sacco di Roma del 1527 e l'immaginario collectivo* (Rome: Istituto Nazionale di Studi Romani, 1986); Giulia Ponsiglione, *La 'ruina' di Roma: Il sacco del 1527 e la memoria letteraria* (Rome: Carocci, 2010); and Manfredo Tafuri, 'Rome Coda Mundi, The Sack of Rome: Rupture and Continuity', in Tafuri, *Interpreting the Renaissance: Princes, Cities, Architects*, translated by Daniel Sherer (New Haven: Yale University Press and Cambridge, MA: Harvard University Graduate School of Design, 2006), pp. 157–59.

26 The letter is reproduced in Michele Carcani, *Il Tevere: Le sue inondazioni dall'origine di Roma fino al giorni nostri* (Rome: Tipografia Romana, 1875): pp. 45–46: 'noi tutti assediati in le case nostre'; 'Ad una città afflitta e consumata come questa è parso un'altro sacco'; 'un infinità di robe'. For Sanga, see Marcello Simonetta, 'Sanga, Giovan Battista', *DBI*, 90 (2017), 182–83. See also an English translation of the letter--Luis Gómez, *The Floods of the Tiber*, trans. by Chiara Bariviera, Pamela O. Long and William L. North (New York and Bristol: Italica Press, 2023), pp. 133–35. Eight palms equals about 1.79 metres.

27 Carcani, *Il Tevere*, pp. 46–47: 'ha lasciate le strade e le case così deformate, che è spaventevole l'andar per Roma'; 'ha dato a tutta la città grandissimo terrore'.

28 Anonymous, *Diluvio di Roma che fu a VII d'ottobre, l'anno M.D. XXX... con ordinata discrittione di parte in parte*, ed. by Benvenuto Gasparoni in *Arti e Lettere: Scritti Raccolti da Francesco e Benvenuto Gasparoni*, 2 vols. (Rome: Tipografia delle Scienze Matematiche e Fisiche, 1865), vol. 2, pp. 81–98, 106–31, text at 87–91: 'pareva che le cataratte tutte dil cielo fossero aperte;' 'in un momento fu convertita la terra ferma ivi in un ampio Mare'. For an English translation, see Gómez, *Floods of the Tiber*, trans. by Bariviera, Long and North, pp. 124–132, Only one of the ancient Roman aqueducts, the *Acqua Vergine*, was (barely) functioning at this time, but there were a number of sewers and drains which would have immediately filled with water.

29 *Diluvio di Roma*, p. 9: 'Indistintamente chi a piedi e chi a cavallo cercava luoco da conservarsi da si horrenda furia dacqua: il padre non espettava il figliolo: ne il figliolo di curava dil padre: nel fratello dil fratello: nel amico pensava di salvare altri che se stesso'; 'lamenti, cridi, strida, urli, pianti, battere di palme: stracciare de vestimenta; graffiarsi de volti: percuotersi de petti: che riempievano laere'.

30 He is also known as Ludovico Gomesius. See especially *Archivo Biográfico de España, Portugal e Iberoamérica*, vol. 1, pp. 327–36, vol. 2, pp. 352–61; https://wbis.degruyter.com/biographic-document/S048-869-9; Joseph Folliet, 'Gómez, Luis', in *Dictionnaire de Droit Canonique*, ed. by Raoul Naz (Paris: Librairie Letouzey et Ané, 1953), vol. 5, cols. 974–75; and A. García y García, 'Gómez, Luis', in *Diccionario de Historia Eclesiástica de España* (Madrid: Instituto Enrique Flores, 1927), vol. 2, p. 1026.

31 Luis Gómez, *De prodigiosis Tyberis inundationibus ab orbe condito ad annum 1531...* (Rome: Francesco Minizio Calvo, 1531). The treatise was summarised in Italian by Andrea Bacci in the 1599 edition of his treatise on Tiber flooding, Andrea Bacci, *Del Tevere dell'eccell. dottore medico e filosofo Andrea Baccio Libro Quarto ...* (Rome: Stampatori Camerali 1599), pp. 17–39. For an English translation, see Gómez, *Floods of the Tiber*, trans. by Bariviera, Long, and North, pp. 16–84.

32 Gómez, *Floods of the Tiber*, trans. by Bariviera, Long, and North, pp. 22–38.

33 Gómez, *Floods of the Tiber*, trans. by Bariviera, Long, and North, pp. 38–39.

34 Gómez, *Floods of the Tiber*, trans. by Bariviera, Long, and North, pp. 39–74 for the history of the floods, citation on p. 42; and see Livy, *Ab urbe cond.*, 4.49.3.

35 Gómez, *Floods of the Tiber*, trans. by Bariviera, Long, and North, pp. 58–61, citations on p. 60.

36 Gómez, *Floods of the Tiber*, trans. by Bariviera, Long, and North, pp. 61–74 (for the 1530 flood), here especially pp. 61–62, citations on p. 61.

37 Gómez, *Floods of the Tiber*, trans. by Bariviera, Long, and North, pp. 63–71, citation on p. 63.

38 Gómez, *Floods of the Tiber*, trans. by Bariviera, Long, and North, pp. 71–73, citation on p. 71.

39 Gómez, *Floods of the Tiber*, trans. by Bariviera, Long, and North, pp. 74–84, citation on p. 75. , For the theory of miasma, see especially Renato Sansa, 'L'odore del contagio: ambiente urbano e prevenzione delle epidemie nella prima età moderna', *Medicina e Storia* 2 (2002): 83–108.

40 Aldrete, *Floods of the Tiber*, p. 244, for the masl. See also Di Martino, Di Martino, and Belati, *Huc Tiber Ascendit*, pp. 66–71; Frosini, *Il Tevere*, pp. 166–70; Long, *Engineering the Eternal City*, pp. 19–41.

41 See Mario Crespi, 'Bacci, Andrea', *DBI*, 5 (1963), 29–30.

42 Andrea Bacci, *Del Tevere: Della natura et bonta dell'acque & delle inondationi Libri II* (Rome: Vincenzo Lucino, 1558); Andrea Bacci, *Del Tevere [...] libri tre [...]* (Venice: [Aldo Manuzio], 1576); and Andrea Bacci, *Del Tevere dell'eccell. Dottore medico e filosofo Andrea Baccio libro quarto* (Rome: Stampatori Camerali, 1599).

43 Bacci, *Del Tevere* (1558), fols. 67^r–87^r (causes); 87^r–89^r (flood of 1557).

44 Bacci, *Del Tevere (1558)*, fols. 89^r–100 ^{v}.

45 See Antonio Trevisi, *Fondamento del edifitio [...] sopra la innondatione del fiume [...]* (Rome: Antonio Blado, 1560). For Trevisi and flood control, see Long, *Engineering the Eternal City*, pp. 28–32.

46 For the Bufalini map, see especially Jessica Maier, *Rome Measured and Imagined: Early Modern Maps of the Eternal City* (Chicago; London: University of Chicago Press, 2015), pp. 77–118. For a reprint of Trevisi's treatise, and of the Bufalini map with his attached letters, see Paolo Agostino Vertrugno, *Antonio Trevisi: architetto pugliese del Rinascimento* ([Fasano (Puglia)]: Schena Editore, [1985]).

47 For these topics, including Trevisi's contract to repair the Acqua Vergine and its disastrous results, see Long, *Engineering the Eternal City*, especially pp. 19–41 and 73–78.

48 Aldrete, *Floods of the Tiber*, p. 244, for the masl. See also Di Martino, Di Martino, and Belati, *Huc Tiber Ascendit*, pp. 73–81; D'Onofrio, *Il Tevere*, pp. 155–60; Frosini, *Il Tevere*, pp. 171–75; and Segarra Lagunes, *Il Tevere e Roma*, pp. 76–80.

49 Giacomo Castiglione, *Trattato dell' inondatione del Tevere di Iacomo Castiglione Romano* (Rome: Guglielmo Facciotto, 1599). It is dedicated "All'Ill[ustrissi[mo] et Rev[erendissi[m] Sig[nore] Il Signor Cardinale Aldobrandini." The congregation is mentioned on pp. 1–2. Thus far, I have not been able to discover Castiglione's life dates.

50 Castiglione, *Trattato dell' inondatione*, p. 7: 'stavannole persone in casa attonite vedendo si horribile spettacolo'; 'quelle povere persone e famiglie colete sotto la ruina di alcune cassette, che non potendo resistere àtanto impeto caderono'.

51 Castiglione, *Trattato dell' inondatione*, pp. 18–71 and 73–76 for the Castel Sant'Angelo rescues.

52 Bacci, *Del Tevere [...] libri tre [...]* (1576), especially p. 293; and see Long, *Engineering the Eternal City*, pp. 36–38 for a more detailed discussion of this treatise.

53 Bacci, *Del Tevere* (1599), especially pp. 42–82 for the topics mentioned here.

54 Margherita Fratarcangeli, 'Giovanni Fontana e la sua stirpe: edifice d'acque e inondationi del Tevere', and Silvia Mangiasciutto, 'Ramo dei Fontana di Melide: Giovanni Fontana', in *Studi sui Fontana: una dinastia di architetti ticinesi a Roma tra Manierismo e Barocco*, ed. by Marcello Fagiolo and Giuseppe Bonaccorso (Rome: Gangemi Editore, 2008), pp. 339–54 and 419–20, respectively.

55 Giovanni Fontana, *Mesure raccolte da Giovanni Fontana architetto, dell'accrescimento che hanno fatto li fiumi, torrenti, e fossi che hanno causato l'inondatione à Roma il Natale 1598* (Rome: Stampatori Camerale, 1599). I warmly thank Cesare Maffioli for providing me a digital copy of this very rare text. Fontana's methodology was much criticised in the next generation by the founder of modern hydraulics, Benedetto Castelli (ca. 1577–1643), who noted

among other things that Fontana failed to consider or measure the velocity of the water flow. See Cesare S. Maffioli, *La via delle acque (1500–1700): appropriazione delle arti e trasformazione delle matematiche* (Florence: Olschki, 2010), especially pp. 151–215.

56 See especially Anna Bedon, 'Della Porta, Giacomo', *DBI*, 37 (1989), 160–70; Federico Bellini, *La basilica di San Pietro da Michelangelo a Della Porta*, 2 vols. (Rome: Argos, 2011); Long, *Engineering the Eternal City*, especially pp. 82–83; Katherine Wentworth Rinne, 'Fluid Precision: Giacomo della Porta and the Acqua Vergine Fountains of Rome', in *Landscapes of Memory and Experience*, ed. by Jan Birksted (London: Spon, 2000), pp. 183–201; and Vitaliano Tiberia, *Giacomo della Porta: un architetto tra Manierismo e Barocco* ([Rome]: Bulzoni, 1974).

57 The document is in the Biblioteca Apostolica Vaticana, Cod. Chigi H. II 43, fols. 163^r–165^v. It was transcribed and published by D'Onofrio, *Il Tevere*, pp. 339–40.

58 See Aldrete, *Floods of the Tiber*, pp. 247–52.

59 Recent work includes David Karmon, *The Ruin of the Eternal City: Antiquity and Preservation in Renaissance Rome* (New York: Oxford University Press, 2011); Long, *Engineering the Eternal City*; Elizabeth McCahill, *Reviving the Eternal City: Rome and the Papal Court, 1420–1447* (Cambridge, MA; London: Harvard University Press, 2013); and Katherine Wentworth Rinne, *The Waters of Rome: Aqueducts, Fountains, and the Birth of the Baroque City* (New Haven; London: Yale University Press, 2010).

60 See especially Maurizia Cicconi, 'E il papa cambiò Strada: Giulio II e Roma. Un nuovo documento sulla fondazione di via Giulia', *Römisches Jahrbuch der Bibliotheca Hertziana*, 41 (2013/14), 227–59; Luigi Salerno, Luigi Spezzaferro, and Manfredo Tafuri, *Via Giulia: una utopia urbanistica del 500* (Rome: Stabilmento Aristide Staderini, 1973); and Nicholas Temple, *Renovatio Urbis: Architecture, Urbanism and Ceremony in the Rome of Julius II* (London: Routledge, 2011), pp. 34–93.

61 For the 'Little Ice Age', see John Aberth, *An Environmental History of the Middle Ages: The Crucible of Nature* (London: Routledge, 2013), pp. 49–51; Brian Fagan, *The Little Ice Age: How Climate Made History, 1300–1850* (New York, NY: Basic Books, 2000); and for the global picture, Jean M. Grove, *Little Ice Ages: Ancient and Modern*, Vol. 1, 2nd ed. (London: Routledge, 2004), esp. pp. 564–90. For 'trading zones', see esp. Pamela O. Long, 'Trading Zones in Early Modern Europe', *Isis*, 106 (December 2015), 840–7.

Part 3

Interventions

10 Flood, war and economy

Leonardo da Vinci and the plan to divert the Arno River

Emanuela Ferretti

And I discussed [with Piero Alamanni, Florentine Ambassador to Milan] the plan that I had with the magnificent Piero [de' Medici] your father to make the canal and greatly benefit the country. ... In sum, it was concluded that the canal should begin at the Molina d'Ogni Santi then, by necessity, end at Signia. But the bed of the Arno presents two obstacles. The first is that routing the Arno into a canal is impossible without weirs to direct its course. The other obstacle is that, if the Arno is put in a canal without a weir, it is reckoned that it will not hold because there would be much more current than at present, and boats will not be able to go there [...]. And for this reason it was decided to make a canal from Prato to Signa into which we would allow only a certain quantity of water to enter and not let other streams flow into it. In the plain below it would be very useful to make a canal in which there could be a more copious flow of water, because the Arno could take more below Florence. This would achieve two things: the first is that it reclaims a lot of land since the embankments and excavations defend us [...]; the other is the ease of navigating, while at each weir all the water can be directed to mills and other buildings.[1]

In this letter of Luca Fancelli to Florence's *de facto* ruler Lorenzo de' Medici (1449–1492), dated 19 December 1487, the subject of changing the course of the river Arno is presented with the discussion of the construction of a canal crossing the lower Arno valley downstream of Florence. This plan was meant to meet a variety of needs, such as improving river transit from Florence to Pisa and allowing for water-powered mills and workshops. At the same time, it aimed to create a new way for managing the waters of a complex river network and improve the defence of the territory from flooding of the Arno and its numerous tributaries: the fifteenth century witnessed five disastrous floods.[2] Lorenzo de' Medici (il Magnifico), the informal 'signore' of Florence from 1469–92, put in effect modifications to water drainage systems in the Pisan countryside and planned others, though without definitively solving the precarious hydrogeological situation which constantly threatened these areas with floods or swamp conditions.[3] In this context, and therefore on the subject of Leonardo's participation in the

DOI: 10.4324/9781003029823-14

circle of Lorenzo,[4] it is worth recalling Giorgio Vasari's hints in the biography of Leonardo in the Giuntina edition of the *Lives of the Artists* (1568), where he writes that the artist 'was the first who, as a youth, discussed putting the river Arno in a canal from Pisa to Florence'.[5] This emphasises the tie between the early project designs of the Medici in the second half of the Quattrocento and the major hydraulic works on the river promoted by Duke Cosimo I dei Medici (1537–1574)[6], and well known to Vasari and to his friend Vincenzio Borghini. Also noteworthy is the interest of Fancelli and the Florentine government in the problem of river regulation and the navigability of the Arno, understood as a system, a concern that stretches back to the thirteenth century.[7]

This chapter revisits a specific hydraulic project that falls between the fifteenth-century interventions by Lorenzo de' Medici and the major works of the early duchy under Cosimo I. This was the well-known attempt to reroute the Arno near Pisa in 1503–1504, a failed project that nonetheless represents a key intersection in the history of Renaissance Italy of hydraulic engineering ideas and military-political imperatives. As it does so, the essay also reconsiders the nature of Leonardo's involvement in this endeavour. As mentioned above, Vasari's remarks on Leonardo's precocious reflections on managing the Arno with corrections to its course suggest Leonardo had developed specific skills in this field early on (as was common, indeed, for other Renaissance artists and architects)[8]. Indeed Leonardo himself refers to this in his well-known letter of presentation, c1482, to Ludovico il Moro, duke of Milan:

> in peacetime I believe that I satisfy very well the comparison to all others in architecture, in construction of buildings public and private, and in moving water from one place to another capable of offence and defence.[9]

The use of techniques developed in the field of hydraulic engineering for military architecture – with the exchange of practical and management methods – was, in fact, an established practice in this period, and this borrowing would have been reinforced at the beginning of the sixteenth century, with the evolution of fortifications tied to the introduction of new and powerful firearms, widespread after Charles VIII's descent into Italy (1494–1495).[10] Meanwhile, humanistic culture conferred a specific 'aura' to this link, through the recuperation and the diffusion of classical texts that celebrated hydraulic endeavours in the context of narratives about the martial exploits of mercenaries and emperors. The organisation of the army in ancient Rome, in fact, included the presence of technicians specialised to meet the needs of long military campaigns, in terms of housing troops, laying siege or creating infrastructure in general. Not only the construction of roads and bridges, but digging canals were among the tasks of the engineers who followed the troops.[11] For example, in *De bello civili,* Julius Caesar relates how the army engineered the deviation of the river Sicori (Segre) in Spain[12]. Or as Tacitus discusses, Corbulo, one of Emperor Claudius' generals, dug the *Fossa Corbulonis*, an extremely long canal (34 km) in current-day Holland.[13] These literary works were well known in Florentine humanistic circles, their diffusion ensured by their translation into the vernacular and by printed versions.[14] Given the cultural depth of Florentine chancery officials in the fifteenth century and the first decade of the sixteenth century these observations do not appear out of place.[15]

It is within this context of Florentine reflections on the Arno problem over a period of decades that plans by Piero Soderini, Florence's new lifetime Gonfaloniere or president, and Niccolò Machiavelli took shape between 1503 and 1504. The goal was to overcome decisively the rebellious city of Pisa by deviating the course of the Arno upstream to deprive Pisa of its access to the sea, through the excavation of a new river bed.[16] A significant historiography has hypothesised a central role for Leonardo in this military plan on the basis of the documented survey and measurement of the Arno carried out by the artist for the Republic in July of 1503.[17] I have recently revisited the question of Leonardo's participation on-site in 1504 (typically assumed in the historiography) and shown that if the artist had a role in the operation against Pisa, it should be understood as applying exclusively to the preliminary phases of the work. On the other hand, Leonardo dedicated himself at length to the theme of the construction of canals[18] along with various questions connected to the Arno,[19] leaving observations and drawings both on changes to put in place to mitigate flood destruction, and the project to design a canal downstream of Florence.[20] These are well-known subjects within the historiography of Leonardo, yet unequivocal chronological attachment points are often lacking, and we still await an objective verification of the technical contents of the drawings of the artist (using new digital tools), which may allow for clarification on the feasibility of these 'designs'.[21]

In a nutshell, we can see that hydraulic engineering, and the theme of water more generally, appears in Leonardo's thoughts diachronically: it beginsin the first Florentine period and was enriched during his time in Milan.[22] In his months in the service of Cesare Borgia (1502), moreover, Leonardo must have acquired an even greater awareness of the associations connecting military techniques and hydraulic engineering.[23] This strengthened the practical aspect of his work with water, which had taken on a structured form as early as 1492 in a specific treatise that was, then developed in the Leicester Codex, a project which remained unfinished.[24]

In recent times, in connection with the commemoration of the five-hundredth anniversary of death of Leonardo (2019), new studies re-addressed the plan drawn up by the artist for a canal to be created downstream of the city that would cross the plains between Florence, Prato and Pistoia and then re-join the Arno at the level of Vicopisano, crossing over the difficult Serravalle saddle pass. This ambitious plan, which takes shape in an early form in folio 127[v] of the Codex Atlanticus (ca. 1490),[25] reappears later in a group of drawings on the course of the Arno, traditionally dated 1503[26] and usually connected to the military initiative against Pisa for the deviation of the river.[27] Regarding this grandiose plan imagined by Leonardo, Filippo Camerota – who, like other scholars, relates the revival of the idea of this canal to the planned diversion of the Arno during the Pisa conflict – wrote: 'the project was developed on several occasions over many years but it never reached a final stage because, in fact, there was never an official appointment for its drafting'.[28] Numerous navigable canals had been built in Italy in the Middle Ages, but in flat areas or with scarce differences in height in the territories crossed by these waterways. Leonardo's imagined project, on the other hand, was applied to an orographically very complex area and therefore his proposal represented a technical challenge, and even utopian engineering. The danger here is the risk of a forced interpretation based on scarce documentary sources. It is not possible to find in Leonardo's ideas – as many scholars have done – a fully worked out technical strategy; it is better instead to acknowledge the generative and 'visionary' component of his thinking. Leonardo's waterway – an alternative to navigating the lower course of the Arno – should, according to a common interpretation in the historiography, also have contributed to solving the problem of the destructive flooding of the river and of the network of lesser watercourses in the valley between Florence and Pistoia, collecting and draining excess water and helping to avoid the creation of swamps. Added to this was the idea of creating new massive river embankments from the gates of the city to Signa, prefiguring a systematic and thorough plan.[29] Such a canal may have found funding from the Arte della Lana (wool merchants' guild) as it would have had the economic benefit of improving river transportation, which in the summer was rendered impractical due to the low water levels in the Arno.[30] It can be seen, in any case, that these topics were already present in the letter of Fancelli to Lorenzo de' Medici, quoted at the beginning.

It is plausible to argue that when the Florentine government's plan in 1503 to divert the Arno to bring Pisa to heel took shape, Leonardo resumed the ideas and plans he elaborated during his first stay in Milan, the navigable canal alternative to the river just discussed. These reflections now intersected with the government's new project to change the course of the Arno in order to bend the Pisans to their will, which led Leonardo to return to study the course of the river and the entire hydraulic system between Florence and the sea, and thus to elaborate drawings and notes on a territorial scale. Setting aside the feasibility of Leonardo's envisioned canal (no evidence so far establishes a real intent to carry out the project), the depth of his reflections along with his past experiences made him an ideal interlocutor for the Florentine government. In the following pages, therefore, we will re-examine the project to deviate the Arno, retracing its various phases from its conception to its ultimate failure. As I do so, it will be clear that the copious documentary sources support Leonardo's role only as a consultant in the Pisan military enterprise, without indicating a working role in the decisive operational phases between August and October 1504.

The plan to deviate the Arno against Pisa

On 8 November 1494, the Pisans welcomed Charles VIII of Valois (1483–1498) with festivities. A day later, having thrown the *marzocchi*, heraldic stone lions that symbolised Florence, into the river, the city proclaimed its freedom and the end of Florentine dominion. So began an extremely tough and bloody conflict between Pisa and Florence that would be drawn out over more than 15 years. In this context, in 1504, the plan for deviating the Arno upstream from Pisa, in the making since the year before, was set in motion. The aim was to leave Pisa without its river, thus depriving it of that essential communication route that had assured its survival through the long years of war.

For scholars of Leonardo and Machiavelli (and Piero Soderini), this is a well-known undertaking.[31] The documents that speak to the early formulation of the idea in July 1503 regard Leonardo himself, while the operational phases in 1504 place Niccolò Machiavelli at the front line.[32] In fact, a payment from the government records expenses for a survey carried out by Leonardo to 'level the Pisan Arno to remove it from its bed', or rather to outline the changing elevation of the river bed[33] (Figure 10.1).

According to a broad consensus, a group of Leonardo drawings that represent the course of the river from Florence to the sea belong to this chronological context. Of major importance, in particular, is the Windsor RL 12683 drawing on which several themes overlap.[34] Here a grandiose plan for a canal is plotted, conceived to definitively resolve the problem of the navigability of the entire Arno; but in addition, there is a sketch of a plan for a new length of river bed towards the sea, which might be linked to the military endeavour against Pisa. This design incorporates, furthermore, the three different points where the river waters change direction, with the most western almost coinciding with the place chosen the

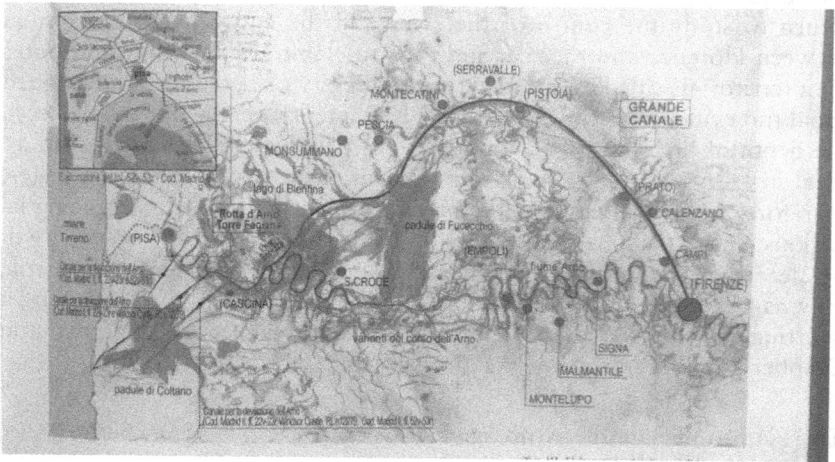

Figure 10.1 Pietro Ruschi, *Graphic remarks on possible new canal routes for the Arno diversion on the basis of Windsor, RL 12279* (from Benigni, Ruschi, 'Brunelleschi e Leonardo: l'acqua e l'assedio', p. 115). Courtesy of Pietro Ruschi.

following year for the work site, the Torre del Fagiano in the outskirts of Pisa (Figure 10.2).[35]

Machiavelli's rich correspondence allows us to follow the operational phases of the attempt to deviate the Arno almost day by day and is an important source for understanding the endeavour as a whole. Recently, the participation on site of the Ferrarese architect Biagio Rossetti (1441–1517), in the service of Ercole I d'Este (1471–1505), has been examined along with Master Alessandro Doria.[36] Rossetti became famous for supervising the expansion of Ferrara, the so-called *Erculean Addition*.[37] Ercole I d'Este most likely also played a significant role in the Arno project, by virtue of the centuries-old hydraulic engineering culture that had been established in the Ferrarese territory,[38] plus the military experience gained in person by the duke. Ercole I had also acquired a thorough knowledge – directly and then through his son Ferrante – of Tuscany.[39] As for Biagio Rossetti, he was called to the banks of the Arno for his expertise in hydraulic engineering – shared with other architects of the time[40] – to reveal a side of his work that has only recently received attention.[41] Rossetti's arrival at the site, in fact, was a last resort before abandoning the extremely expensive and unlucky – or perhaps it would be better to say reckless – work of creating a new path for the Arno. This was a very expensive undertaking, both in economic terms and in terms of city prestige. Moreover, in one of the meetings of Florence's governing bodies,

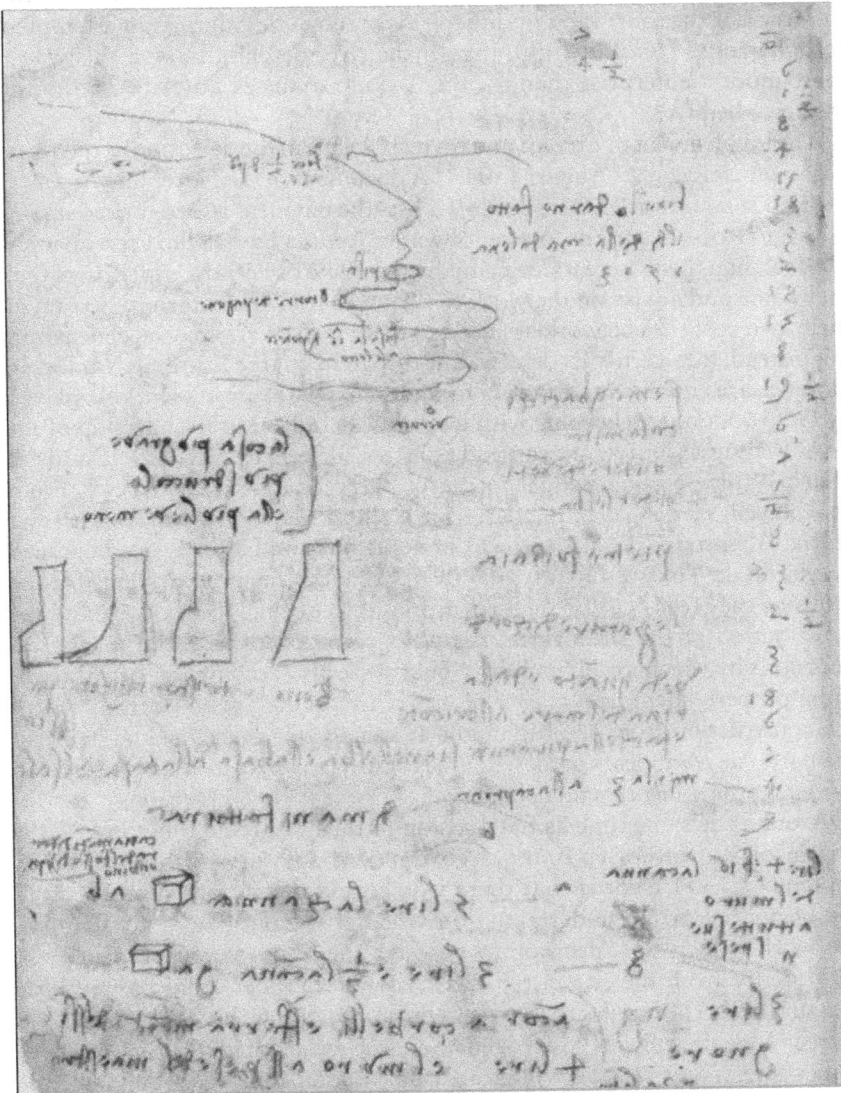

Figure 10.2 Leonardo da Vinci, *Schematic drawing of the Arno from Vicopisano to the sea*, 22 July 1503 (Madrid, Biblioteca Nacional de España, Madrid II, fol. 1*v*, CC BY 4.0).

as a *memento*, they noted the failure of Filippo Brunelleschi (1377–1446) to carry out the same kind of difficult river re-routing during an attempted action against Lucca in 1430: that of the river Serchio, though this was a more modest enterprise than for the Arno in terms of flow and section of the river bed.[42]

The involvement of Ercole I and his technicians in the deviation of the Arno began officially on 22 August 1504.[43] A document in the Florentine Archivio di Stato, a record of 21 October 1504, lists the payment issued for the survey of Biagio Rossetti and the 'master of water' Andrea Doria who accompanied him, sending them 'here to the camp by the Duke of Ferrara with orders that they come and advise on the work to turn the Arno and return to Ferrara at their own cost'.[44] The work assigned to them is to try to save the endeavour, but instead, they certify its demise. The dispatches of the Este ambassadors in Florence are equally important.[45] In these emerge the prominent role played by Piero Soderini, in keeping with his political and cultural leadership of the city.[46] To these sources we can add numerous letters by Machiavelli and the *Diario* of Biagio Buonaccorsi, coadjutor of the Florentine chancery.[47] From Machiavelli's letters, in particular, there is a note that along with Rossetti and Doria, Alessandro degli Albizi was present.[48] He had already accompanied Leonardo da Vinci in July of 1503 in working on the measurements for the levels of the river bed for the first proposal to deviate its course[49] and his contribution at the site of the digging is recorded as essential on several occasions by Machiavelli himself.[50] They were also accompanied by the yet-to-be-properly identified 'master of water' Colombino, who coordinated all the construction from the beginning. It should be noted that the name of Leonardo never appears in these sources and, again, his direct involvement in executing the work remain conjectural.[51]

Around the same time as the decision to turn to Ercole I, but before the arrival of Rossetti, a *consulta*, a government advisory committee, of 28 September 1504 records that there was heated debate one month before the beginning of the undertaking on whether to continue or to abandon the work.[52] Despite a decision to proceed, the frequent reports sent back to Florence from the site in the early days of October 1504 offered little reassurance. The conflicts between those in charge became worse and worse. Machiavelli acknowledged this and wrote that 'they spoke with some supervisors and we will clear it up in good time'.[53] A day later the arrival of two masters from Ferrara was announced. They must have been summoned at least a few weeks earlier: the Ferrarese envoy, in fact, in a letter of 30 September 1504, mentioned having sent to Ferrara 'the drawing of the digging that we were doing to turn the Arno' and asked the duke to decide about 'finding those water engineers' that Piero Soderini had asked for.[54] Moreover, Ercole I, in Florence from 6 to 8 July 1504,[55] might have already granted his support for the project in the months preceding the effective beginning of the work, promising, in case of need, to send experts. The presence of the Este technicians may have been initially considered useful precisely so as to resolve the management conflicts that had already arisen in early September.[56]

The hopes of the Florentine authorities, pinned to the arrival of the two master builders from Ferrara, quickly waned, however, giving way to bitter disappointment.[57] The construction site was rapidly dismantled and the only concern expressed by Machiavelli was that the work already carried out might damage the Florentine forces if the Arno should flood, as the first rains had arrived, swelling the river.[58] Francesco Guicciardini in his *Storia d'Italia* records the episode of the attempt to deviate the Arno in an openly negative way.[59]

The story of a failure

If we consider the array of artistic commissions that Piero Soderini promoted to reinforce the city's military pride and prestige between 1503 and 1504, the symbolic value of the plan to deviate the Arno cannot be overlooked: it echoes an analogous episode related by Ceasar in *De bello civile*, discussed earlier. More immediately, on 25 January 1504, the discussion for choosing the most representative public place for Michelangelo's *David* began, as it had been recently completed and immediately invested with attributes representative of the strength and virtue of the Florentine Republic.[60] Concurrently, Leonardo and Michelangelo had been charged with celebrating the military glories of Florence with a great wall painting in the Salone dei Cinquecento, a project that remained unfinished.[61]

It was in this atmosphere that discussion and debate for the plan to deviate the Arno proceeded. On 24 February a session of the consulting bodies for the Comune on 24 February, Piero Guicciardini deferred to the magistracy of the *Dieci*, the Ten of War, while others warned about the onerous economic cost. The openly negative positions of Francesco Gualterotti and Lorenzo Morelli stand out. The first recalled the failed siege of Lucca: even declaring 'that it is hard to judge things of water' underscoring that 'Pippo di ser Brunellescho [Filippo Brunelleschi], finding our camp at Lucca, wanted to flood Lucca, and instead flooded our camp; and in short, [Gualterotti] would let nature take care of it'[62]. In the war against Lucca, this initiative differed in manner and in principle from the Arno proposal, but was similar from the perspective of military strategy – to use water as a weapon, through massive digging works, construction of embankments and the application of technical and organisational methods. For these kinds of operations it was, in fact, necessary to have precise control of the elevation values over long distances and bring in specific engineering expertise in order to make massive locks and dikes, not to mention recruiting and coordinating thousands of men to dig the new river bed. Questions and doubts about the economic sustainability of the work were raised by Lorenzo Morelli.[63]

> As for the waters, turning the Arno to Stagno will give little trouble to the Pisans; and as he [Morelli] said before the fruits of this will not be as expected. And the cost would be greater than planned: where you plan for 5000 florins, he reckons it would cost more than 10000: but if nature did it that would help, etc.[64]

The reference to *nature* has a precise meaning: observation of the Arno floods and the historical record indicated, in fact, that the river naturally overflowed in the area of Stagno, and more than once had created a long and deep channel to the sea.[65] The endeavour, thus, proposed to artificially recreate a phenomenon that had occurred repeatedly over the centuries: by redirecting all the water from the main river bed of the Arno, probably taking advantage of the floods tied to the autumn rains. It is this aspect of the plan that is most likely attributable to Leonardo, with his unsurpassed capacity to observe natural phenomena and his experience in hydraulic engineering developed in Milan in the service of the Sforza.[66]

In the spring of 1504, the debate revolved around the need to make a decisive push in the war, and once the necessary funding was found, to bring Pisa finally back under Florentine domination. Tempers flared and Tommaso Pucci, for example, asserted that a major new campaign, bringing the entire army under the wall of the rebel city, must be carried out because 'Pisa is the soul of our body'.[67] The plan to deviate the Arno, however, disappears from the records of the *consulte* and reappears only in mid-August of 1504.[68] In this August *consulta*, Giovanni Berardi, a member of the Ten of War also participates, called to testify about the status of the sites during a feverish moment of the operations for the deviation of the river. Berardi, in fact, had gone in the previous weeks to the area of Pisa.[69] It is worth pausing to reflect on Berardi's mission, which appears much more important than that given to Leonardo the previous year (which is always emphasised in the historiography), because it was aimed at giving the go-ahead to a concrete and workable plan. Furthermore, from Machiavelli's letters, Berardi appears to be a key figure in the project not only in terms of general coordination, but also from a technical point of view, along with the engineer Colombini and the official Alessandro degli Albizi. Berardi's report opens a new phase of debate inside the bodies of the government of the Republic, and, while the go-ahead is given, conflicting positions emerge once more on the endeavour.[70]

Giuliano Lapi was chosen to be in charge of the operation[71] with the technical direction in the hands of the aforementioned Master Colombino, an engineer in the service of the *Dieci*.[72] Four hundred men of the vicariate of Valdinievole were called to the site and they were to bring with them '200 spades and 200 shovels, hoes, and billhooks'.[73] Other workers were requested from the nearby vicariate of San Miniato al Tedesco and had to present themselves at the site with the same tools.[74] Workers were also called from Certaldo and joined other men from Prato, Pistoia and the Mugello, though without ever arriving at the number of 2,000 workers deemed necessary for the success of the work.[75] The problem of recruiting manpower comes out in the correspondence of Machiavelli both because the men requested failed to arrive in the expected timeframe, and because the work site conditions pushed them to defect or be expelled, to the extent that diggers were requested from the distant vicariate of San Giovanni Valdarno (near Arezzo) and from Fivizzano in Lunigiana. They worked in mud after the rains of mid-September and the difficulties mounted.[76] In light of this, it should be noted that the grandiose machines for digging canals drawn by Leonardo in the *Codex Atlanticus* would have greatly facilitated the work. These designs have been associated with the Arno project, but there is no mention of them in the source documents.[77] However, in Leonardo's papers, there are numerous references to the excavation capacity of individual workers in the creation of ditches and canals and this suggests once again the role of the artist as an external consultant to the project.[78] As for the workers, their discontent must have been significant if Lapi had to ask Florence to send four supervisors and directors for the work.[79] The day wages of each digger[80] were 1 carlino and on 24 September, Giuliano Lapi received from the *Dieci* the conspicuous sum of 5,000 florins – cash for roughly 60,000 man-days – to pay those working on the Arno.[81] The government did not, therefore, ever fail to provide economic support for the endeavour.

The initial project foresaw the creation of three parallel channels, probably to have narrower sections to excavate, and at the same time, increase the speed of the water from the Arno at the mouths of the respective canals, thus ensuring a greater flow. The excavation progress immediately appears too slow, but the work proceeded. Machiavelli demanded to be informed daily on the amount of earth excavated, while at the camp, specialised craftsmen like stone cutters were requested.[82] On 7 September, Machiavelli writes:[83]

A channel two thousand *braccia* long (1160 meters) and wide 25 (14.6 meters) has already been made. They have to do two others like that: within 10 days, whatever working hours are needed, it must be done; and then water can be allowed in and you will see what effect it has.

The three channels were reduced to two as the work progressed. In reality, as time passed, the whole plan was resized, especially the sections of the canals in correspondence to the mouths that opened onto the river.

Despite the difficulties recorded daily, on 16 September 1504 work was begun on a second canal. By the time the construction site was shut down, the two ditches were, respectively, 30 braccia (17.5 metres) wide and 6 (3.5 metres) deep, and 18 braccia (10.5 metres) wide by 7 (4 metres) deep.[84] Machiavelli, furthermore, to reinforce the intention to continue with the work, expressed the idea of being able to use at least one as a waterway, such that 'it would be possible to carry our merchants that way'.[85] This important objective would be repeated again by Albizi.[86] The idea of taking financial advantage of the canal can, as observed earlier, also be found in the reflections of Leonardo, that is both in the project of the artificial canal that would have created an alternative waterway to the Arno downstream of Florence and within his observations on the straightening of the river bed with the elimination of the numerous meanderings to improve its navigability.

A central element of the military campaign against Pisa was the construction of a large weir or barrier dam to direct the Arno waters into the channels being dug. For the construction of the weir, which began on 10 September 1504, three master builders experienced in this kind of river work were sent to the site.[87] The Torre al Fagiano, which gives its name to the place, was destroyed by the Florentines scavenging construction material to make this weir.[88] For the construction of this structure, the cornerstone of the operation, shafts and withies (*ritorte*)[89] were sent for from Bientina suggesting a kind of barrier with masonry shoulders that extended into the river with mixed materials, using the traditional technique of gabions (modular cages of wood filled with aggregates of various weights and bundles, anchored with poles driven in the ground). Structures of this kind are part of a technical culture that finds examples in the drawings of Leonardo, those of Francesco di Giorgio and all the way up to pre-contemporary treatises.[90] In one of the *consulte* mentioned above, Giovanni Vettori emphasised the significant cost of building the weir and the quantity of soldiers necessary for guarding it, since it could be used to cross the river.[91] Water filled one of the two channels on 3 October 1504, taking advantage of high levels in the river, but it was not enough to guarantee the success of the enterprise.[92]

In 1976, Denis Fachard published an important iconographic clue to this work he found in the *Diario* of Biagio Buonaccorsi[93] (Figure 10.3). This shows how at the place chosen to carry out the work, the Arno forked, creating an island: the weir was made in the diverticulum of the river towards Stagno, so that there would be less distance to cover with the artificial barrier. Buonaccorsi's drawing shows the unfinished weir, that is

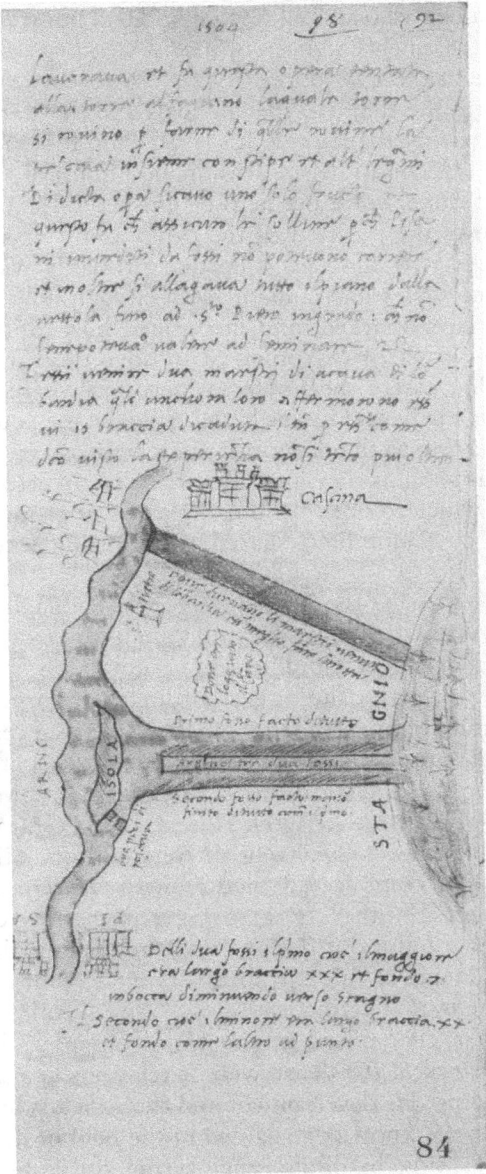

Figure 10.3 Biagio Buonaccorsi, *Drawing of the area of the cut of the Arno at the Torre al Fagiano with the two channels started on a design by Colombino and the channel proposed by the Ferrara engineers*, October 1504 (Florence, Biblioteca Riccardiana, Ricc. 1920, fol. 84r). Courtesy of the Biblioteca Riccardiana.

the two barriers that were started at the opposite banks are not joined: in effect, in a letter of 24 September we read that Colombino had not wanted to finish the weir until 'both channels had been opened',[94] lending weight to the iconographic evidence of the *Diario*.

The same drawing captures the shape of the channels and the state of the work before the work site was abandoned. It says, 'Primo fosso fatto di tutto'/first ditch all done'; 'secondo fosso fatto, manco finito di tutto come il primo'/'second ditch done, but it needs to be finished off like the first'. The drawing also shows the position of the camp and a massive embankment that divides the two channels, one that probably would have been smaller if the Arno water had come in with enough flow and force. Furthermore, Buonaccorsi marked, and mentioned in a caption, another canal located to the west of the other two: 'dove dicevano li maestri venuti di Lombardia, essere meglio fatto la rotta'/'where the masters who came from Lombardy said it was better to make the route'. Buonaccorsi was clearly referring to the proposals made by Rossetti and Doria, who had identified, as soon as they arrived at the site, an error in having chosen that point at which to open the digging and had proposed an alternative to the plan by Colombino shared with the Florentine authorities.

In conclusion, the endeavour to cut the Arno at the Torre del Fagiano, which predates the later hydraulic engineering and canalisation of the river carried out by Cosimo I dei Medici and his son Ferdinando I in the second half of the Cinquecento, was a site of extraordinary importance for the political, military and technical history of the Renaissance. It opens a window on the design process and on the methods for executing this kind of work, and along the way furnishes a missing piece of the puzzle for an understanding of relations between Florence and Ferrara at the outset of the century. However, Machiavelli's letters and our other sources establish an extremely complicated picture with overlapping accounts that often contradict each other. In these discrepancies, we can see the primary reasons for the failure of the project. The copious documentation provides a vivid image of this extraordinary endeavour that goes beyond – for the content that it incorporates – this 'Leonardo-Machiavelli' project as it has crystallised in the historiography. The numerous and varied sources reveal the dense web of relationships and conflicts that marked the genesis, the development and the conclusion of the project. In this enterprise, the knowledge gained in the field of flood defence and the practice of using hydraulic engineering for military purposes converge, along with its potential economic benefits. It was an ambitious dream, to which Leonardo may have contributed at least conceptually, and which crumbled when faced with technical and organisational difficulties.

Notes

1 "Ed io li ho ragionato [con Piero Alamanni, ambasciatore fiorentino a Milano] la praticha ch'io avi cholla magnificentia di Piero vostro padre, di fare il chanale e di bonifichar molto paexe. Credo avere anchora in nota tutti e provedimenti. En suma si choncruxe che el canale chominciasse alle Molina d'Ogni Santi per fino a Signia di neciesità però che per lo letto d'Arno è due chontrarietà. La prima si è che derizando Arno in chanale, egli è imposibile ch'egli vi stia sanza pescare [pescaie] che gli toghino el chorso [...]; l'altra chontrarietà si è che, se Arno si mette in canale sanza pescaia, si stima non vi si possa tenere perché arà tanto più chorente che non v'à ora, e non che vi vada navi [...]. E per questa chagione fu terminato si faciexe un chanale dal Prato a Signia dove si metteva una cierta quantità d'acqua la quale d'ogni tempo entrava per una medexima mixura e provvisto agli altri fiumi che nonne intrasino in detto chanale [...] (p. 62). Nel piano di soto anche sarà utilissimo fare un chanale, el quale può essere più chopioso d'aqua perché Arno gli può porgere più per i fiumi metono da Firenze in giù e questo fa due efeti buoni: el primo si è che bonificha un gran paese, perché chogli argini e per lo chavamento si difendono [...]; l'altro è la facilità del navichare, mulini e edicifi: ad ogni challa tuta l'aqua si da a li dificiˮ. Corinna Vasić Vatovec, *Epistolario Gonzaghesco* (Florence: Uniedit, 1979), pp. 60–62. In particular, Fancelli refers to having worked on ideas for the river earlier under the charge of Piero dei Medici the Gouty (1416–1469), and specifically, in this long letter, mentions that the canalisation of the water course from Florence to Signa is foreseeable only using locks, while in the area of Empoli, it is easier, albeit with the unknown variations of the flow and dangerous annual floods. In fact, there is a provision of the Republic dated 1458 to create a canal to connect Florence with Pisa: see Carlo Starnazzi, *Leonardo cartografo* (Florence: Istituto Geografico Militare, 2003), p. 90. See also Mario Baratta, 'Leonardo da Vinci negli studi per la navigazione dell'Arno', *Bollettino della Società geografica italiana*, 10–11 (1905), 893–921.
2 Emanuela Ferretti, Davide Turrini, *Navigare in Arno: acqua, uomini e marmi da Firenze al mare* (Pisa: Pacini, 2010), p. 27.
3 Patrizia Salvadori, *Dominio e patronato: Lorenzo dei Medici e la Toscana del Quattrocento* (Rome: Edizioni di Storia e Letteratura, 2000), pp. 171, 173, 176; Ferretti, Turrini, pp. 12–3.
4 Eliana Carrara, 'L'Adorazione dei Magi e i tempi di Leonardo', in *Il restauro dell'Adorazione dei Magi di Leonardo: la riscoperta di un capolavoro*, ed. by Marco Ciatti, Cecilia Frosinini (Florence: Edifir–Opificio delle Pietre Dure, 2017). pp. 51–62.
5 "Fu il primo ancora che, giovanetto, discor[r]esse sopra il fiume d'Arno per metterlo in canale da Pisa a Fiorenza" Giorgio Vasari, *Le Vite*, [Life of Leonardo da Vinci] (http://vasari.sns.it/cgi-bin/vasari/Vasari-all?code_f=print_page&work=le_vite&volume_n=4&page_n=17). This note, absent in the Torrentiniana edition, does not appear either in the *Anonimo Gaddiano* or in the so-called *Libro di Antonio Billi*, manuscript sources held to be essential to the bibliography of Leonardo as related by Vasari: see Eliana Carrara, 'Biografi e biografie di Leonardo fra Rinascimento e prima età moderna', in *Leonardo da Vinci: disegnare il futuro*, ed. by Enrica Pagella, Francesco Paolo Di Teodoro, Paola Salvi, exhibition catalogue from Turin, Musei Reali – Galleria Sabauda, 16 April –14 July 2019, (Cinisello Balsamo, Silvana Editore, 2019), pp. 157–81.

6 Emanuela Ferretti, '"Imminutus crevit": il problema della regimazione idraulica dai documenti'. gli Ufficiali dei Fiumi di Firenze (1549–1574)', in *La città e il fiume. Conference Proceedings, Rome, 2001*, ed. by Carlo Travaglini (Rome, École Française de Rome, 2008), pp. 105–28. Among the many sources for ducal hydraulic works, see *Architettura e politica da Cosimo I a Ferdinando I*, edited by Giorgio Spini (Florence: Olschki, 1976); Paolo Santini, 'Arno Vecchio e il "taglio" di Limite: Storia e vicenda di una grande opera medicea', *Quaderni d'Archivio: Rivista dell'Associazione Amici dell'Archivio Storico di Empoli*, 6 (2016), 6, 49–60.

7 Francesco Salvestrini, *Libera città su fiume regale: Firenze e l'Arno dall'Antichità al Quattrocento* (Florence: Nardini, 2005), p. 28.

8 Carlo Pedretti, *Leonardo architetto*, (Milan, Electa, 1978), pp. 309–10; Daniela Lamberini, 'Giuliano da Maiano e l'architettura militare', in *Giuliano e la bottega dei da Maiano*, conference proceedings, Fiesole 13–15 June 1991 (Florence: Octavo, 1994), pp. 3–27; Alberto Carlo Carpiceci, 'Il cantiere di Leonardo prima del Cinquecento', in *Studi vinciani in memoria di Nando de Toni* (Brescia: Geroldi, 1986), pp. 145–76.

9 "[...] in tempo di pace credo satisfare benissimo ad paragone de omni altro in architectura, in compositione di aedificii et publici et privati, et in conducer aqua da uno loco ad uno altro acto ad offendere e di difendere": the famous letter can be read in: Edoardo Villata, 'Scheda IV.23', in *Leonardo da Vinci: La vera immagine,* catalogo della mostra (Florence, Archivio di Stato, 19 October – 28 January 2006), ed. by Vanna Arrighi, Anna Bellinazzi, Edoardo Villata, (Florence, Giunti, 2005), pp. 141–42; cf. also, Pietro C. Marani, *L'architettura fortificata negli studi di Leonardo da Vinci*, (Florence, Olschki, 1984), p. 12–15. Cf, for example, Codex Atlanticus (from now on CA.), f. 38v.

10 Enrico Guidoni, Angela Marino, *Storia dell'urbanistica: Il Cinquecento*, (Rome–Vari, Laterza, 1991), pp. 9–29; Nicholas Adams, 'L'architettura militare in Italia nella prima metà del Cinquecento', in *Storia dell'architettura italiana del Cinquecento*, ed. by Arnaldo Bruschi (Milan, Electa, 2002), pp. 546–61.

11 Graham Webster, *The Roman Imperial Army*, (London, Adam & Charles Black, 1969); Yhann Le Bohec, *L'esercito romano: Le armi imperiali da Augusto a Caracalla*, (Rome, La Nuova Italia Scientifica, 1989).

12 Cesare, *De bello civili*, Book I, 61–62: 'Quibus rebus perterritis animis adversariorum Caesar, ne semper magno circuitu per pontem equitatus esset mittendus, nactus idoneum locum fossas pedum XXX in latitudinem complures facere instituit, quibus partem aliquam Sicoris averteret vadumque in eo flumine efficeret [...]'. 'While the souls of the adversary were drawn to these events, Caesar, so as not to have to send the cavalry over the bridges on a long circuit, found an appropriate place, had built quite a few ditches thirty feet wide with which to deviate part of the river Sicori and create a ford in this river'.

13 Tacitus, *Annales*, Book XI, 20. The episode of the construction of the long canal is recorded by Dione Cassio (79, XX) as well. With the creation of this canal, 34 km long, the boats that transported merchandise would have been able to move from the Rhein to the Meuse in total safety. It seems that the Romans had to dig only a few meters of depth, using the bed of what was then the river Gantel. Recent archeological excavations have established the canal dates from 50 AD, a channel 12–13 meters wide and about 3 meters deep: cf. Guus Besuijen, Rodanum. *A Study of the Roman Settlement at Aardenburg and Its Metal Finds* (Leiden Sidestone Press, 2008), pp. 21–3.

14 L'*editio princeps* of the works of Tacitus, containing books XI–XVI of the *Annales*, I–V of the *Historiae* and *De situ moribus et populis Germaniae libellus* and *De oratoribus dialogus* were printed in 1470, at the Venetian press of Vindelino da Spira, see Kenneth C. Schellhase, *Tacitus in Renaissance Political Thought*, (Chicago, University of Chicago Press. 1976). For the first vernacular editions of the works of Caesar from the mid-fifteenth century for the Sforza: Paolo Ponzù Donato, 'Pier Candido Decembrio editore di Cesare', *Italia medioevale e umanistica*, VII, s. III, (2018), 59, pp. 165–81. The incunabulum *Commentarioum de bello Gallico*, (Milano, tip. Antonio Zarotto) is dated to 1477, and also containsil *De bello civili*.

15 Nella vasta bibliografia sul tema, Cf. Roberto Cardini, Paolo Viti, *I cancellieri aretini della Repubblica di Firenze*, (Firenze, Polsitampa, 2003); Giulio Ferroni, *Storia della letteratura italiana. Dalle origini al Quattrocento*, (Torino, Einaudi, 1991), pp. 332–335.

16 Emanuela Ferretti, 'Fra Leonardo, Machiavelli e Soderini. Ercole I d'Este e Biagio Rossetti nell'impresa "del volgere l'Arno" da Pisa', *Archivio Storico Italiano*, 177, (2019), 2, pp. 235–72. See the following paragraph.

17 See note 33.

18 See note 78.

19 Among the most recent contributions, see: Francesco Paolo Di Teodoro, 'L'Arno a Firenze entro le mura: note e toponomastica, dalla pescaia della Giustizia a quella d'Ognissanti', in *Leonardo & Firenze: fogli scelti dal Codice Atlantico*, exhibition catalogue (Florence 29 March – 24 June 2019), ed. by Cristina Acidini (Florence, Giunti, 2019), pp. 166–69. See also in the Leonardo's *corpus*: CA, fol. 404^v, 785^r; Arundel, fol. 271 ^{r}, 278 ^{v}, 508 ^{r}; Madrid II, fol. 1 ^{v}; Ms L, fol. 31 ^{r}; Leicester, fol. 8 ^{v}, 9 ^{r}, 15 ^{r}, 18 ^{v}, 22 ^{v}, 31 ^{v}. For the other Leonardo's Arno drawings in Madrid e Royal Library, cf notes 27 and 36.

20 See here below.

21 These are striking new techniques from the digital humanities used in the research of Enrica Caporali, Ignazio Becchi, Matteo Isola, Nicodemo Parrilla, Tiziana Pileggi, 'Un naviglio per rilanciare l'economia Toscana: considerazioni sul progetto di Leonardo di deviazione del corso dell'Arno', and Emanuela Ferretti, Michela Chiti, 'Il corso dell'Arno da Firenze al mare nella cartografia Leonardiana fra rilievo, progetto e proiezioni immaginifiche: Nuovi approcci metodologici per una effettiva multidisciplinarietà', both in *Lo sguardo territorialista di Leonardo: il cartografo, l'ingegnere idraulico, il progettista di città e territori, 14–15–16 novembre 2019, Empoli*, conference proceedings, forthcoming.

22 *Leonardo e le vie d'acqua*, exhibition catalogue from Milan in 1983 (Florence, Giunti, 1983); Piero C. Marani, 'Leonardo e le acque in Lombardia: Dal "primo libro delle acque" ai Diluvi', in *La Civiltà delle acque tra Medioevo e Rinascimento* (Florence, Olschki, 2010), pp. 329–46.

23 Enrico Ferdinando Londei, 'I progetti leonardiani di macchine scavatrici per il canale di Cesena per Cesare Borgia', in *Leonardo, Machiavelli, Cesare Borgia. Arte, storia e scienza in Romagna (1500–1503)*, exhibition catalogue (Rimini, 1 March –15 June 2003), (Rome, De Luca, 2003), pp. 55–72. Note, furthermore, that in April of 1500 Leonardo presented the Venetian Senate with a plan for a new system of river defence along the river Isonzo (not executed) that envisioned water barriers to prevent the advance of the enemy or to push them back with the sudden release of waters. (CA, fol. 234^v). See Antonio Cassi Ramelli, 'Leonardo da Vinci geniere', *Bollettino dell'Istituto storico e di cultura dell'Arma del genio*, 47 (1981), 21–9.

24 Ms. A, fol. 55ᵛ. Among the many sources, Augusto Marinoni, 'Il Codice Leicester', in *"Che chosa è l'acqua"*: *Atti del Simposio su Leonardo da Vinci e l'idraulica*, Milan, Congressi Cariplo, 1 December 1995, ed. Costantino A. Fassò, (Milan, Fondazione Cassa di Risparmio delle Provincie Libri Scheiwiller, 1997, pp. 22–38; Francesco Paolo Di Teodoro, 'L'architettura idraulica negli studi di Leonardo da Vinci', in *Acque tecniche e cantieri nell'architettura rinacimentale*, ed. by Claudia Conforti, Andrew Hopkins, (Rome, Nuova Argos, 2002), pp. 259–78; Martin Kemp, '"E questo fia un racolto sanza ordine": Compilazione e caos nel Codice Leicester', in *L'acqua microscopio della natura*, pp. 23–41

25 CA fol. 127ʳ. *Questo [canale] bonificherà il paese; e Prato, Pistoia e Pisa insieme con Firenze fia l'anno di meglio dugento mila ducati, e porgeranno le mani a spesa a esso aiutorio, e i Lucchesi il simile.*

This [canal] will reclaim the country; and Prato, Pistoia and Pisa, together with Florence, will make the year better than two hundred thousand ducats, and they will help with the expenses, and the Lucchesi the like.

26 *I disegni geografici di Leonardo da Vinci conservati nel Castello di Windsor*, ed. by Mario Baratta (Rome, Libreria dello Stato, 1941). Also of significance and very well known is the drawing of the Arno valley in the Codice di Madri, II, fols 52ᵛ–53ʳ and 22ᵛ–23ʳ.

27 On the latter: Filippo Camerota, 'Leonardo and the Florence Canal. Sheets 126–127 of the Codex Atlanticus', *Substantia*, 4.1 (2020), 37–50; Filippo Camerota, 'La scienza delle acque e i suoi "giovamenti": le carte idrografiche della Toscana', in *L'acqua microcosmo della natura*, pp. 99–115; also, Alessandro Vezzosi, 'Il canale di Firenze. Scienza, utopia e land art', in *Leonardo e Firenze*, pp. 55–63.

28 Camerota, p. 38.

29 In the same conceptual environment but without a tight chronological connection with the military operation against Pisa are the observations of Leonardo on the necessity to eliminate the numerous meanders of the Arno, through excavating canals where the river would run, bypassing the curves and narrowing the bed of the river with new embankments: CA, fol. 256ʳ; CA, fol. 785ʳ: Carlo Starnazzi, *Leonardo cartografo*, (Florence, Istituto Geografico Militare, 2003), p. 101.

30 Starnazzi, p. 90. CA, fol. 1107ʳ.

31 The bibliography on this topic is vast and multifaceted. For essential references, see: Pasquale Villari, *Niccolò Machiavelli e i suoi tempi: illustrati con nuovi documenti*, (Florence, Le Monnier, 1877), I, pp. 478–479; Edmondo Solmi, 'Leonardo e Machiavelli', *Archivio Storico Lombardo*, s. 4, 39, (1912), 34, pp. 209–44: 226–30; Francesco Bausi, *Machiavelli*, Rome, Salerno, 2005, pp. 46–48; Carlo Vecce, *Leonardo*, (Rome, Salerno, 2006), p. 46; Enrico Tolaini, '"Se pensa levare lo Arno a Pisa": a proposito della Mappa del Pian di Pisa di Leonardo', in *'Conosco un ottimo storico dell'arte…': per Enrico Castelnuovo, scritti di allievi e amici pisani*, ed. by Maria Monica Donato, Massimo Ferretti, (Pisa, Edizioni della Normale, 2012), pp. 223–26; Marco Versiero, *Il dono della libertà e l'ambizione dei tiranni: L'arte della politica nel pensiero di Leonardo da Vinci*, (Naples, Istituto italiano per gli studi filosofici, 2012), especially pp. 174–77, 405–06.

32 Among others, see: *Scritti inediti di Niccolò Machiavelli risguardanti la storia e la milizia (1499–1512)*, ed. by Giuseppe Canestrini (Florence, Barbèra, Bianchi e comp. tipografi-editori, 1857), pp. VI–VIII; Fredi Chiappelli, 'Machiavelli as Secretary', *Italian Quarterly*, 14 (1971), 27–44.

33 'Livellare Arno in quello di Pisa per levallo dal leto suo' ASF, *Camera del Comune, Depositario dei Signori, Entrata e uscita*, 15, fol. 52ᵛ, 26 luglio 1503, published in various places among which: Villata, doc. 181, p. 161 with preceding bibliography. Francesco Guiducci, on 24 July 1503, refers furthermore to the *Dieci* having carried out a survey with Alessandro degli Albizi and Leonardo 'et veduto el disegno insieme con el ghoveratore, doppo molte discussioni et dubbi conlusesi che l'opera fussi molto ad proposito, o sì veramente Arno volgersi qui, o restarvi con un cabale, che almeno vieterebbe che le colline da nimici non potrebbono essere offese' [and saw the drawing together with the governor, after much discussion and doubts he concluded that the work involved was susbtantia, or if truly the Arno were to turn here, or stay with a subterfuge, that at least it would prevent the hills from being attacked by the enemy]: Giovanni Gaye, *Carteggio inedito d'artisti dei secc. XIV-XVI pubblicato ed illustrato con documenti inediti*, (Florence, Molini, 1839–1840), II, p. 62.

34 Paola Benigni, Pietri Ruschi, 'Brunelleschi e Leonardo: l'acqua e l'assedio', in *Leonardo e l'Arno*, ed. by Roberta Barsanti, (Ospedaletto-Pisa, Pacini Editore), 2015, pp. 99–129. See also, Madrid II, fol. 1v.: 'Livello d'Arno fatto il dì della Maddalena 1503 '[level of Arno done on the the day of the Magdalene (22th july)1503]. Cf. also, Windsor RL 12279: Benigni, Ruschi, p. 115.

35 http://stats-1.archeogr.unisi.it/repetti/includes/pdf/main.php?id=1856 (30 October 2020).

36 Emanuela Ferretti, 'Fra Leonardo, Machiavelli e Soderini. Ercole I d'Este e Biagio Rossetti nell'impresa "del volgere l'Arno" da Pisa', *Archivio Storico Italiano*, 660, (2019), 2, pp. 235–72.

37 Francesco Ceccarelli, 'Alla ricerca di Biagio Rossetti', in *Biagio Rossetti e il suo tempo. Architettura e città*, atti del convegno (Ferrara, 25 novembre 2017), ed. by Allessandro Ippoliti, with extensive bibliography (Roma, GBE/ Ginevra Bentivoglio editoria, 2018), pp. 23–35.

38 Teresa Bacchi, 'Il territorio ferrarese orientale nel medioevo', in *La grande bonificazione ferrarese: Vicende del comprensorio dall'Età romana alla istituzione del Consorzio (1883)*. I, (Ferrara, Consorzio della Grande Bonificazione Ferrarese), 1987, pp. 69–102 (92–94); Franco Cazzola, 'Difficili riforme: i Lavorieri del Po nella Ferrara pontificia', in *Cultura nell'età delle Legazioni. Atti del convegno, Ferrara, 20–22 marzo 2003*, ed. by Franco Cazzola, Ranieri Varese, (Florence, Le Lettere, 2005), pp. 201–31

39 Enrica Guerra, *Soggetti a ribalda fortuna: gli uomini dello stato estense nelle guerre dell'Italia quattrocentesca*, (Milan, Franco Angeli, 2005). For Ferrante d'Este and the Pisan territory, Sergio Mantovani, '*Ad honore del signore vostropatre et satisfactione nostra': Ferrante d'Este condottiero di Venezia*, (Modena–Ferrara, Deputazione di storia patria per le antiche provincie modenesi, 2005).

40 From the vast bibliography, see Paolo Galluzzi, 'Introduzione', in *Gli ingegneri del Rinascimento da Brunelleschi a Leonardo*, exhibition catalogue (Florence, 22 June 1996–6 January 1997) ed. by Paolo Galluzzi (Florence, Giunti), pp. 44–45; Cesare S. Maffioli, 'Saper condurre le acque', in *Enciclopedia italiana di scienze, lettere ed arti: Il contributo italiano alla storia del pensiero. Ottava appendice*, (Rome, Istituto Italiano della Enciclopedia Italiana, 2013), pp. 79–90 (82–83).

41 For the biography of Rossetti, see Francesca Mattei, 'Rossetti, Biagio', in *Dizionario Biografico degli Italiani*, LXXXVIII, 2017, https://treccani.it/ enciclopedia/biagio-rossetti_%28Dizionario-Biografico%29/ (accessed 30 October 2020).

42 See note 68.

43 Luca Landucci, *Diario fiorentino dal 1450 al 1516, continuato da un anonimo fino al 1542, con annotazioni di Iodoco Del Badia*, anastatic reprint, (Florence, Studio Biblos, 1969, [facsimile edition, Florence, 1883]), p. 271.

44 'Qui in campo dal Signor Duca di Ferrara con ordine che veghino et consiglino circa alla opera del volgere Arno et tornarsene a Ferrara a loro spesa'. ASF, *Dieci di Balia, Condotte e stanziamenti*, 52, fol. 63^v: Luigi Napoleone Cittadella, *Notizie amministrative, storiche, artistiche relative a Ferrara ricavate da documenti*, II, (Ferrara, Tipografia di Domenico Taddei, 1868), p. 256, note 1 and again in, Bruno Zevi, *Biagio Rossetti architetto ferrarese: Il primo urbanista moderno europeo*, (Turin, Einaudi, 1960), p. 661.

45 These documents are fully transcribed for the first time in Ferretti, pp. 267–72.

46 J. Jacques Marchand, 'Il carteggio semiufficiale inedito del gonfaloniere perpetuo Piero Soderini: prime indagini', *Interpres*, 34 (2016), 143–71; Nicoletta Marcelli, 'Pier Soderini, Leonardo da Vinci e la Battaglia di Anghiari', *Interpres*, 36 (2018), 191–210; for the political context of 1503–1504, see, *Machiavelli e il mestiere delle armi: Guerra e potere nell'Umbria del Rinascimento*, ed. by Alessandro Campi, Erminia Irace, Francesco F. Mancini, Maurizio Tarantino, (Passignano, Aguaplano, 2014).

47 Niccolò Machiavelli, *Legazioni. Commissarie. Scritti di governo. IV (1504–1505)*, ed. by Denis Fachard, Emanuela Cutinelli-Rendina, (Rome, Salerno editrice, 2006); Denis Fachard, *Biagio Buonaccorsi*, (Bologna, M. Boni, 1976).

48 Machiavelli, *Legazioni*, doc. 169, p. 227: Machiavelli a Tommaso Tosinghi, 9 October 1504.

49 ASF, *Dieci di Balia, Responsive*, 73, fol. 290^r, 24 July 1503, published in several places including *Leonardo da Vinci, I documenti*, doc. 180, p. 160, with preceding bibliography.

50 Machiavelli, *Legazioni* doc. 79, p. 121: Machiavelli to Giuliano Lapi, 28 August 1504; ivi, doc. 102, p. 141: Machiavelli to Giuliano Lapi, 4 September 1504. His authority is also noted in this contemporary source: 'Sulla vita di Antonio Giacomini e l'Apologia de' cappucci di Iacopo Pitti', ed. by Cirillo Monzani, *Archivio Storico Italiano*, 4 (1853), 2, 99–270 (p. 208).

51 A different opinion (that is a decisive role for Leonardo) is presented in Roger D. Masters, *Fortune is a River: Leonardo da Vinci and Niccolò Machiavelli's Magnificent Dream to Change the Course of Florentine History*, (New York, Free Press, 1998); Pascal Briost, *Léonard de Vinci, l'homme de guerre*, (Paris, Alma, 2013); and again in, Patrick Boucheron, *Leonardo e Machiavelli: vite incrociate*, (Rome, Viella, 2014). Again, an active role for Leonardo in 1504 is assertively presented in the text of Alessandro Vezzosi, Agnese Sabato in https://brunelleschi. imss.fi.it/itinerari/itinerario/StudiDeviazioneArnoRiglionePisaAlloStagnoLivorno. html (accessed 30 October 2020). Important observations on the shakiness of the claims about the friendship between Leonardo and Machiavelli especially in relationship to the events surrounding the deviation of the Arno are found in Romain Descendre, *Leonardo da Vinci*, in *Machiavelli: Enciclopedia machiavelliana*, [2014]. https://treccani.it/enciclopedia/leonardo-da-vinci_%28Enciclopedia-machiavelliana%29/ (accessed 30 October 2020). A more prudent approach to Leonardo's role is in Dora D'Errico, '"Se e da fare opera da volgere Arno". Leonard au service du projet de detournement de l'Arno", *Cromohs*, 19, (2014), pp. 79–97.

52 *Consulte e pratiche della Repubblica Fiorentina, 1498–1505*, ed. by Denis Fachard, (Geneva, Droz, 1993), p. 1017, 28 September 1504.

53 'Parleronno con qualche intendente et a bell'agio ci risolveremo'. Machiavelli, *Legazioni*, doc. 168, p. 227: Niccolò Machiavelli to Giuliano Lapi, 8 October 1504.
54 'Ritrovare quelli ingigneri da acqua'. Manfredo Manfredi to Ercole I d'Este, from Florence on 30 September 1504: ASMo, *Carteggio Ambasciatori con i principi Estensi*, Florence, n. 10. See Ferretti, p. 269.
55 Thomas Tuohy, *Herculean Ferrara: Ercole d'Este, 1471–1505, and the Invention of a Ducal Capital*, (Cambridge, Cambridge University Press, 1996), p.152, note 29. Also, ASMo, *Carteggio ambasciatori, Roma*, 15: Beltrame Constabili to Ippolito d'Este, from Rome on 14 July 1504.
56 Machiavelli, *Legazioni*, doc. 102, p. 141: Niccolò Machiavelli to Giuliano Lapi, 4 September 1504. In a letter of 11 September 1504 written to the Serenissimo Venetian Ambassador to Rome, perplexities on the success and usefulness of the operation emerge: 'I Fiorentini attendono a svolgere da Pisa il corso dell'Arno, per impedire l'arrivo delle vettovaglie per via di mare ai Pisani; ma è impresa tentata anche altre volte senza effetti e dalla quale non si caverà alcun frutto; giacché se tale diversione avesse effetto, i Pisani, per quanto dicesi, potebbero approvigionarsi per un'altra acqua che fa foce in mare' [The Florentines expect to turn the course of the Arno from Pisa to impede the arrival of provisions by sea to the Pisans; but it is an undertaking attempted other times too without effect and from which nothing is gained; for if that diversion had been done, the Pisans, like they say, could supply themselves from another water that opens on the sea]: *Dispacci di Antonio Giustinian. Ambasciatore veneto in Roma dal 1502 al 1505*, ed. by Pasquale Villari, (Florence, Le Monnier, 1876), III, p. 228; The Este ambassador, in these same days, described to Ercole I the intense difficulties that marked the progress of the operation, at a few weeks from the beginning of digging: Manfredo Manfredi to Ercole I d'Este, da Firenze il 13 September 1504; Ferretti, p. 269, doc. 3f.
57 Machiavelli, *Legazioni*, doc. 173, pp. 231–232: 231; Niccolò Machiavelli to Tommaso Tosinghi, 12 October 1504.
58 Machiavelli, *Legazioni*, doc. 176, p. 234: Machiavelli to Tommaso Tosinghi, 16 October 1504.
59 Francesco Guicciardini, *Storia d'Italia*, edited by Silvana Seidel Menchi, (Turin, Einaudi, 1971), I, p. 604.
60 Among the many sources, see especially John T. Paoletti, *Michelangelo's David: Florentine history and civic identity*, (Cambridge, Cambridge University Press, 2015), pp. 141–74.
61 See, recently on this theme, *La Sala Grande di Palazzo Vecchio e la Battaglia di Anghiari di Leonardo: Dalla configurazione architettonica all'apparato decorative*, ed. by Roberta Barsanti, Gianluca Belli, Emanuela Ferretti, Cecilia Frosinini (Florence, Olschki, 2019).
62 'Che delle cose delle acque può dare poco iudicio'; che Pippo di ser Brunellescho, trovandosi il campo nostro a Luccha, volle allagare Luccha, et allaghò il nostro campo; et insomma [Gualterotti] lascerebbe fare alla natura' *Consulte e pratiche della repubblica fiorentina*, doc. 441, p. 990.
63 *Consulte e pratiche della repubblica fiorentina*, doc. 441, p. 990. Giovanni Cavalcanti, *Istorie fiorentine*, (Florence, Tipografia all'insegna di Dante, 1838–1839), I, p. 328. For this episode from Brunelleschi's biography, Paola Benigni, Pietro Ruschi, 'Il contributo di Filippo Brunelleschi all'assedio di Lucca', in *Filippo Brunelleschi, la sua opera e il suo tempo*, conference proceedings, Florence, 1977, (Florence, Centro Di, 1980), II, pp. 517–33.

64 'Quanto alle acque, che voltando Arno allo Stagno darà poca noia a' Pisani; et come altra volta ha decto, non stima faccia fructi si sono disegnati, et dixene più ragioni. Et che la spesa sarebbe maggiore non si disegna, et che dove si disegna fiorini 5000, stima sarebbono più di 10000: ma facendo la natura aiuterebbe etc.'.

65 Giuseppe Caciagli, 'Rettifiche e varianti del basso corso dell'Arno in epoca storica', *L'Universo*, 49 (1969), 1, 133–62 (p. 148).

66 This theme is central to Leonardo studies and is explored in an extensive bibliography. For the purposes of this contribution, it is important to note, at least for those reading with a historiographic interest on the subject, Fabio Frosini, 'Leonardo da Vinci on Nature: Knowledge and Representation', in *Leonardo da Vinci on Nature. Knowledge and Representation*, edited by Fabio Frosini, Alessandro Nova, (Venice, Marsilio, 2015), pp. 11–31.

67 'Pisa è l'anima del corpo nostro'. *Consulte e pratiche della repubblica fiorentina*, doc. 451, p. 1006, 31 May 1504.

68 Ivi, doc. 456, p. 1013, 14 August 1504.

69 Machiavelli, *Legazioni*, doc. 53, p. 91: Machiavelli to Antonio Tebalducci Giacomini, 2 August 1504.

70 *Consulte e pratiche della repubblica fiorentina*, doc. 456, p. 1014.

71 Giuliano Lapi is mentioned, with great familiarity, in a letter from Machiavelli to Niccolò Valori on 30 October 1501 'E vi priego tocchiate la mano al nostro Juliano Lapi che è gentile cosa' [And pray clasp the hand of our Juliano Lapi who is so kind]: Niccolò Machiavelli, *Opere. II Lettere. Legazioni. Commissarie*, ed. by Corrado Vivanti, (Turin, Einaudi, 1999), p. 42. When the work begins, Lapi is the Florentine *commissario* at Cascina: Machiavelli, *Legazioni*, p. 112, note 8.

72 Machiavelli, *Legazioni*, doc. 71, p. 113: Machiavelli to Antonio Giacomini, 21 August 1504.

73 '200 vanghe et dugento fra pale, zappe e ronconi'. Ivi, doc. 72, p. 115: Machiavelli to the Vicario della Valdinievole, 22 August 1504.

74 Ivi, doc. 73, pp. 116: Machiavelli to the Vicario di San Miniato, 24 August 1504.

75 ivi, doc. 118, p. 161: Machiavelli to Antonio Giacomini and to Giuliano Lapi, 11 September 1504.

76 Ivi, doc. 125, p.168: Machiavelli to Giuliano Lapi, 17 September 1504.

77 C.A. fol. 1va; C.A., fol. 1vb: Carlo Pedretti, *Leonardo architetto*, (Milan, Electa, 1978), pp. 180–181; Galluzzi, p. 71.

78 Andrea Bernardoni, Alexander Neuwal, 'Lavoro manuale e soluzioni tecnologiche nello scavo dei canali', in *L'acqua microscopio della natura: Il codice Leicester di Leonardo da Vinci*, exhibition catalogue, ed. by Paolo Galluzzi, Florence, 30 October 2018–20 January 2019, (Florence, Giunti, 2018), pp. 134–53.

79 Machiavelli, *Legazioni*, doc. 118, pp. 162: Machiavelli to Giuliano Lapi, 11 September 1504.

80 Giuliano Pinto, 'L'organizzazione del lavoro nei cantieri edili (Italia centro-settentrionale)', in *Artigiani e salariati: Il mondo del lavoro nell'Italia dei secoli XII–XV*, conference proceedings (Pistoia, 9–13 October 1981), (Pistoia, Centro Italiano di Studi di Storia e d'Arte, 1984), pp. 69–101: 87.

81 ASF, *Dieci di Balia*, 52, fol. 60^v.

82 'Di già è fatto un fosso lungo dumila braccia e largo 25. Debbesene fare du' altri così: e' quali fra 10 dì, servendo el tempo come serve, doverebbono essere fatti; e di poi si darà loro l'acqua e vedrassi che effetti partorirà'. Machiavelli,

Legazioni, doc. 85, p. 126. Machiavelli to Antonio Giacomini Tebalducci and Giuliano Lapi, 31 August 1504.

83 Machiavelli, *Legazioni*, doc. 111, pp. 153–154: 154, Machiavelli to Giovanni Ridolfi, 7 September 1504.

84 'Ricordi di Ser Perizolo da Pisa dall'anno 1422 sino al 1510', ed. by Francesco Bonaini, *Archivio Storico Italiano*, 6 (1845), 2, 385–96: 395, cited in Tolaini, p. 223.

85 'Si potessi per quella via condurre le mercantie nostre'. Machiavelli, *Legazioni*, doc. 117, p. 159: Machiavelli to Giuliano Lapi, 10 September 1504.

86 Machiavelli, *Legazioni*, doc. 142, p. 190: Machiavelli to Tommaso Tosinghi, 24 September 1504.

87 Machiavelli, *Legazioni*, doc. 116, p.158: Machiavelli to Giuliano Lapi, 9 September 1504.

88 *Ricordi di Ser Perizolo da Pisa*, p. 223. See also the letter to the *Dieci* requesting '10 boni scarpellini che bisognano per tagliare la torre di Fagiano che resta sulla riva del fosso verso Pisa; et li sassi serviranno alla pescaia', [10 good stone cutters for cutting the tower of Fagiano which is at the bank of the channel toward Pisa; and the stones will serve for the weir]. Antonio Giacomini to the *Dieci*, from camp 30 August 1504: 'Documenti per servire alla storia della milizia italiana dal XIII al XVI secolo', ed. by Giuseppe Canestrini, *Archivio Storico Italiano*, XV, (1851), p. 302.

89 Machiavelli, *Legazioni*, doc. 128, p. 173. Machiavelli to Giuliano Lapi, 18 September 1504. The *ritorte* are used to tie the shafts together. In the *Vocabolario della Crusca*, at the entry 'ritorta' it reads: 'Vermena verde, la quale attorcigliata, serve per legame di fastella, e di cose simili; [green shoots or branches which wound together can tie bundles and similar things] See *Vocabolario degli Accademici della Crusca*, (Venice, Alberti, 1612), *ad vocem*.

90 Ferretti and Turrini, Navigare in Arno, pp. 9–26 '[...] parci che sia d'avanzare tempo di cominciare a condurre pali, et fascine et sassi in su l'opera della pescaia'[it seems that there is time to being to bring shafts and bundles and stones for this work on the weir], writes Machiavelli to Giuliano Lapi on 10 September 1504: Machiavelli, Legazioni, doc. 372, p. 427.

91 *Consulte e pratiche della repubblica fiorentina*, doc. 456, p. 1014.

92 For the arrival of the water: Machiavelli, *Legazioni*, doc. 175, p. 213: Machiavelli to Tommaso Tosinghi, 3 October 1504.

93 Fachard, plate III, [*Diario*, fol. 84^r]. The image, after Fachard, was published after I brought it to the author's attention, in Benigni, Ruschi, p. 128.

94 'Tutti a dua e' fossi non sono sboccati'. Machiavelli, *Legazioni*, doc. 142, p. 191: Machiavelli to Tommaso Tosinghi, 24 September 1504.

11 The making of a transnational disaster saint

Francisco Borja, patron saint of earthquakes from the Andes to Europe

Monica Azzolini

In 1627, an image of Francisco Borja (1510–1572), 4th Duke of Gandìa and prominent member of the Jesuit Order, started to perspire profusely under the eyes of a young boy in the small Andean town of Tunja, in modern-day Colombia. All attempts to dry up the picture were futile: the moisture kept resurfacing over and over. If this fact was not wondrous enough, with the passing of time the expression on Borja's face became increasingly sad and doleful, 'as if to indicate a forthcoming calamity'.[1]

Borja died in 1572, after devoting nearly 25 years of his life to serving God as a member of the Jesuit Order. During this period, he had raised through the ranks of the Order to the highest level, that of Superior General (the third member to take on that role). His earlier life as a Jesuit, however, was not without controversy and ambivalence, so much so that very early on he had a brush with the Inquisition due to his highly unorthodox ideas.[2] Despite all of this, a few years before the events in Tunja, in 1624, Urban VIII had beatified him. By the seventeenth century, several *Lives* celebrated his heroic virtues and his miracles both when he was alive and after his death. Unsurprisingly, many of these works were written by Jesuits: they were clearly aimed at boosting the cult of the former Duke of Gandìa to pave the way to the next step: his canonisation.[3] It would take over 35 years, however, before Clement X canonised him in 1670; 35 years in which both the Jesuits and the House of Borja worked tirelessly to plead Francisco Borja's cause.

The story that will be recounted in this essay constitutes a neglected part of this saint's life and his canonisation, the latter of which took many years, involved numerous people, demanded great resources and produced a remarkable amount of documentation.[4] The facts sketched briefly at the beginning of this essay constitute the rough contours of a story about the creation of a disaster saint cult in the Andes, a cult that then travelled back to Rome, Naples and other cities in Italy and Europe in the wake of other

DOI: 10.4324/9781003029823-15

calamities. It is one of many stories that illustrate the way in which early modern men and women made sense of catastrophes and connected them to the divine.[5] At a broader level, moreover, it is a story that exemplifies how earthquakes are never quite simply natural phenomena; they are cultural constructions that acquire social, political and cultural significance. In short, much like the case of the Magdeburg earthquake with which Grégory Quenet's monumental study of France opens, an earthquake is never just an earthquake. The 'curious history' of how the King of Prussia, Friederick William II forbid everybody from talking or even acknowledging the natural event for fear of being thrown in prison is indicative of the political and social import of these events. Small (like the Magdeburg earthquake) or big (like the famous Lisbon one), these events were never inconsequential, no matter the intensity or the damage.[6]

Finally, this is also a story of how cult-making is dependent on place and space, on power dynamics and on historical conditions that may determine both its development and its lasting success (or lack thereof, as in this case). The birth of this cult in the Andes, as we shall see, may not have been entirely accidental: it was Borja who had sent the first Jesuits to the Americas with the aim of converting the indigenous population through peaceful means. For this reason, both within and outside the order, Borja's name was strongly associated with the Jesuits' evangelisation of the New World.[7]

In this essay, I shall attempt to situate the case of St Francis Borja within the broader contours of political colonial power, natural disasters and the relationship between the Jesuits and the political establishment in the Iberian Empire, in Rome and in the Kingdom of Naples in the early years of the order. In doing so, I rely on incomplete evidence: while the miracle of Tunja is significant enough to have been transmitted in several sources, it seems that some of the documentation may have not made it as far as the Roman Curia: so far, I have been unable to locate this particular miracle among the numerous documents that constitute Borja's canonisation process.[8]

In exploring these historical events, my aim is threefold: to draw out the implications of this complex web of political and religious connections stretching across the Iberian Empire from Bogotá to Rome and the rest of Italy; to examine how conceptions of nature, sin and salvation in the early modern period shaped and were shaped by local religious and political interests; and, finally, to shed some light on the rich historiographical and textual tradition that sustained this cult from the moment it emerged until the dawn of the Enlightenment. In doing so, this case study represents an additional lens through which we can explore the cultural nature of disasters and the way they generated new forms of religiosity besides sustaining old ones.

A rebours: Seeking protection against earthquakes in Enlightenment Bologna

When the city of Bologna was struck by an earthquake in June 1779 its civic, religious and intellectual community was seemingly unprepared. Tremors were felt already on 1 June 1779: the first earthquake stroke the city on the 4th of June (MCS VII), followed by another of similar intensity on the 10th, and another on the 14th. Two more earthquakes followed, on 23 November 1779 and 6 February 1780, leaving Bologna's citizens shaken and confused.[9] Being unusual for its intensity in and around Bologna, this sequence of earthquakes left the community wondering about the most appropriate response. Copious correspondence between members of the Academy of Sciences of the Institute of Bologna and members of the Academy of Sciences in Paris as well as other documentation in the State Archive bear witness to the ample echo that the tremors had within the city and beyond.[10] One of the most typical responses in these cases was to create collective moments of expiation and community building in the form of religious processions and cults of intercession, and Bologna was no different from other places. Not without tensions and disagreement, for lack of a local cult devoted to these kinds of events, the city turned to the Virgin Mary and to St Emygdius for intercession.[11] As we shall see, however, other cults seem to have emerged within more localised communities.

The prolific Bolognese printer Stamperia di San Tommaso d'Aquino was quick to exploit the dramatic event for commercial purposes, publishing a series of works related to the theme of earthquakes. Three short books appeared in quick succession: a short history of earthquakes by the Olivetan monk Michele Augusti; a pamphlet detailing the meteorological conditions and the atmospheric phenomena surrounding the shocks and aftershocks of 1779–1780; and a religious text written by an anonymous Spanish priest entitled *Notizie della protezione speciale di S. Francesco di Borgia contro i terremoti*. These were followed by *De effectibus terraemotus in corpore humano* (1784) by the physician Vincenzo Domenico Mignani a few years later, indicating the lasting impact of the events among the Bolognese medical and scientific community.[12]

The variety of approaches presented by these publications suggests that they were targeting different audiences, but it is the religious text documenting the special protection against earthquakes offered by St Francis Borja that is most relevant here. While Bologna had no direct historical connection with the House of Borja, the city attracted numerous Spanish students and had long-standing connections with Rome and the Iberian Empire. Loyola had been a guest of the famous Real Colegio de España in Bologna.[13] A visit by Francis Xavier to Bologna in 1537/1538, moreover, prepared the ground for the establishment of a Jesuit College in the city, which officially opened its doors in 1551. The Bolognese Jesuit College was always in stiff competition with the *Studium* to attract students, and this meant the Jesuits did not always have an easy life in Bologna; yet the Order was able to build a privileged relationship with one important historical figure, Cardinal Gabriele Paleotti (1522–1597), bishop and archbishop of Bologna (from 1567 and 1582, respectively) and a great patron of the arts and sciences. Paleotti looked at the Jesuits favourably and chose a Jesuit, Father Francesco Palmio (1518–1585), as his trusted confessor.[14] In short, even in a city like Bologna, long-standing ties with both Spain and the Jesuits can explain the genesis of such a unique publication. Despite its distance – geographical as well as temporal – from the New Kingdom of Granada, Bologna was not an utterly outlandish place where to trace the cult of Borja.

The *Notizie* marks the end point of a long-lasting cult that started in the early seventeenth century with the miracle with which I opened this essay. To the best of my knowledge, this ephemeral publication survives in only four copies, all presently housed in or around Bologna. Its impact, therefore, may have been very limited. The story that it recounted, however, had a venerable pedigree that spanned the long swath of land and sea stretching from Tunja and Santa Fé de Bogotá, in the New Kingdom of Granada, to the Kingdom of Naples, Portugal and Spain (Figure 11.1). At some point after Borja's death, this cult consolidated into a set of events that celebrated Borja's ability to intercede with God in times of catastrophe.

The anonymous author of the *Notizie* explained how the cult of Borja emerged in Bologna: after the tremor of the 4th of June, a relic of the Spanish saint was exposed in the church of St Sigismund, right in the heart of the city, together with those of St Catherine of Saragozza (thus reinforcing the hypothesis of a Spanish origin of Borja's cult in Bologna). After a second, milder tremor, on 4 March (presumably of the following year), prayers were directed to Borja, together with St Joseph, Ignatius Loyola and Francis Xavier (this time bringing out the evident Jesuit connection).[15] The anonymous Spanish priest also recounted how he had accidentally discovered a copper plate with the image of Borja in ecstasy in front of the consecrated host. The image was accompanied by the phrase: 'St. Francis Borgia, elected defender against the earthquakes first by the New Kingdom of Granada, then by the Kingdom of Naples under the authority of Innocent XII, then also by other cities, especially in Italy, Spain, and Portugal'.[16] Bologna was only the last of a series of cities that had prayed for the saint's intercession.

In the opening pages of the work, the printer – possibly under the guidance of the author, possibly independently – included an image of the

Figure 11.1 Detail from the frontispiece of *Notizie della protezione speciale di S. Francesco di Borgia contro i terremoti raccolte da un sacerdote spaguolo divoto del santo* (Bologna, Stamperia di San Tommaso d'Aquino, 1780), showing the towns that had elected Francis Borja patron saint of earthquakes. Fondo Piancastelli. Courtesy of the Biblioteca Aurelio Saffi, Forlì.

saint and the eucharist and a map of the places that elected him as patron saint (Figure 11.2). As historians of religion have pointed out, the iconography of Borja praying in front of the eucharist was based on the accounts of some biographers who reported how Borja was particularly devout to this practice and had addressed a letter to Loyola asking if the deep piety felt after communion, which engendered copious tears in his eyes, was appropriate. The picture thus functioned as a reminder of Borja's deepest moral qualities of humility and piety.[17]

The image also encloses another iconographic detail that allows us to identify the saint, namely, the crowned skull, a reference to the moment of Borja's 'conversion' upon seeing the decomposing corpse of Queen Isabel of Portugal, Charles V's wife, a poignant representation of the transience of life and an apt metonymy for the Christian concept of *vanitas*.[18] Borja's popularity in the Americas can be also evinced by the many colonial representations of the saint in salient moments of his life that are still

Non timebimus dum turbabitur terra

Portogallo · Madrid · Napoli · Valenza · Muro · Gandia · Teruel · Regno di Granata · Seviglia

S. Francesco di Borgia, eletto Patrono, o Protettore da più Regni e Città nel flagello de' Terremoti

G.B.f.a B.

Figure 11.2 Saint Francis Borja, Patron Saint of Earthquakes, venerating the Eucharist, in anon., *Notizie della protezione speciale di S. Francesco di Borgia contro i terremoti raccolte da un sacerdote spaguolo divoto del santo* (Bologna: Stamperia di San Tommaso d'Aquino, 1780), frontispiece. Fondo Piancastelli. Courtesy of the Biblioteca Aurelio Saffi, Forlì.

extant.[19] According to the author of the *Notizie*, it was the discovery of this particular image that encouraged him to write about Borja and his special powers of intercession: for this, as he indicated in the text, he drew both on Cardinal Alvaro Cienfuegos' *Vita* of Borja and on various other sources, both in manuscript and in print (including printed images, paintings and medals he had access to).[20]

A patron saint of earthquakes is made in the New Kingdom of Granada: The evidence

The core of this Bolognese text articulates in chronological order events in the saint's life that God set in his path to 'train' (*esercitare*) him in his role of intercessor against earthquakes. From his illness at the age of 12 during an earthquake in a small town in the Kingdom of Granada, to the miraculous escape from the destruction of the convent dedicated to him (set up by his uncle, Ercole II d'Este, upon his visit in 1550) when the city of Ferrara was hit by a devastating earthquake in 1570, the author listed a series of events that hinted at how this special gift had punctuated Borja's life.[21] These powers of intercession, however, were only fully understood after his death, and especially after the miracle of the New Kingdom of Granada with which I opened this essay.[22] The second chapter of the *Notizie* emphasises how this event, at a time when Borja had just been made *beato*, finally revealed to the faithful of the New World his powers of intercession; according to the anonymous writer and his source (once again Cienfuegos), this event was instrumental to his canonisation. Drawing heavily on Ciensfuegos' account, but also on other sources in Spanish and Portuguese he did not openly name, the anonymous author traced the broad contours of how the miracle happened in Tunja, and then explained how the cult spread, slowly but steadily, across southern Europe, first to the Kingdom of Naples, then to the Kingdom of Valencia, then to Madrid and Seville and, finally, to Lisbon and Portugal when the region was hit by the ruinous earthquake of 1755.[23]

It is not clear when the story of Borja's apotropaic powers against earthquakes crossed the ocean, moving from Tunja and Santa Fé to Europe. The process of canonisation was put in motion in 1607, well before the Tunja event, when a relic of Borja was credited with having miraculously healed Mariana de Padilla Manrique, daughter-in-law of Francisco Goméz de Sandoval y Rojas, Duke of Lerma and one of Borja's grandchildren.[24] The time seemed ripe to pursue Borja's canonisation and this also meant collecting testimonies of miracles. The Jesuit order was relatively new to the practice. Indeed, Loyola's first process of canonisation under Clement VIII in 1599 failed on the grounds that too few miracles had been attributed to the founder of the Order.[25] While Pedro de

Ribadeneyra, who wrote Loyola's first *Vita* on behest of Borja, and Claudio Acquaviva, the fifth Superior General of the Order, were reluctant to overplay Loyola's miracle-making qualities,[26] the seven bound volumes of Borja's canonisation material, which together count thousands of pages, bear clear testimony that a hard lesson had been learned. The Jesuits and everybody else involved had a clearer understanding of what was needed to succeed beyond holy virtues: plenty of miracles. Borja's was a long-drawn case, but one that had a better chance of succeeding.

The petitions for the opening of Borja's process in 1611 were signed by the Duke of Lerma himself, but also by the Spanish monarch and high members of the Jesuit Order, clearly revealing that a strong partnership between the Borja family, the Crown and the Jesuits was there from the very beginning.[27] The Congregation of Rites headed by Cardinal Roberto Bellarmino approved these related processes in 1615. On 3 April 1617, Juan Esterlic, bishop of Drago, carried the official papers from Madrid to Rome, thus officially opening the second phase of the canonisation proceedings, the 'apostolic' phase.[28] The *processus ordinarius* material collected mainly in the Iberian Peninsula dates to 1611–1617, but the *processus apostolicus* did not come to a close until 1650–1651, and a further phase collecting new miracles continued until 1668, two years prior to his official canonisation under Clement X. This created an enormous amount of documentation in three languages: Spanish, Italian and Latin. I could count 163 witnesses that were interviewed just for the apostolic phase. Over a century after his death, Francis Borja had made it to sainthood and these witness reports, and many others collected before, were crucial to its success.

Attempts to locate traces of the Tunja miracle in the huge amount of documentation about Borja's canonisation in the Archivio Apostolico Vaticano have so far come to nothing, but I have been able to locate other significant sources that corroborate and enrich Cienfuegos' account. The first is a short manuscript note in a miscellaneous printed collection held at the Biblioteca Casanatense in Rome entitled *Divotione a S. Francesco Borgia contro i Terremoti*.[29] As it turns out, this anonymous note is a translation of a passage in Pedro de Mercado's *Historia de la Provincia del Nuevo Reino y Quito de la Compañia de Jesús*, a Jesuit history of the Order in the New Kingdom of Granada which was not printed at the time, but that clearly circulated in manuscript well beyond the New World at the end of the seventeenth century.[30] As both the note and Mercado's original text explain, the cult of Borja patron saint of earthquakes started in Bogotá shortly after the sumptuous celebrations of 1624–1625 that took place following his beatification. Both Mercado and our anonymous scribe state that it was around this time that Borja was elected patron saint of earthquakes, as – the note vaguely explains – he had already proven himself to be efficacious in protecting the city of Santa Fé from tremors.[31]

What the manuscript note failed to include (but Mercado had fully documented) is that the lavish celebrations that took place in Bogotá were sponsored by no other than Borja's grandchild, Juan Buenaventura de Borja y Armendia, who at the time was holding the most powerful post in the region as General Captain of the New Kingdom of Granada and head of the Real Audiencia de Santa Fé de Bogotá, the high tribunal of the Spanish crown in the New Kingdom of Granada (a post that he held for twenty years).[32] The other sponsors mentioned, unsurprisingly, were the top religious authorities in the region, especially the Archbishop of Santa Fé de Bogotá, Bogotá-born Ferdinando Arias de Uguarte, son of a local *encomandero*.[33] It was under the newly elected archbishop of Santa Fé de Bogotá, Julian de Cortézar, however, that progress was made to document Borja's intercessory powers against earthquakes. And this is where the events of 1627 in the small town of Tunja become especially relevant: the miracles of Tunja became a catalyst for renewed efforts to advocate for Borja's special powers. It is under Cortézar's leadership that the local authorities collected vital testimonies to support the efforts of the Borja family and the Jesuits to further Francis Borja's canonisation.[34]

A second, much more significant document records the numerous testimonies that Cortézar had collected and had planned to send to Rome.[35] At the start of this extraordinary document, Cortézar indicates that he had heard news of how within Sebastián de Mojica Buitrón's estate in Chitagoto, near the city of Tunja, there was a chapel that housed the image of St Francis Borja 'through which Our Lord operated some marvels and miracles' on the 6th of May of that year.[36] In order to shed light on these unusual events, which were witnessed by many people over the course of several days, Cortézar decided to send a commission to Tunja to investigate the facts and entrusted the vicar of Tunja, Sancho Ramírez de Figueredo, with the task of ascertaining the truth 'with due care and diligence' and then write a report about the events.[37] This report – as the notary Martin de Velasco, who collected the testimonies states – contained declarations (*autos*) about the miraculous events that occurred in Tunja and was to be sent to the archbishop 'closed and sealed through a trusted person' to make sure it reached him safely. The document was dated 13 September 1627.[38]

Among the testimonies included, we can read that of the Franciscan brother, Adriano de Ribera, who visited the chapel in Tunja to give mass on the day of St John before the Latin Gate (the 6th of May) and witnessed how the image of St Francis Borja in that chapel exuded profusely and mysteriously. Initially, the sweat exuded from the face, hands and body. De Ribera and another Franciscan friar cleaned it and dried it up and then moved it to the centre of the altar, but the image started to exude sweat drops once again. This time only the face became wet and, Ribera testified, it was as if the saint was crying. The two friars cleaned the image from the tears and sweat once again and left. When they returned two days later, however, the image was once again wet. As they witnessed this phenomenon for eight to nine days in a row, news of the miraculous occurrence spread, and many people came to see Borja's exuding image.[39]

But the miracles, unsurprisingly, did not end here: when Ana de Oquendo, the wife of Juan Gómez, head servant of Sebastian de Mojica Buitrón, was ill and crippled by pain, a cloth with the droplets collected from the exuding painting was applied locally and the woman was miraculously healed. Likewise, the daughter of Sebastian de Mojica Buitrón, who was unwell, was treated similarly and recovered. In his testimony, de Ribeira also mentioned how the colour of the face of the saint changed from a moment to the other, first being very pale and then turning red like if the saint had exercised or was suffering. Moreover, his hand, which was holding a crucifix, opened and closed. De Ribeira asserted how he had seen these things with his own eyes, and he believed they were miraculous and supernatural. He concluded his testimony swearing to have told the truth and declaring his age (he was 37).[40]

Subsequent testimonies corroborate the story of the first witnesses: they all confirmed that the image miraculously exuded droplets which, if collected on a cloth and applied, healed various illnesses. Among the testimonies, there is also that of Sebastian de Mojica Buitrón, the owner of the image. His account adds some details to the story. It was his children, he tells us, who saw the image exude for the first time. Sebastian had sent them to clean the chapel for the celebration of the Feast of St John before the Latin Gate, and it was his youngest, Luis, who was eight years old, who run back home to tell him that the image of Francis Borja was exuding profusely.[41]

The hazy details of these testimonies become much clearer once we read Cienfuegos' *La heroyca vida, virtudes, y milagros de el Grande San Francisco de Borja*, the key source of our eighteenth-century Bolognese text. Here Cienfuegos dedicates a chapter to the 'portentous image of Borja in the New Kingdom of Granada'.[42] His account enriches our understanding of the events with key details. To start with, Cienfuegos mentions once again that Francis Borja was elected patron saint of earthquakes after earth tremors in the Province of New Granada. Then he continues by introducing the story of the painting: he adds the interesting detail that the first owner of the painting was a Jesuit, who used to travel with it from town to town, until one day he lost it. A strong Jesuit connection is thus established. Then we are told that the painting was found by an indigenous man, and this man sold it to Sebastian de Mojica Buytron, who decorated his chapel with it. We also learn another important detail: Buytron's celebrations in honour of St John the Evangelist on 6 May were made to ask him to intercede with God and send away the locusts that had been plaguing his fields. We can only speculate that Buytron was already aware of Borja's powers of intercession against natural calamity, but the events clearly made the connection visible to all.

Cienfuegos then recounted how Buytron had sent his three children to the chapel to prepare it, and it was then that Luis de Mojica Buytron, Sebastian's youngest son (who, Cienfuegos specified, was 'particularly innocent', and thus, we may assume, an extremely reliable witness) observed Borja's image crying and sweating, with the expression of somebody fainting. Luis run to call his father, who verified the miracle. Tears were coming down from the portrait's eyes and forehead, and he had drops of blood on his left hand as if he had been crucified. Buytron sent two servants to call Father Pedro Zavaleta, the priest in charge of the chapel. Even the weather reflected the events happening within the confined space of the altar: the sky was ominously dark, and a storm was raging. Zavaleta tried to dry up the streams of water oozing off the painting, unsuccessfully. Zavaleta gave mass and prayed God to send an explanation for such an event; then he dried the painting a third time and left.[43] When they returned and found the image still oozing, they decided to remove it from the altar to check that the water did not have a direct source; but the wall was dry. Their conclusion was, therefore, that the water was coming from the canvas itself and that it was due to the 'aching heart' of Borja's portrait. They dried up the painting once again, they locked the church, and when they returned the same phenomenon had taken place. The image was exuding copious water. The same phenomenon repeated itself for 22 or 24 days.

The major of the neighbouring town of Duitama (also in modern-day Colombia), Don Martin de Berganza Gamboa, travelled to Tunja to see the miracle.[44] During his visit, once again, Borja's portrait exuded profusely, his hand holding the crucifix and opening and closing it as it had happened before. His complexion, moreover, changed colour from pale to red to dark. In Cienfuegos' words, everybody was astonished at the sight of what looked like a living portrait of a dying man.[45] Gamboa decided to write to the Archbishop of Santa Fé to ask for witness accounts to be collected to document the saint's posthumous *fama*.[46] But to make things particularly momentous, the earth started to tremble and shake the mountains around Tunja: 'the beautiful machine of nature' started to quiver.[47] Cienfuegos saw a strong connection between the catastrophic natural events and the miraculous vision of the suffering Borja: around the same days, the dead body of Borja was being transferred to Spain to be buried in the Jesuit professed house in Madrid built for this very purpose by the 1st Duke of Lerma, Francisco Goméz de Sandoval y Rojas. The process, however, had been a contested one, and Cienfuegos connected openly the pain expressed by the painting with the treatment of Borja's remains and with the death of Borja's grandson, Juan Borja y Armendia, the

President of the Real Audiencia de Santa Fé de Bogotá.[48] But there were also some positive effects: the drops and tears exuded by the image could heal: this was the case of Sebastiana, the daughter of Sebastian de Mojica Buytron, who was healed miraculously when her life was in peril; and also of Ana de Oquendo, one of Buytron's servants. The miraculous water treated also both deafness and blindness.[49]

As noted, the new archbishop of Santa Fé, Julian de Cortézar, initiated the investigation, and many of the towns in the area declared Borja their patron saint to protect them from calamities like droughts and earthquakes. The major authorities involved, including the rector of the Jesuit College and the Provincial superior, wrote to Urban VIII pledging for Borja's canonisation. Yet, not all the cities in the region had initially elected the saint as their patron and intercessor. However, when in 1641 the earth trembled again, their people run to Santa Fé to pray at the Church of its patron, St Francis Borja, and vow their devotion to him – and the earthquake stopped. Borja was seen as the intercessor between the inhabitants of the Kingdom and the natural powers of air and earth: it was he who could placate the elements in the name of God. Not too far away from where these events took place, in present-day Ecuador, another Borja, Francisco Borja y Aragon, Principe de Esquilache (Squillace), Viceroy of Lima, funded the city of San Francisco de Borja extending 'the fame of his grandfather from place to place'.[50] Political power, the religious authorities and local personalities all conspired to make St Francis Borja a disaster saint of the Andes.

The cult crosses the Atlantic: Borja's cult in Rome and the Kingdom of Naples

The birth and development of the cult of Borja in the Andes reflects the power of the Borja family within the New Kingdom of Granada, but also of the Jesuits as a religious order deeply linked with Spanish political power. But the story of St Francis Borja, patron saint of earthquakes and calamities, does not end here. As stressed also by Ida Mauro, who focused on documents that trace the cult in the Kingdom of Naples, all branches of the family worked collectively towards Borja's canonisation, and not just in the Americas. It is not by chance that the rise of Borja's cult within the Kingdom of Naples coincided with a series of strategic marriages between important local aristocratic families and the Borja throughout the sixteenth century, while the Jesuits were establishing themselves more firmly in the region.[51] The earliest trace of Borja as patron saint of earthquakes in Italy seems to date to the years around 1688, when, following a devastating earthquake near Benevento, Francis Borja was elected patron saint of Massa Lubrense – a town that had had two Borgias (a Girolamo and a Giovan Battista) as bishops – and Naples. The sponsor of the cult was

Francisca de Aragón y Sandoval (1647–1697), wife of the Viceroy, distantly related to Borja via both the Aragón and Sandoval blood lines.[52] The links with the Kingdom of Naples and its cities, however, predated the terrible events of 1688; they go back as far as 1624, the year of Borja's beatification, when the wife of the then Viceroy was Catalina de la Cerda y Sandoval (1580–1648), daughter of Francisco de Sandoval y Rojas, 1st Duke of Lerma, and great-granddaughter of Borja. With a charitable act, Catalina bequeathed a considerable sum of money to the Jesuit College dedicated to St Francis Xavier that was being built but she asked that it be dedicated to both Xavier and 'the Blessed Francisco Borgia'.[53] Despite the fact that the construction of the Jesuit College was severely delayed, the building was finally inaugurated in 1665. In 1671, the Neapolitan Jesuits were thus able to celebrate Borja's canonisation with great pomp.[54]

As Mauro highlights, the accounts of the celebrations of Borja's canonisation are revealing of substantial synergies between the Spanish political power, represented by the Viceroy, Pedro Antonio de Aragón (whose brother was once married to Marianna de Sandoval y Rojas, Duchess of Lerma and thus, a Borja) and the Jesuits. The procession's path across the city was typical of the celebrations dedicated to Naples' patron saints like San Gennaro.[55] It is not surprising that the staging of these celebrations included an image of Borja praying in front of the Eucharist. The catalyst of Borja's success in the Kingdom may have been the earthquake that hit the region around Naples on 8 September 1695: the people of the kingdom turned their gaze to Borja pleading for his intercession (who, in the meantime, had also acquired the additional power of intercession against tertian fever!).[56] It was the Viceroy at the time, the Count of Santisteban, and his wife, Francisca de Aragón y Sandoval, who elected Borja patron saint of Naples, as Cienfuegos documented copying a passage from a Spanish publication that reminded his readers that Catalina de la Cerda y Sandoval had been instrumental in introducing the cult.[57] In this way, Borja was admitted officially to the pantheon of Neapolitan patron saints – over 50 – responsible for protecting Naples from calamity.[58]

As it has been pointed out, the cult was sponsored by the highest levels of the Neapolitan aristocracy, but lacked the fervour characteristic of other, more popular cults. It is for this reason, probably, that it is not still present within the religious social fabric of Naples and neighbouring cities, unlike others, and particularly St Januarius. Yet, the cult of Borja left traces well beyond Naples, demonstrating that even when not directly connected with the political parties that had created the compelling narrative of his intercessory powers, the Jesuit order continued to promote his cult, through devotional practices as well as textual transmission.[59] The Suppression of the Order in the eighteenth century, however, must have weakened the cult of Jesuit saints sufficiently to erase it from more recent memory.

Conclusions

Why did communities as far apart as Bogotá, Naples and Bologna elect a disaster saint? And why Francis Borja? Fear (of God), it has been pointed out by numerous historians, is an emotion so deeply rooted in Catholicism and in other Abrahamic religions to make it impossible to separate fear and guilt from these faiths. Medieval practices of confession and penance had only made this theology of sin, both individual and collective, more articulate and refined over time. Indeed, according to French historian Jean Delumeau, this emphasis on death, the wrath of God, and the inescapability of divine judgement is a key trait of the Western modern self.[60] At the same time, natural disasters, with their disruptive force and with their unpredictability, naturally engender deep uncertainty and fear. Therefore, to the social and political crises that formed the backdrop of Delumeau's 'Western guilt culture' – plague epidemics, protracted wars, riots and rebellions – we can certainly add natural disasters such as earthquakes.[61] The disruptive nature and sudden onset of an earthquake could provoke fear, both mortal and soteriological.[62] If we then add the fact that pre-moderns had a very imprecise understanding of the cause of earthquakes that veered between the divine and the natural, it is not hard to understand why in a Catholic worldview, natural disasters could be interpreted as the action of an angry God.[63]

A rich array of historical figures populates the pantheon of Christian saints, and some of these, as in Borja's case, came to be associated with natural disasters. The use of these intermediaries is as old as Christianity, and this practice was both powerfully reaffirmed and expanded after the Council of Trent.[64] The use of saint cults was clearly functional to the process of evangelisation of the Americas, and the cult of Borja described in this essay is just one example among many of how the process of beatification and canonisation could intersect with evangelisation within the framework of Jesuit expansion and conversion. This does not mean, however, that the symbols and the figures that acted as mediators between God and its people were always the same across time and space. The choice of a specific intermediary to venerate was not casual. The factors that contributed to this choice could be multiple. As Borja's case amply illustrates, the election of patron saints was often a political affair: the elevation of Borja as patron saint of earthquakes in the New Kingdom of Granada saw the concerted effort of the Jesuits and the Borjas, and it was dictated by the need to support the cult as much as possible through the collection (one could even say 'creation') of miraculous stories that would amplify the cult locally to the point of generating documentation that could be sent to Rome through the channel of local religious hierarchies. Once established, moreover, this cult could be transferred to faraway places: once Borja's intercessory powers against earthquakes and other calamities were known, other calamitous events could be associated with his name, promulgating the cult, but also potentially multiplying the examples that could end up in the hands of the Roman religious tribunals in charge of his canonisation. This was indeed the case in Naples, where the presence of various descendants of the Saint in the capital of the Kingdom made it possible by 1695 for St Francis Borja to become one of the patron saints of the city.[65] But even this act, which was generally driven by popular devotion, was orchestrated by the Viceroy, thus revealing the deep connections between the Borjas, other Spanish and Neapolitan families and the Jesuits.[66] To make a saint, and especially a disaster saint, nature needed to be read both politically and theologically. Calamities could be signs of God's wrath, but they were also powerful instruments of social containment and control to be deployed for the political advancement of both the aristocracy and the clergy. Through rituals of expiation and collective demonstrations of contrition, disasters could restore order and promote the interests of the clergy and local governments. Within a society that ultimately rested on the authority of God, the destructive power of nature could be appeased through the intercession of old and new figures, all of whom served the purpose of abating fear and restoring faith.

Acknowledgements

In writing this essay, I incurred many debts: I am very grateful to Rodrigo Cacho for kindly lending his precious expertise in Spanish palaeography at a crucial time in the development of this project. I wish to thank the students and colleagues who attended the Shelby Cullom Davis Center Seminar at Princeton University in April 2017 for listening to a very early incarnation of this story. My deepest thanks go also to Simon Ditchfield, Jonathan Greenwood, Domenico Cecere and the colleagues of the York Workshop 'Saints and Science', who discussed with me some of this material as I was working through it in 2018. A final thank you goes to the editors, Ovanes Akopyan and David Rosenthal, for inviting me to present a shorter version of this paper at the RSA in 2019. The anonymous reader's comments and the editors' constructive criticism have improved this essay substantially. All errors remain my own.

Notes

1 *Notizie della protezione speciale di S. Francesco di Borgia contro i terremoti* (Bologna: Stamperia di S. Tommaso d'Aquino, 1780), pp. 17–8. The earliest printed account of these events I could find is in Àlvaro Cienfuegos, *La heroyca vida, virtudes, y milagros del grande S. Francisco de Borja*, available in various editions (1702, 1717, 1726, 1754). I have used Cienfuegos, *La heroyca vida* (Madrid: Bernardo Peralta, 1726), bk. 7, chap. 8, pp. 590–95.

2 Stefania Pastore, 'Francisco de Borja, santo', in *Dizionario storico dell'Inquisizione*, 4 vols, ed. by Adriano Prosperi et al. (Pisa: Edizioni della Normale, 2010), vol. 2, pp. 622–24. On Borja and his time, see *Francisco de Borja y su tiempo: política, religión y cultura en la Edad Moderna*, edited by Enrique García Hernán and Maria del Pilar Ryan (Valencia-Rome: Albatros Ediciones and Institutum Historicum Societatis Iesu, 2012); Enrique García Hernán, *Francisco de Borja, Grande de España* (Valencia: Ediciones Alfonso el Magnánimo, 1999); Maria del Pilar Ryan, *El Jesuita secreto. San Francisco Borja* (Valencia: Biblioteca Valenciana, 2008) with earlier bibliography.

3 On early biographies, see now Henar Pizarro Llorente, 'Política y Santidad: los bíografos de San Francisco de Borja durante el Barocco', in *La Corte del Barroco. Textos literarios, avisos, manuales de la corte, etiquetas y oratoria*, edited by Antonio Rey Hazas, Mariano de la Campa Gutíerrez, and Esther Jiménez Pablo (Madrid: Ediciones Polifemo, 2016), pp. 685–712. See also below.

4 Amparo Felipo Orts, 'La actitud institucional ante el proceso de canonización de san Francisco de Borja', in *Francisco Borja y el suo tiempo*, pp. 59–78; Archivio Apostolico Vaticano (AAV; formerly Archivio Segreto Vaticano), Congr. Riti, Processus NN 2443–2447.

5 For a wealth of examples and references, see now Rienk Vermij, *Thinking on Earthquakes in Early Modern Europe. Firm Beliefs on Shaky Ground*, London-New York, Routledge, 2021, pp. 22–5, and part 2 ('Early modern confessionalized science'). For some eloquent case studies exploring the connection between natural disasters, religion and politics that have shaped my approach, see the essays of Gerrit Schenk, Elaine Fulton and Grégory Quenet in *Historical Disasters in Context: Science, Religion and Politics*, edited by Andrea Janku, Gerrit Jasper Schenk, and Franz Mauelshagen (London-New York, Routledge, 2012).

6 Grégory Quenet, *Les tremblements de terre aux XVIIe et XVIIIe siècles: La naissance d'un risqué* (Seyssel: Champ Vallon, 2005), pp. 7–12. Quenet remains the most impressive and thorough cultural study of earthquakes in early modern Europe.

7 Borja promoted three separate missions, one in 1567, another one in 1569, and a third one in 1572. Sergi Doménech García, 'La imagen de San Francisco de Borja y el discurso de la Compañía de Jesús', in *Francisco der Borja y su tiempo*, p. 325.

8 AAV, Congr. Riti, Processus NN 2443–2447. I checked these volumes swiftly in the summer of 2017, and I have been unable to return to the AAV since to conduct more thorough research. The processual documents amount to thousands and thousands of pages.

9 Historical data and further details can be found in Emanuela Guidoboni et al., "CFTI5Med, the New Release of the Catalogue of Strong Earthquakes in Italy and in the Mediterranean Area," *Scientific Data*, 6, 80 (2019), doi: https://doi.org/10.1038/s41597-019-0091-9. Details about the documents identified so far about this particular event are at: http://storing.ingv.it/cfti/cfti5/quake.php?02388IT

10 These documents will be the topic of a separate article.

11 Archivio di Stato di Bologna (henceforth ASBo), Governo misto, Senato, Filze, vol. 105 (1779), Preghiere pubbliche per la disgrazia del terremoto, Bologna, 5 giugno 1779 (fol. 246*r–v*); Voto pubblico per la disgrazia del terremoto, Bologna, 11 giugno 1779 (fols. 266*r–267v*); Voto pubblico per la disgrazia del terremoto, Bologna 19 giugno 1779 (fols. 302*r–304v*); Governo misto, Senato,

Filze, vol. 106 (1780), Elezione di Sant'Emidio Patrono, Bologna, 8 e 16 agosto 1780 (fols. 305*r*–307*v*); Decretum super electione sanctorum in Patronos a Sac. Rit. Congreg. de ordine SS.mi D.ni N.ri, Urbani VIII (fol. 309*r*–*v*); Documentazione relativa alla costruzione nella piazza di Castel S. Pietro di una statua dedicata alla Beata Vergine del Rosario per la liberazione dal terremoto, Bologna, 22 agosto 1780 – 4 dicembre 1780 (fols. 594*r*–596*r*); Governo misto, Senato, Partiti, reg. 42 (1778–1783), Provvedimento deliberativo del Senato di Bologna relativo alle messe da celebrare in onore della Madonna in seguito ai terremoti, Bologna, 19 giugno 1779 (fols. 66*v*–67*r*); Provvedimento deliberativo del Senato di Bologna relativo alla concessione a favore dell'arciconfraternita del Santissimo Rosario per erigere una statua dedicata alla Beata Vergine per la liberazione dal terremoto, Bologna, 18 dicembre 1780 (fol. 126*v*). Governo misto, Senato, Vacchettoni, reg. 80 (1778–1779), verbale di seduta del Senato di Bologna relativo alla esposizione di immagini sacre e alla proibizione degli spettacoli pubblici in seguito ai terremoti, Bologna, 5 giugno 1779; verbale di seduta del Senato di Bologna relativo alle cerimonie da officiare in onore della Madonna in seguito ai terremoti, Bologna, 19 giugno 1779; verbale di seduta del Senato di Bologna relativo alle messe da celebrarsi in onore della Madonna in seguito ai terremoti, Bologna, 26 novembre 1779. For the controversy that arose between the magistrate of the Anziani Consoli and the Senate regarding the religious processions for the Virgin's intercession, see ASBo, Archivio Salina-Amorini-Bolognini, b. 526, Atti del Senato di Bologna per il terremoto del 1779.

12 *Dei terremoti di Bologna. Opuscoli di d. Michele Augusti monaco olivetano* (Bologna: Stamperia di San Tommaso d'Aquino, 1780) (a second, augmented edition appeared later in the same year); *Lettera risponsiva ad altra in cui richidevasi, che diligentemente si notasse quanto accadeva in Bologna in occasione de' terremoti dello scorso anno 1779* (Bologna: Stamperia di San Tommaso d'Aquino, 1780); *Notizie della protezione speciale di S. Francesco di Borgia contro i terremoti* (Bologna: Stamperia di S. Tommaso d'Aquino, 1780); Vincenzo Mignani, *De effectibus terraemotus in corpore humano* (Bologna: Stamperia di San Tommaso d'Aquino, 1784).

13 Originally founded in 1364, the institution received the royal title in 1530, during Charles V's momentous visit to the city. Together with the Collegio dei Fiamminghi, it is the only remaining college to represent the *nationes* of the University of Bologna and it is still fully operating.

14 Paul Grendler, *The Jesuits and Italian Universities, 1548–1773* (Washington: The Catholic University of America Press, 2017), pp. 282–83.

15 *Notizie*, p. 6. The triad Loyola-Borja-Xavier had become a shorthand for the Jesuits' evangelisation of, respectively, Europe, the Americas and Asia. Doménech García, 'La imagen de San Francisco de Borja y el discurso de la Compañía de Jesús', pp. 325–32. On the Jesuits in Bologna, dated but still useful are Natale Fabrini S.J., *Lo Studio pubblico di Bologna ed i gesuiti* (Bologna: Parma, 1941); idem, *Le congregazioni dei gesuiti a Bologna* (Roma: Stella Matutina, 1946). An overview is offered in *Dall'Isola alla Città. I gesuiti a Bologna*, edited by Gian Paolo Brizzi and Anna Maria Matteucci (Bologna: Nuova Alfa Editoriale, 1988), with bibliographical references. On the Jesuits' expulsion from Spain and Portugal and Spanish overseas territories in 1759 that led to many of them living in the Papal States, including Bologna, see *La presenza in Italia dei gesuiti iberici espulsi. Aspetti religiosi, politici, culturali*, edited by Ugo Baldini and Gian Paolo Brizzi (Bologna: CLUEB, 2010).

16 *Notizie*, p. 1. Antonio Pignatelli, later Pope Innocent XII, may have had an interest in supporting the cult of an earthquake saint as he probably experienced the Sannio earthquake of 1688. See my 'Coping with Catastrophe: St Filippo Neri as Patron Saint of Earthquakes', *Quaderni Storici*, 52, 3 (2017), pp. 727–50.

17 Doménech García, 'La imagen de San Francisco de Borja', p. 321. See Pedro de Ribadeneira, *Vida del Padre Francisco de Borja* (Madrid: En casa de P. Madrigal, 1592), pp. 320–22; Ivan Eusebio Nieremberg, *Vida del santo padre, y gran siervo de Dios el B. Francisco de Borja* (Madrid: Maria de Quiñones, 1644), pp. 45–7. The image of the Spanish saint in ecstasy in front of the eucharist is common to other sources of the saint, like the *Epitome de la vida de San Francisco de Borja* (Naples: Parrino & Mutio, 1695). In the two editions I consulted, the images (which are different in the two copies but portray the same iconography) were not accompanied by the writing that attributed to Borja the election of patron saint of earthquakes. The two copies are, respectively, in Biblioteca Nazionale Centrale, Roma, and at the Complutense in Madrid.

18 This is also a common iconographic theme, possibly the most common as it is sometimes presented alone, sometimes in composite iconographies. See Doménech García, 'La imagen de San Francisco de Borja' and Wilfredo Rincón García, 'Iconografía de San Francisco de Borja en el arte español', in *Francisco der Borja y su tiempo*, pp. 415–37 (pp. 417–23).

19 See Doménech García's 'La imagen de San Francisco de Borja' as well as the many images available in PESSCA, *Project on the Engraved Sources of Spanish Colonial Art* under 'Francis Borgia' at: https://colonialart.org/archives/subjects/saints/individual-saints/francis-borgia#c478a-1695b (last accessed 3 November 2021).

20 *Notizie*, p. 7.

21 *Notizie*, p. 7.

22 *Notizie*, pp. 9–14. On the Ferrarese earthquake, see Craig Martin, *Renaissance Meteorology: Pomponazzi to Descartes* (Baltimore: The Johns Hopkins University Press, 2011), pp. 60–79; Vermij, *Thinking on Earthquakes*, pp. 115–19.

23 *Notizie*, pp. 23–25 (on Naples); pp. 26–33 (on Valencia); pp. 33–36 (on Aragona); pp. 36–9 (on Madrid); pp. 39–45 (on Seville); pp. 45–8 (on Lisbon).

24 Pedro Suau, S.J., *Historia de S. Francisco de Borja, Tercer General de la Compañia de Jesus (1510–1572)* (Zaragoza: Hechos y Dichos, 1963), p. 445. On these early phases of the process, see also Orts, 'La actitud institucional ante el proceso', p. 65.

25 Simon Ditchfield, 'Coping with the *Beati moderni*: Canonisation Procedure in the Aftermath of the Council of Trent', in Ite inflammate omnia: *Selected historical papers from conferences held at Loyola and Rome in 2006*, ed. by Thomas McCoog (Rome: Institutum Historicum Societatis Iesu, 2006), pp. 413–40; Miguel Gotor, *I beati del papa. Santità, Inquisizione e obbedienza in età moderna* (Florence: Olschki, 2002), pp. 57–65, Bradford A. Bouley, *Pious Postmortems. Anatomy, Sanctity, and the Catholic Church in Early Modern Europe* (Philadelphia: University of Pennsylvania Press, 2017), p. 23.

26 Ditchfield, 'Coping with the *Beati moderni*', pp. 415–23.

27 Bouley, *Pious Postmortems*, pp. 26–7. See also Carmen Sanz Ayán, 'La canonizacion de Francisco de Borja: una lectura política', in *Francisco de Borja en su tiempo*, pp. 73–92.

28 Suau, *Historia de S. Francisco de Borja*, p. 445.

29 Biblioteca Casanatense, Rome, Vol. Misc. 739, *Divotione a S. Francesco Borgia contro gli Terremoti*, n.p.

30 Pedro de Mercado (1620–1701) was enviably placed to recount the history of the Order in the New Kingdom of Granada. He was born in Riobamba, Equador, and died in Bogotá. He entered the Order in 1636 and was rector and teacher at the Jesuit College of Tunja, among other places. For a short biography, see Hernán Rodríguez Castelo, 'Mercado, Pedro', at https://dbe.rah.es/biografias/20672/pedro-mercado (last accessed 21 November 2021); for a richer account of Mercado's life and his *Historia*, see José del Rey Fajardo, *Los Jesuitas en Venezuela*, 2 vols (Caracas; Bogotá: Universidad Católica Andrés Bello; Pontificia Universidad Javeriana, 2006), vol. 1, pp. 247–69.

31 Pedro de Mercado, S.J., *Historia de la Provincia del Nuevo Reino y Quito de la Compañía de Jesús*, 4 vols (Bogotá: Empress Nacional de Publicaciones, 1957), vol. 1, p. 86: "En este con un acuerdo y con afectuosa devoción se eligió a San Francisco de Borja por patrón y abogado contra los temblores que hacían estremecer esta tierra y la ponían en peligro de asolar las casas y matar a sus moradores. Tomó este santo grande a su cargo el patrocinio como lo ha experimentado esta ciudad de Santa Fé en la cesación de sus terremotos."

32 Don Juan de Borja y Matheu was born in Gandia, Spain, his father was Fernando de Borja y Castro and his lover, Violante Matheu de Armendia. On the Borjas in the Americas, including Juan Bonaventura, see Manuel Garcia Rivas, 'Los Borjas americanos: su contribución al mundo de la cultura', *Revista Borja. Revista de l'Institut Internacional d'Estudis Borgians*, 5 (2016), 1–15.

33 Mercado, S.J., *Historia de la Provincia del Nuevo Reino y Quito*, vol. 1, p. 86: "Acerca de las fiestas con que la ciudad de Santa Fé celebró la beatificación de nuestro padre San Francisco de Borja en el año de seiscientos y veinte y cinco, no me han dado las *Annuas* más que unas noticias en común diciendo que se hicieron unas reales y suntuosas fiestas. Y para entender en particular cuán suntuosas y reales fueron, no es necesario más que saber que las hizo el señor don Joan de Borja nieto del Santo Padre y presidente de todo este Nuevo Reino de Granada."

34 Ferdinando Arias de Uguarte was moved to Caracas and died in 1628.

35 Biblioteca Nacional de Colombia (BNC), Ms RM 183, fols. 50–83: "Nos el doctor don Jullian de Cortazar electo arcobipo de el Nuevo Reyno de Granada de el Consejo de su Magestas [*sic* Majestas] etc, habemos saver a el Vicario de la cividad de Tunja como por nos seproveyo un auto de el thenor siguiente [...]" (fol. 50*r*). I wish to express my deepest gratitude to the staff of the BNC, in particular Drs. Camilo P. Jaramillo and Paola Londoño for locating, scanning, and sending me the document free of charge after my initial inquiry. See also Juan Manuel Pachecho, S.J., *Los Jesuitas en Colombia, Tomo 1 (1567–1654)* (Bogotá: San Juan Eudes, 1959), pp. 174–75.

36 BNC, RM 183, fol. 50*r*: "En questa una ymagen de San Francisco de Borja por la qual asido Nuestro Señor servido de obrar algunas maravillas y milagros."

37 BNC, RM 183, fol. 50*r*: "[...] y para que se averigue y sepa la verdad [...] mando se despache commision a el vicario de la cividad de Tunja para que personalmente con el cuyda doy diligencia que el caso pide haga ynformacion de todo lo contenido."

38 BNC, RM 183, fol. 50*r*: "La relacion de [fusso?] y haga los demas autos que coniungan para que conto da claridad seben siguen los dechos milagros y la zerticumbre de ellos y hecha la dicha ynformacion cerrada y sellada con quenta

y rrazon y numeracion de hojas orginalmente con persona de confianza." The signatures and dates are on fol. *50v*.

39 BNC, RM 183, fols. *52v–53v*.

40 BNC, RM 183, fols. *52v–54v*.

41 BNC, RM 183, fol. *58v*. Sebastian's testimony is one of the longest and can be found at fols. *58v–62v*.

42 Cienfuegos, *Vida*, p. 590.

43 Cienfuegos, *Vida*, p. 591.

44 Gamboa's witness account is in BNC, RM 183, fols. *54v–56r*.

45 Cienfuegos, *Vida*, p. 592.

46 Cienfuegos, *Vida*, pp. 591–92.

47 Cienfuegos, *Vida*, 592.

48 Cienfuegos, *Vida*, p. 592.

49 Cienfuegos, *Vida*, p. 593.

50 Cienfuegos, *Vida*, pp. 592–93.

51 Ida Mauro, 'La diffusione del culto di San Francesco Borgia a Napoli tra feste pubbliche e orgoglio nobiliare', *Revista Borja. Revista de l'Institut Internacional d'Estudis Borgians*, 4 (2012), 449–560 (549–51).

52 Mauro, 'La diffusione del culto', pp. 557–58. See also ARSI, Post. Gen. 69, "Flagellata da spessi terremuoti la nobilissima Città di Napoli s'atterri sopra modo nelli 8 di Settembre del 1694" (on the 1694 earthquake), which mentions also the 1688 earthquake.

53 Saverio Santagata, *Istoria della Compagnia del Gesú appartenuta al Regno di Napoli*, 4 vols (Naples: Vincenzo Mazzola, 1755–1757), vol. 4, p. 319, cited in Mauro, 'La diffusione del culto', p. 554.

54 Mauro, 'La diffusione del culto', pp. 554–55.

55 Mauro, 'La diffusione del culto', p. 555.

56 Cienfuegos, *Vida*, p. 593.

57 Possibly from the *Relación de los Milagros que Dios nuestro señor ha obrado por una imagen del glorioso Padre San Francisco de Borja, en el Nuevo Reino de Granada, sacada de los procesos originales de la información y aprobación que dello hizo el ilustrísimo señor don Iulián de Cortázar, arzobispo de Santafé*, 1629.

58 Cienfuegos, *Vida*, pp. 556–57. On the rich array of patron saints of the city, see the classic study of Jean-Michel Sallman, *Naples et ses saints à l'âge baroque (1540–1750)* (Paris: Presses Universitaires de France, 1994).

59 Abundant documentation about the cult and seventeenth-century celebrations in Lecce, Messina, Paola, Cremona, Mantova, Ascoli, but also Regensburg exist in ARSI, Postulazione Generale, 69. On its presence in Sicily, see now Valeria Enea, 'Seeking the Protector Saint: Cults and Devotions in Palermo after the 1693 Earthquake' 1693', in *Heroes in Dark Times. Saints and Officials Tackling Disaster (16th-17th Centuries)*, ed. by Milena Viceconte, Gennaro Schiano and Domenico Cecere (Rome: Viella 2023), 287-303 (on Borja, see 297-301). I wish to thank Valeria Enea for kindly sharing her article with me before publication.

60 See Jean Delumeau, *Sin and Fear: The Emergence of a Western Guilt Culture, 13th–18th centuries* (New York: St Martin's Press, 1990). Fear is, of course, a thoroughly human emotion shared by different cultures and civilizations. Its social and political dimensions, therefore, are not unique to Catholicism. Similarly, the link between religious deities and punishment is a theme that runs through various cultures and civilizations from antiquity onwards.

61 See Delumeau's influential, *Le péché et la peur: La culpabilisation en Occident. 13–18. Siècles* (Paris: Fayard, 1983).

62 Early modern Catholics were terrified of dying suddenly, as this threatened their salvation by hindering repentance and the chance of receiving the last rites. For an exploration of this theme, see Maria Pia Donato, *Sudden Death: Medicine and Religion in Eighteenth-Century Rome* (Burlington, Vt.: Ashgate, 2014).

63 See Rogelio Altez, 'Historias de Milagros y temblores: fe y eficacia simbólica en Hispanoamérica, siglos XVI–XVIII', *Revista de Historia Moderna. Anales de la Universidad de Alicante*, 35 (2017), 178–213 (179–80). On natural disasters and the history of emotions, see *Fear in Early Modern Society*, ed. by William G. Naphy and Penny Roberts (Manchester; New York: Manchester University Press, 1997); and especially, *Disaster, Death and the Emotions in the Shadow of the Apocalypse, 1400–1700*, ed. by Jennifer Spinks and Charles Zika (London: Palgrave Macmillan, 2016). On pre-modern responses to disasters, see 'Calamità/paure/risposte', special issue of *Quaderni Storici*, nuova serie, vol. 19, n. 55 (1984), ed. by Giulia Calvi and Alberto Caracciolo.

64 Clare Copeland, 'Sanctity', in *The Ashgate Research Companion to the Counter-Reformation*, ed. by Alexandra Bamji et al. (Ashgate: Aldershot, 2013), pp. 212–24.

65 Mauro, 'La diffusione del culto', 449.

66 Mauro, 'La diffusione del culto', 559–60.

12 Dikes, ships and worms

Testing the limits of envirotechnical transfer during the Dutch shipworm epidemic of the 1730s

Adam Sundberg

In late September of 1730, Dutch dike authorities discovered a hitherto little-known species of mollusk (*Teredo navalis*) burrowed into the wooden piles, revetments and groynes that protected coastal dikes. The wood-boring bivalves carved cavities into the wooden structures, which fractured and broke following a recent minor storm. Further investigation revealed 'shipworm' infestation across three provinces. By 1732, the threat of widespread dike failure prompted a crisis in water management that assumed existential proportions. This was the first large-scale outbreak of shipworms in Dutch waters. Dike officials labelled the shipworm a novel threat and descriptions of the shipworm 'plague' in popular media emphasised its 'strange' and 'previously unknown' character. This language of novelty was an important motivation to adapt dike designs.[1] Lacking time-tested solutions and fearing disaster if dikes failed, authorities enacted capital-intensive dike reconstruction programmes, in many cases replacing wooden components with imported stone. This transformation is well known in Dutch water history and considered a pivotal moment in dike modernisation.[2]

The history of Dutch dike adaptation in the 1730s embraces the catalytic role of shipworms, yet little scholarship explores the process of dike adaptation, the broader context of Dutch experience with the mollusk, or their influence on the envirotechnical systems the Dutch fashioned to manage life on or near the sea. *Teredo navalis* appeared an unprecedented threat to flood security, yet the animal was far from unknown. Shipworms were a primary hazard of oceanic travel throughout the early modern period and mariners had coped with them for centuries. Although dike authorities emphasised shipworm novelty, they nevertheless proposed, tested and implemented adaptations derived from maritime knowledge. These trials operated at the nexus of two largely distinct sets of envirotechnical systems – maritime shipbuilding and coastal water management.

This chapter explores the challenges shipworms presented to both dikes and ships, the potential of translating maritime adaptations to the coasts,

DOI: 10.4324/9781003029823-16

and dike authorities' ultimate decisions to accept or reject these strategies. It argues that the shipworm 'plague' of the 1730s was an envirotechnical disaster that encouraged the transfer of technologies among dike authorities, but also bridged flood management and shipbuilding systems. In the process, dike officials rediscovered the limitations of adaptations long accepted in shipbuilding. Dike authorities in the province of Zeeland opted to incorporate maritime techniques as a central strategy in their shipworm adaptations, whereas authorities in Holland refined pre-existing dike knowledge. The incremental and uneven character of response belies the radical characterisation of shipworm adaptation. *Teredo navalis* was a shared threat, but the differences between envirotechnical systems in Zeeland and Holland incentivised diverging responses.

Shipworms and envirotechnical disaster

The shipworm epidemic of the 1730s is an ideal subject of envirotechnical analysis both because it emphasises key contributions of the field and moves the discussion into new territory. The synthesis of environmental history and the history of technology has encouraged richer interpretations of the ways that technologies reflect and mediate human–environmental interactions.[3] Technologies facilitated environmental manipulation, motivated practical and scientific evaluations of environments, influenced environmental decision making and were in turn shaped by those dynamic environments. This synthesis has encouraged greater interest in the role of environments, and more recently animals, as dynamic agents of change within large technological systems.[4] Human actors and institutions co-produced these systems through historically inscribed envirotechnical regimes. More recently, envirotechnical analysis has moved into the realm of disaster history. In her case study of the 2011 Fukushima-Daiichi disaster, historian Sara Pritchard argued that the incident revealed the convergence of natural and sociotechnical processes in both the creation of a high-risk system, but also the response to its catastrophic failure.[5]

The shipworm epidemic of the 1730s showcased each of these insights. Early modern flood infrastructure and shipbuilding were composed of multiple envirotechnical systems that emerged out of their diverse social and physical environments. Technological changes in each responded to a wide array of human and non-human influences, whether property relations, social and political institutions, or shifting dunes and rapacious mollusks. The shipworm epidemic of the 1730s was also an envirotechnical disaster that reflected the long history of technological and institutional innovation in Dutch flood defences, emerged from the transformation of coastal environments and influenced later adaptations in dike building regimes.

The shipworm epidemic also revealed the important role that environments played in the transfer of technology. Since all technologies emerge in specific social and physical environments, dissemination often demands adaptation accommodate new conditions. Environments, thus, act as a buffer limiting transfer. In his essay on the evolution of large technical systems, Thomas Hughes acknowledged the close relationship between technological transfer and adaptation, yet dynamic environments did not play a central role in his theory.[6] Other scholars have probed deeper into the role environments played in technological transfer. Karel Davids, for instance, investigated the sensitivity of hydraulic expertise to environmental differences between European and colonial environments. The different degrees of transfer between these envirotechnical systems arose out of an array of social, political and environmental factors. 'Water', he concludes, 'was also a powerful non-human actor in its own right'.[7] Different environments presented geographically specific challenges that fostered divergent envirotechnical systems as a result.

Focusing on technological transfer amidst disaster affords a unique perspective on the interaction between envirotechnical systems. Although often presented as a barrier, environmental components of distinct envirotechnical systems could encourage or facilitate interaction. The common threat of the shipworm animal was a case in point. In ordinary conditions, maritime shipbuilding and coastal dike engineering in Holland and Zeeland operated as largely separate systems composed of distinct social and ecological elements. The crisis moment of the early 1730s presented a unique opportunity for shipworms to bridge that gap.

Construction of an envirotechnical disaster

The discovery of *Teredo navalis* burrowed into broken wooden piles must have been a jarring experience for Edualdus Reynvaan. Reynvaan was upper commissioner for the island of Walcheren and he oversaw the water board charged with maintaining its sea dikes. The Westkapelle dike was arguably its most important sea defence. It protected the westernmost edge of Walcheren, which jutted like a thumb into the North Sea. The island was home to the provincial capital and among the most densely populated regions of the Dutch Republic.[8] Little land buffered the outer face of the dike, so it was protected by wooden piles, which were arranged either parallel or perpendicular to the shore as wave-breaking groynes called *hoofden* or wood and stone buffers called *staketwerken*. At its northern and southern edges, the dike connected to long chains of dunes, which were equally important flood barriers. The significance of these defences ensured vigilant appraisal of any important changes.

The discovery of broken piles may not have initially surprised Reynvaan. Piles had limited lifespans and regularly required repair. What triggered his alarm was that 'most of the piles were not driven out of the ground' as was typical, 'but instead broken above the ground'. Upon closer inspection, he noted peculiar, small holes 'full of worms'.[9] Subsequent investigation identified infestation across much of the island. By December, Walcheren's officials characterised the molluscan outbreak as a disaster. 'It cannot be seen as anything' other than an event that is of 'the utmost consequence, if not total ruin of the island'.[10] News spread quickly and by the fall of 1731, dike inspectors discovered shipworms in the provinces of Holland and Friesland as well (Figure 12.1).

Although early reports emphasised the sudden, unexpected character of the outbreak, the appearance of shipworms in the 1730s resulted from the conjuncture of long-term technological and environmental changes along Dutch coastlines. Dutch communities had modified coastal landscapes to suit their needs since the earliest settlement. As populations increased and land use intensified, flood protection grew more vital. By the medieval period, the Dutch had constructed complex envirotechnical systems of flood and water management that included seawalls, groynes, dams and sluices, but also coastal dunes, beaches and saltmarshes.[11] The resulting mosaic blurred the distinction between the natural and technical. Dune landscapes in Zeeland and the foreshores of West Friesland in Holland were natural flood barriers and important buffers during storm surges. Communities thus managed and regulated their use. Dikes, sluices and groynes altered natural erosion and sedimentation patterns and fashioned novel ecosystems.[12] An assortment of water management institutions, loosely defined as water boards, governed these envirotechnical land-scapes. Flood management was diverse and institutions decentralised, but they all contributed to a common set of goals: promoting security, fostering productivity and stabilising a dynamic coastline.

The history of Dutch flood engineering was not a linear story of progress, however. Maintaining the balance between land and water was a constant, often pyrrhic struggle. Until the sixteenth century, as much land was lost as gained from the sea. Dike breaches and flood disasters sometimes catalysed important changes in dike design, but more often they did not.[13] Dikes were 'thick' objects that reflected political interests, cultural imperatives and anxieties and the organisation of labour and capital.[14] They also influenced and were subject to shifting environmental conditions. As a result, these relationships evolved slowly, incrementally and unevenly. The resulting envirotechnical systems were anchored by flood management technologies yet perpetually in flux.

Several environmental and technological transitions converged in the early eighteenth century. In Holland, centuries of drainage and peat excavation in

Legend

- • Major Ports
- ≈≈≈ Shipworm Infestations

Harlingen

Texel

NORTH SEA

West Friesland • Enkhuizen

Hoorn

SOUTHERN SEA

Amsterdam

Rotterdam

Delfshaven

RHINE/MEUSE DELTA

Walcheren

Middelburg

Figure 12.1 Shipworm infestations in the Netherlands between 1730–1732. The initial inspections took place along the Westkapelle Sea Dike on the island of Walcheren and the Drechterland Northern Dike in West Friesland.

Sources: Boundaries adapted from O.W.A. Boonstra (2007): NLGis shapefiles. DANS. http:// dx.doi.org/10.17026/dans-xb9-t677 using Moll, Herman, d. 1732, Bowles, John, Bowles, Thomas, d. 1767, and King, John. 'A new and exact map of the United Provinces, or Netherlands &c.' Map. 1710. *Norman B. Leventhal Map & Education Center*, https:// collections.leventhalmap.org/search/commonwealth:cj82kt31g (accessed December 06, 2021). First published in Adam Sundberg, *Natural Disaster at the Closing of the Dutch Golden Age: Floods, Worms, and Cattle Plague* (Cambridge, UK; New York: Cambridge University Press, 2022).

already low-lying landscapes resulted in subsidence.[15] This lowered the level of the land relative to the sea and increased the risk of catastrophic floods. At the same time, natural shifts in tides and storm surges eroded beaches and the foreshores that fronted dikes (*voorland*), which shrank these important buffer zones.[16] Water boards responded by developing a suite of technologies to protect dikes immediately fronted by coastal waters. In Zeeland, these included the *hoofden* and *staketwerken* described by Reynvaan, as well as wicker and straw mats (*rijswerken*) stapled to the foot of Zeeland's characteristically shallow dike slopes. Beaches and dune ridges protected Walcheren's eastern and northern coasts. Centuries of coastal erosion and dune shrinkage required the gradual lengthening of the Westkapelle dike to compensate for these changes. Walcheren also relied on *hoofden* to anchor its beaches and stabilise its dunes. A ring of low dikes inside the dunes formed a last line of defence.[17] In Holland, a very different dike tradition (*wierdijken*) emerged that employed boxy wooden frames that supported cushions made from compacted eelgrass, which was locally available.[18] These were then protected with a combination of wooden piles and stones. Other dikes in Friesland and the Diemer dike near Amsterdam were protected by wooden palisades (*paaldijken*).[19] These wooden barriers and buffers were expensive to maintain and required specialised technical skills to construct. By the early eighteenth century, many water boards distributed these costs to communities living inland from the dike. Others shared the expense of 'post and pileworks' with the provinces. The construction of wooden technologies necessitated new expenditures and political arrangements, but they were necessary to stabilize the coast and protect its fertile polderlands.[20] They also reordered coastal ecologies, opening new habitats for wood-boring species (Figure 12.2).

Shipworms required more than wooden habitats to survive and proliferate, however. Shipworm growth and fecundity depended on a variety of environmental conditions. The most important limiting factors for *Teredo navalis* are temperature and salinity. Their optimal living conditions are water temperatures between 15° and 25°C and salinity levels of 10% and higher.[21] Both conditions grew increasingly favourable by the eighteenth century. Between the fifteenth and the eighteenth centuries, the channels that connected the North Sea to the Southern Sea gradually eroded, elevating the salinity of the latter. At the same time, river modification along the Rhine increased sedimentation and slowly silted up of the mouth of the IJssel River, reducing its discharge of fresh water into the Southern Sea. Finally, a series of warm springs and droughts between 1729 and 1737 lowered water levels along the Rhine and Meuse and elevated coastal temperatures.[22] Although drought likely triggered the shipworm explosion in a proximate sense, the stage had been set long before. The reorganisation of the coast was an envirotechnical process that

Figure 12.2 Print depicts three pieces of broken dike piles infested with shipworms.

Sources: Jan Ruyter, 'Drie Stukken Eiken hout van het Paalwerk aan de Zee-Dyken, getekent naar het leven zoo als het zelve van de Wurmen doorboord is', Print, 1731–1735, Rijksmuseum, Amsterdam, http://hdl.handle.net/10934/RM0001.COLLECT.505096.

mobilised the power of currents, sediments and subsiding landscapes, but also technical skill, political will and the cultural imperative to gain and retain dry land. The same structures water boards constructed to preserve security and prosperity opened pathways for a novel biological agent that threatened its very foundations.

Shipworms as an existential disaster

The discovery of *Teredo navalis* on the beaches of Walcheren in 1730 provoked fear in Zeeland, but anxieties broadened a year later when shipworms were discovered in Holland. Dike inspections revealed that shipworms had colonised wooden structures that protected coastal dikes on the northern tip of the province as well as the low-lying polderlands of West Friesland. An October 1731 inspection of one affected dike in the West Frisian water board Drechterland found shipworms in nearly 95 percent of dikes.[23] A second investigation in a neighbouring water board called the Vier Noorder Koggen revealed a 'certain unknown sort of Worm' in large sections of its own dikes. Inspectors marvelled at the speed of the infestation and warned of further consequences if God did not mercifully spare them from this 'plague in our territory'.[24]

The discovery of shipworms in Holland transformed the geography of shipworm risk. Unlike the island of Walcheren, dike breaches in Holland risked affecting neighbouring territories. During large coastal floods, water that breached West Friesland's coastal dikes could spill into adjacent polders, especially those that lay below sea level due to subsidence. Inland dikes mitigated this scenario by compartmentalising the landscape, but they were usually low and not built to sustain extended periods of inundation.[25] Once broken, floodwaters would then slowly spill from polder to polder. This had occurred in West Friesland as recently as 1675 when a December storm initiated a series of dike failures that left much of West Friesland below water for months.[26] Later, a series of river floods in 1726 inundated vast territories in the interior of southern Holland, threatening several large towns and cities. These and other inundations birthed a doomsday scenario where inundations in West Friesland spread from this northern region into the economic heart of the Republic, sweeping over valuable farmland and into its largest and most important urban centres, including Amsterdam, Leiden and The Hague. Water boards in West Friesland, thus, considered their seawalls a 'barrier and fortress that protects the entirety of Holland against their shared, restless enemy'.[27] By late 1731, *shipworms* had breached this first line of defence and dire consequences appeared imminent.

The shipworm panic assumed broader existential proportions in 1732 when dike inspectors discovered mollusks in the Diemer dike near Amsterdam. Up until this point, shipworm anxieties remained largely confined to water boards. Their discovery near Holland's largest and most prosperous city vaulted shipworms into the public sphere. A flood of pamphlets and articles in newspapers, periodicals and news digests reported the extent and severity of the disaster.[28] The Dutch news digest *Europische Mercurius*, for instance, published a 'Report on the Plague of Worms in the Pileworks of Dikes in Zeeland and West Friesland' in 1732, drawing explicitly on early dike inspection reports to outline the severity and extent of the disaster.[29] The *Mercurius* described what was then known, and not known, about the threat. At this early stage, the latter was far more worrisome.

It was in this context that moralising pamphlets, books and sermons latched onto the shipworm narrative. Ministers argued that shipworms were a punishment from God and a direct result of Dutch malfeasance. This providential interpretation attributed the disaster to a variety of sins, including atheism and blasphemy.[30] Others linked the outbreak of shipworms to an ongoing and deadly series of sodomy trials and executions.[31] According to this moral calculus, shipworms and sodomy were each symptomatic of a diseased and decadent Republic in decline.[32] Regardless of diagnosis, moralists advocated a similar suite of treatments, including repentance and prayer. The Estates General of the Dutch Republic, thus, called for a day of prayer across the Republic to combat the 'unusual plague of destructive worms in the piles and woodwork'.[33] In the meantime, new outbreaks appeared in Holland, Friesland and Zeeland. By 1733, shipworms had become a state-wide emergency.

The expansive scope and existential implications of shipworm risk demanded an immediate response from water managers as well, especially as public awareness of the molluscan threat grew to near panic. Yet the novelty of the shipworm outbreak proved its most troubling attribute. Reports invariably characterised shipworms as 'strange', 'uncommon' and 'previously unknown'.[34] This was problematic because water managers could not rely on time-tested solutions. The envirotechnical systems Dutch communities depended upon for flood control were never static, but the social and environmental agents of change were broadly understood. Shipworms introduced disturbingly unfamiliar challenges. In the early months following discovery, water boards tended to seek input from within their ranks. When no solutions appeared, they broadened their network to include formally trained hydraulics experts. In 1732, for instance, the Drechterland water board in West Friesland appealed to the Leiden professor and polymath Willem Jacob 's Gravesande for assistance. 's Gravesande replied that he could offer little useful advice, having 'no experience relative to the matter, and having found nothing in the books that can help me'.[35] The shipworm infestation seemed to grow with each passing season. As expenses mounted and risk intensified, water boards expanded their search yet again. They quickly confirmed that although shipworm disaster was unprecedented, the animal was in fact far from unknown.

Shipworms and maritime technical change

European shipwrights and mariners had centuries of experience with shipworms by the time they appeared along Dutch coasts in the 1730s. Sailors returning from months at sea brought back harrowing stories of compromised hulls and, in the worst cases, the loss of entire vessels. Most accounts associated shipworms with the tropics. As the Dutch empire expanded in the sixteenth and seventeenth centuries, the geography of shipworm risk widened. Dutch sailors lost vessels in the Caribbean, the coasts of Africa and Brazil, and Dutch East India Trading Company (VOC) records noted woodborers in southeast Asia. Dutch notarial records, maritime journals, company accounts and shipbuilding manuals all noted the destructive impact of 'worms'.[36] Wherever ships sailed in the tropics, they seemed to encounter new wood-borer habitats.

Dutch adaptation to shipworm hazards proceeded slowly and asymmetrically. Like flood management systems, Dutch shipbuilding was decentralised. Shipbuilding was a highly localised craft governed by guilds in cities, builders in the industrial *Zaanstreek* region of Holland, or large institutions like the West and East India Companies and the Admiralty. By the mid-seventeenth century, ships built for overseas voyages were increasingly constructed by these large institutions, especially the VOC.[37] The VOC was itself decentralised and each of its six chambers maintained its own shipyards. The VOC laid out general specifications for its ships in shipbuilding charters, but local variation persisted. Shipworm adaptations likely varied in part due to these decentralised operations.[38]

Numerous other human and non-human factors influenced changes in VOC hull design and shipworm adaptations. Warfare, for instance, incentivised improved defensibility, commercial imperatives emphasised profit maximisation and technical innovation and diffusion provided the tools to meet these needs. Environmental conditions likewise influenced ship design, whether the sand banks and prevailing winds of the North Sea, or the very different demands that foreign environments presented ships as they navigated an increasingly global empire. By the late sixteenth century, these included shipworms.[39] Dutch ships also reorganised the environments they connected. Ships shuttled entire marine communities around the world, whether in ballast or attached to hulls as 'biofouling'.[40] Indeed, European ships may have even introduced *Teredo navalis* to Dutch coastal ecosystems, although conclusive evidence remains lacking. By the eighteenth century, people and environments (including shipworms) had transformed companies like the VOC into envirotechnical systems that spanned continents and oceans.

Between the late sixteenth and the early eighteenth century, the Dutch experimented with a host of shipworm adaptations, which evolved as shipbuilders and crews acquired a greater understanding of shipworm risk and the limitations of their technologies. The most widely employed strategy was cleaning. Sailors dry docked, ran aground or careened ships to expose timbers below the waterline. They then scraped and burned the outer hull (called 'breaming'), replaced degraded timbers and applied new coats of pitch and tar.[41] By the early seventeenth century, Dutch mariners relied on anti-fouling paints as well. Sailors applied a wide variety of compounds they believed prevented infestation. In his 1671 shipbuilding treatise, Nicolaes Witsen advocated a concoction made from 'refined resin, whale oil, and sulfur'. The Dutch called this compound 'white stuff' and in combination with *harpuis* (a boiled pine resin), Witsen concluded 'it is thought no worm penetrates this'.[42] Witsen's confidence notwithstanding, the efficacy of any anti-fouling paint was questionable. Although some compounds appeared toxic, they degraded quickly and required continuous careening and reapplication.

Careening itself was time and labour intensive, particularly outside the familiar, sheltered harbours of Europe. Cleaning and anti-fouling paints provided at best provisional protection from shipworm damage. Although far from perfect, these strategies remained critical implements in the shipworm toolkit well into the eighteenth century.

Wooden sheathing presented another option to protect hulls from worms. These 'sacrificial hulls' provided an added level of security. All VOC ships in the seventeenth and eighteenth centuries employed a combination of pine or fir sheathing as a second hull. The Dutch fastened sheathing to the hull with iron nails. As the nails degraded in saltwater, they formed an iron oxide barrier thought to prevent biofouling.[43] Wooden sheathing did not replace anti-fouling compositions. More than a century after the first experiments, Cornelius van Yk still advised builders apply *harpuis* and sheath the ship in pine in his shipbuilding manual.[44] Sheathing also suffered limitations, however, the most significant of which was cost. Sheathing required continual maintenance and replacement, which increased in expense when conducted overseas.[45] Yet the widespread use of doubling hulls and rusting nails was testament to their advantages as insurance policies. They were expensive, but far less so than the loss of a ship.

Metal sheathing was an ambitious leap forward in anti-fouling protection. Rather than making ships less palatable or presenting a sacrificial offering to shipworms, metal promised to prevent wood-boring activity altogether. Dutch shipbuilders likely experimented with lead sheathing as early as the 1590s. Lead, however, was costly and prone to disrepair. Although the VOC mandated lead sheathing in 1602, they discontinued the practice four years later.[46] Copper presented a promising alternative because it protected hulls from shipworms and biofouling more generally. Copper exposed to saltwater produces a toxic film that deters biotic accumulation. The Dutch admiral Piet Heyn experimented with copper sheathing as early as the 1620s.[47] Van Yk later promoted its use as well, particularly on ships that sailed south to the Cape or west to the Caribbean.[48] Despite its advantages, the practice never became widespread in the seventeenth century. It was costly and prone to corrosion when fastened to the hulls with iron nails. Its benefits were significant enough that the VOC continued to sheath vulnerable components of ships, such as rudders, throughout the seventeenth and eighteenth centuries.[49]

The Dutch were certainly not the only Europeans experimenting with shipworm adaptations. Many strategies employed on Dutch ships had been pioneered by the Spanish, French, Portuguese or English. Technology transfer between these shipbuilding traditions was complicated by the difficulty of translating designs and techniques in an era when technical knowledge was rarely transferred to paper and shipbuilding guides were intended to elevate the prestige of shipbuilding traditions, more than transmitting reproducible designs.[50] Cross-pollination proceeded despite these barriers. Dutch society was relatively 'open', with few restrictions on the transfer of technical knowledge.[51] Transfer was facilitated by mobile workforces and institutions that competed for their labour in domestic and international markets. Large building sites like the VOC shipyards drew workers from across Europe and shipwrights often sold their services abroad.[52] States sometimes employed spies to learn foreign techniques. The practical knowledge of shipbuilding that passed between ships carpenters and workers, through guilds and corporations, through espionage, or simple observation of ships in ports, gradually reshaped hull design. The broad dissemination of shipworm remedies demonstrated that barriers to technological transfer in early modern Europe could be quite porous. These conditions would likewise smooth the transfer to completely separate envirotechnical systems of flood management.

Shipworms and envirotechnical transfer

After more than a century of effort to wrest control of the ecology of their own ships, mariners and shipworms had achieved a tenuous détente. No technology eliminated risk, but careful maintenance and thoughtful investment transformed shipworms from an existential threat into a manageable cost of life at sea. This process of accommodation played out slowly. By contrast, the introduction of shipworms into Dutch flood infrastructure produced an immediate and disruptive influence. This was the first explosive outbreak and no threat to the integrity of flood management had occurred across such a broad extent of coastline.[53] The novelty of the shipworm disaster and the existential nature of its potential consequences prompted reevaluation of these envirotechnical systems. Relative to slowly evolving maritime systems, the pressing need for solutions accelerated response.

In July of 1732, the Drechterland water board purchased an advertisement in the *Amsterdamse Courant*. They appealed for help from 'anyone who had experimented or developed inventions against the boring of worms into dike piles'.[54] The news digest *Europische Mercurius* republished this appeal in the second volume of its 1732 edition.[55] It also included a rubric to evaluate potential remedies. The most promising proposals would '1. repair [the dikes] in a way that prevents further damage from the worms 2. Protect against the violence of the sea so that people can live peacefully 3. Do so using the least costly methods'.[56] Over the next year and precisely during the period when public perception of the threat grew, a surge of letters, pamphlets and patent requests flooded into Holland from across the Republic and across Europe, proposing solutions to the shipworm problem.

Most recommendations were inspired or directly derived from maritime experience. Proposals frequently called for the removal and cleaning of piles in a manner similar to breaming. One suggestion from a 'Capitain Schrijver' suggested piles be scraped 'just as is done with ships'. This work could be completed by 'a few hundred sailors and their necessary officers', presumably because they were already familiar with the process.[57] Others advised lightly burning piles before scraping. Still others turned to shipbuilders to confirm their expertise. The Prussian inventor Daniel Gabriel Fahrenheit developed a unique anti-fouling paint and called on the equipage master of the Amsterdam Admiralty, another large producer of ships, to test its efficacy.[58] Isack Levi noted that his experience in Dutch Suriname (a shipworm hotspot) lent credence to his proposal.[59] Whether inspiration or confirmation, maritime experience compensated for the dearth of shipworm knowledge internal to water management systems.

The most common proposal was some variation of anti-fouling paint. Suggestions included a diverse array of compounds that usually included *harpuis* or pitch, but also crushed glass or shells to harden piles against attack; hair, wool or hemp to slow wood-boring activity; and compounds of mercury or sulphur, which they believed choked and killed shipworms.[60] Several proposals recognised that the transfer of technology could be a two-way street. Ludovicus Morgenster concluded his appeal by noting his concoction would work equally well on ships. Johan Kiliaan Pflugen meanwhile promised two mixtures. One should be used on piles, the other 'will be of great use to the Seafarer'.[61] Although there is little evidence these techniques were applied to ships, the shipworm outbreak nevertheless expanded opportunities for transfer.

Proposals from across the Republic and across Europe found their way to Holland, where a provincial commission systematically tested them in the shipworm-infested waters of West Friesland. Drechterland began installing test piles in September 1732.[62] By November, they reported that some piles were heavily colonised while others remained partially or completely free of worms. Despite some promising results, they also rediscovered significant limitations to these strategies. Cleaning was labour intensive and costly. Anti-fouling paints quickly washed away, just as they had on ships. Storm surges also posed a risk. Shortly after a November inspection, a storm struck West Friesland and washed away several test piles. Regardless of whether these adaptations decreased molluscan colonisation, they remained useless if dikes could not resist the waves.[63]

Authorities on Walcheren had also begun testing strategies drawn from shipbuilding practices. Shortly after their discovery, authorities ordered that 'piles should be burned, coated in tar and *harpuis* on oak and other wood in those areas most affected by the worms to see after a period of a few months which measures work best'.[64] They also eagerly sought and accepted suggestions from outside their own ranks. Jacob Peijn, a sailor who had worked for the Admiralty of Zeeland, proposed a treatment that relied on a new sort of pitch.[65] He claimed it had 'good success' at limiting wood boring on ships, which he had confirmed with three captains from the East and West India Companies.[66] Walcheren's water boards even considered loading a pair of 'discarded old ships that had been sheathed' with sand and sinking them in front of the vulnerable dike, although there is no indication they carried this out.[67] Walcheren's tests began more than a year before those in Holland, but despite several promising results they were no closer to a solution.

As the full scope of the shipworm disaster resolved, water authorities across the Republic began exchanging information gleaned from these early experiments. In early 1732, Reynvaan sent a letter to Drechterland outlining his findings. 'There is nothing much to say with certainty', he admitted, but maritime compositions including pitch, tar and glass were 'particularly good'.[68] Drechterland also received a letter from Groningen that proposed sheathing piles 'entirely with copper'. The author, Jan Hindrick Eekholt, considered the strategy 'infallible' and argued it would triple the lifespan of piles.[69] Adriaan Bommenee, head of public works for the city of Veere, noted that authorities on Walcheren 'were daily learning of experiments from all of the water boards in Zeeland'. Experts sought advice further afield as well, including from water boards stretching from Antwerp in Flanders to Bergen op Zoom in Brabant.[70] Dike authorities had long shared information amongst one another, but rarely with such speed and across such distance. The shipworm threat had vastly expanded the number of stakeholders, which encouraged this transfer of knowledge.

Despite several promising results, no solution presented the necessary combination of shipworm resistance, dike durability and cost-effectiveness. In its late 1732 update on the state of the outbreak, the *Europische Mercurius* acknowledged these failures.

> Endless designs have been developed and strategies forwarded, from which tests have been conducted, and are still being conducted to protect the piles against the boring of the worms by smearing them with various substances and compositions, several of which have already been found to be fruitless and without any good results; and it is to be feared that any strategies of this nature will not be much better. It is therefore commonly assumed that if this Plague of worms persists (God forbid), it will be necessary to develop changes and improvements in the dikes which, thanks to those destructive worms, with God's help, and with human skill, might enable them to resist, without fear in heavy storm and high tide, serious breaches.[71]

The secretary of Drechterland, Seger Lakenman, voiced similar concerns and conceded that 'entirely new forms of dike construction' might be necessary.[72] As winter turned to spring in 1733 heralding a new season of shipworm risk, shipworm adaptation in Holland and Zeeland began to diverge.

Diverging paths

The hope that a solution might emerge from maritime experience was wearing thin by the end of 1733. Breaming seemed impractical considering the scale of infestations. Anti-fouling paints quickly washed away. The challenge of continually removing and reinstalling piles complicated both strategies. Water boards had experimented with wooden and copper sheathing to a limited extent, but their expense restricted use to sluices.[73] A wave of scientific treatises in 1733 promised new insights by combining maritime knowledge with new empirical investigations. They contributed the first systematic analyses of *Teredo navalis* biology and ecology and several advocated solutions based on their findings, yet none supplied the silver bullet water boards desperately sought.[74] The pace of change was astonishingly quick. In just three years, water authorities had rediscovered the benefits and limits of adaptations developed over more than a century of maritime experience. Despite these efforts, maritime knowledge had seemingly failed to translate to the coast. This opened space to consider new forms of dike construction.

In 1732, a new set of proposals had appeared that shifted focus from maritime knowledge back to water management. These designs called for stone to be laid at the foot of dikes. Water authorities had long employed stone as building material because its weight and durability reinforced wooden infrastructure. These new proposals called for larger stones to shield the dikes. They later added smaller rubble to lower the costs, which incidentally prevented shipworm infestation. Technical and financial conflict between West Frisian water boards and the provincial commission, however, delayed implementation. Holland had been gradually expanding their influence in water management, which sometimes provoked resistance by water boards.[75] These conflicts were largely resolved by 1733 when two authorities from West Friesland, Pieter Straat and Pieter van der Deure, proposed a variation of this basic technique. They promoted their design as the best combination of durability, cost-effectiveness and shipworm prevention. Theirs was a 'new manner of dike building which has never been practiced before' and an 'infallible measure' to protect the dikes. In reality, this was less a radical shift than a refinement of legacy dike technologies, yet Straat and Van der Deure nevertheless employed the language of innovation to promote its use. They understood

Figure 12.3 Pieter Straat and Pieter van der Deure's new dike design for West Friesland. Their design used stone laid on a gradient along the seaward side of the coastal dikes. These designs retained the use of weir, though this would later be phased out.

Sources: Pieter Straat and Pieter van der Deure, Ontwerp tot een minst kostbaare zeekerste en schielykste herstelling van de zorgeliyke toestand der Westfriesche zeedyken, Print, 1733, Zuiderzeemuseum, Enkhuizen, http://hdl.handle.net/21.12111/zzm-collect-4718. CC BY-SA 4.0.

the existential and very public nature of the shipworm threat and cleverly combined this sense of urgency with patriotic appeal. Implementation of their proposal was imperative. 'The prosperity and salvation of our Beloved Fatherland', they argued, 'hangs in the balance'[76] (Figure 12.3).

By 1735, water boards across Holland began implementing variations of Straat and Van der Deure's design. West Frisian water boards started rebuilding dikes as early as 1733. This process was slow because the amount of stone required to line the coastline was immense. Water boards thus prioritised the most vulnerable portions of the dike first. West Frisian coastlines were not equally suited to these changes either. Coastal erosion had taken its greatest toll on the northern foreshores. The deeper waters fronting its dikes required a greater amount of stone.[77] Adding to these challenges, the large stones required for reconstruction were not locally available. Water boards thus relied on merchants to import stone from rural Drenthe in the east of the Republic, and eventually Norway. These supply chains developed quickly and, by 1735, dike reconstruction accelerated.[78]

Elsewhere along the Southern Sea coast, other water boards in Holland followed suit. After several months of consultation with hydraulic experts in 1734 and 1735, the Diemer dike water board began rebuilding their seawall following the pattern pioneered in West Friesland. They added a layer of rubble and broadened its seaward face in a manner reminiscent of Zeeland's dike traditions.[79] At this point, West Friesland's dikes remained relatively steep by comparison, but they gradually adopted gentler slopes as well. The neighbouring Bunschoter dike underwent a similar process beginning in 1737.[80] Regardless of location, the cost of dike adaptation was enormous – usually far more than at-risk communities could finance themselves. Water boards thus appealed to Holland for assistance. Just as it had in West Friesland, the politically fraught nature of dike financing delayed provincial support, but the fear that landholders might abandon their farms (and thus dike maintenance responsibilities) combined with the existential nature of the shipworm crisis to ensure support eventually arrived.[81] Holland ultimately subsidised roughly 66 percent of the more than 6.5 million guilders spent on dike reconstruction between 1732 and 1743.[82]

Holland's willingness to subsidise reconstruction also speaks to the perceived value of these designs. Gently sloping stone-lined dikes reduced the erosion of foreshores and added stability to these shifting coastlines. They effectively resisted storm surges, providing the degree of protection coastal communities had long sought. Although expensive in the short term, stone slopes reduced the costs associated with continuous maintenance of wooden infrastructure. Most importantly, replacing or shielding wood with stone reduced shipworm habitat. These were precisely the requirements laid out in the first *Europische Mercurius* report in 1732. Stone slopes appeared cost-effective strategies to promote flood security and reduce shipworm risk.

Authorities in Zeeland watched these developments in Holland with great interest. The shipworm outbreak had produced a financial crisis just as it had in Holland. By 1732, the province was forced to subsidize cash-strapped water boards on Walcheren.[83] In March 1735, the province sent a delegation to visit with water authorities and inspect water defences in Groningen, Friesland and Holland. They paid particular attention to those regions that responded to the 'plague of sea worms' by developing 'a new manner of replacing pile works with stone'.[84] They were keen to employ similar strategies on Walcheren. The delegates submitted their report in May, and one month later water boards began laying Norwegian stone at the foot of groynes, hoping to prevent shipworm infestation.[85]

Interest in this solution was relatively short lived, however. Walcheren's envirotechnical system differed from those along the Southern Sea because it relied on the beaches and dune landscapes that lined its coasts. Beach erosion and shrinking dunes meant that the sea and sand inched closer to towns and productive landscapes each year. Water authorities considered wooden *hoofden* the best strategy to stabilize these landscapes and invested considerable expense in their construction over the eighteenth century.[86] Shipworms threatened dikes fronted by the sea just as they did in Holland, but they also attacked the structures that protected these beaches and dunes. Shipworms presented a double challenge for Walcheren's envirotechnical system.

Zeeland responded by doubling down of their defence of wooden infrastructure. By 1737, Walcheren's water authorities shifted their attention back to maritime experience. On 29 August, Reynvaan reported that the water board had begun cladding Westkapelle piles with large, flat-faced iron nails (*wormspijkers*).[87] Just as they had on VOC ships, the nails would gradually rust and coat the piles in an iron oxide barrier. Like stone slopes, this solution was enormously expensive. Every pile required 100 pounds of nails and *hoofden* could extend 200 metres out from shore. The water board spent 14,000 guilders on *wormspijkers* in 1737 and nearly double that amount one year later. These costs would mount for the next 60 years, amounting to nearly 3 million guilders.[88] Some questioned the durability of worm nails, which would loosen as wood rotted or could be torn free if struck by ships that moored nearby.[89] Rust also took time to accumulate and, in the meantime, piles risked colonisation. Shipworm protection was far from absolute. Relative to other strategies, however, Walcheren's water boards considered this the most cost-effective solution.

Wormspijkers had other advantages too. They were a relatively simple, battle-tested technology, readily adapted from maritime experience. They did not require dike builders abandon coastal engineering traditions they had developed over centuries. The numerous *hoofden* that protected the coastal landscape of Walcheren were also 'thick' prestige objects and a great source of pride to water authorities. To alter or, worse, abandon their use would have undermined the cultural foundations of this envirotechnical system.[90] Holland's water authorities, by contrast, were already deeply embedded in a paradigm that privileged improvement and innovation.[91] *Wormspijkers* provided neither the security, durability, nor cost-effectiveness of stone slopes, but neither could they have. The Westkapelle dike was merely one part of a complex envirotechnical system that included beaches, dunes and dikes. The implications of shipworm infestation in the very different environmental context of Holland provided at best a partial equivalent. Walcheren adopted a set of solutions that accommodated shipworm risk. In much the same way shipbuilders had learned to live with woodborers, water authorities integrated *Teredo navalis* into their envirotechnical system as a new actor and manageable hazard of life along the coast.

Conclusion

Interpreting the shipworm disaster through an envirotechnical lens challenges important aspects of the dike modernisation narrative. Dikes were merely one component of the envirotechnical systems that Dutch communities depended upon for security and prosperity. A story that focuses exclusively on wier and pile dikes in Holland or the gently sloping Zeelandic dikes protected by *hoofden* and *staketwerken* misses the ways these technologies interacted with dynamic landscapes and were themselves transformed. The slow erosion of foreshores in Holland and Walcheren's beach and dune activity not only set the stage for shipworm colonisation but defined the bounds of potential adaptation.

An envirotechnical approach also emphasises the diverse human factors that transformed these systems. Water boards were important components of envirotechnical systems, and their distinctiveness reflected the diversity of the landscapes they governed. Dikes were objects thick with cultural meaning, which embodied their political power and financial responsibility. Political and social conflict often coloured decision making and certainly influenced the pace and character of shipworm adaptation. The cultural imperatives to foster security, stability and prosperity, however, served as common ground and the emergence of a new biological agent of change strengthened these connections. The outbreak of *Teredo navalis* represented the most widespread challenge to these shared imperatives to date.

The shipworm 'plague' was an envirotechnical disaster. This interpretation emphasises the ecological and social processes that precipitated the disaster and defined the pace and character of the changes that followed. In the centuries preceding the shipworm outbreak, water management institutions had increasingly relied on wooden infrastructure as a technical solution to the increasing flood risk that resulted from the exploitation of coastal environments. These changes widened pathways for shipworm colonisation and transformed flood protection into a high-risk system. Facing existential crisis and in the absence of a reliable solution drawn from dike traditions, water managers rapidly expanded the sphere of technical input.

Over the course of the 1730s, Dutch water authorities tested the limits of transfer across envirotechnical systems. The disaster expanded communication between water boards across the Republic. It also opened new dialogue with maritime shipbuilding, which had long before integrated shipworms into its envirotechnical systems. In the crisis moment of the

1730s, the common threat of shipworms bridged longstanding barriers and by the end of the decade, water authorities had systematically redis-covered the benefits and limits of maritime adaptations. After 1733, they would be weighed against those of stone slopes. The latter entailed the reduction of shipworm habitat. This strategy clearly differed from mari-time experience, which had long accepted shipworms as a manageable cost of life at sea.

Walcheren showcased a divergent trajectory in this story of en-virotechnical transfer. Faced with a similar biological challenge, they retained maritime techniques. The implications of shipworms were no less disastrous than in Holland, but the envirotechnical system was different. Shipworms threatened vulnerable seawalls, but they also colonised the wooden *hoofden* that protected eroding beaches and shifting dunes. Walcheren's water authorities sought techniques that protected these nat-ural barriers. In subsequent decades, Holland would also incorporate *wormspijkers* where stone slopes proved infeasible. Divergent adaptations ultimately reflected Holland and Zeeland's distinct envirotechnical systems, but shipworms successfully bridged an arguably deeper divide with ship-building. The foremost lesson of maritime experience was that shipworms could be limited, though not completely controlled. *Teredo navalis* would never again present an existential threat to the Netherlands, but as subse-quent outbreaks would make clear in the 1770s, 1820s and 1850s, it would remain a permanent player in the dynamic interaction of its coastal tech-nologies and environments.[92]

Notes

1 Adam Sundberg, 'An Uncommon Threat: Shipworms as a Novel Disaster', *Dutch Crossing*, 40, 2 (2016), 122–38. A précis of this work appeared in a special issue of the journal *Water History*. Seohyun Park, Scot McFarlane, Adam Sundberg, et al., 'Water history in the time of COVID-19: cancelled conversations', *Water History*, 12 (2020), 229–249. Springer Nature. The author thanks the editors, Ellen Arnold and Maurits Ertsen, for their feedback.
2 Karel Davids, *The Rise and Decline of Dutch Technological Leadership: Technology, Economy and Culture in the Netherlands, 1350–1800* (Leiden; Boston: Brill, 2008), 67–69; G.P. van de Ven, *Man-made Lowlands: History of Water Management and Land Reclamation in the Netherlands* (Utrecht: Matrijs, 2004), 144–45.
3 Sara B. Pritchard, 'Toward an Environmental History of Technology', in *The Oxford Handbook of Environmental History*, ed. by Andrew Isenberg (Oxford: Oxford University Press, 2014), 228.
4 Dolly Jørgensen, 'Not by Human Hands: Five Technological Tenets for Environmental History in the Anthropocene', *Environment and History*, 20, 4 (2014), 479–89; Etienne Benson, 'Generating Infrastructural Invisibility: Insulation, Interconnection, and Avian Excrement in the Southern California Power Grid', *Environmental Humanities*, 6, 1 (2015), 103–30.

5 Scott Gabriel Knowles, 'Learning from Disaster? The History of Technology and the Future of Disaster Research', *Technology and Culture 55*, 4 (2014), 773–84; Sara B. Pritchard, 'An Envirotechnical Disaster: Nature, Technology, and Politics at Fukushima', *Environmental History*, 17, 2 (2012), 224.

6 Thomas P. Hughes, 'The Evolution of Large Technological Systems', in *The Social Construction of Technological Systems: New Directions in the Sociology and History of Technology*, ed. by Wiebe E. Bijker, Thomas P. Hughes, and Trevor Pinch (Cambridge, MA: MIT Press, 2012), 60.

7 Karel Davids, 'Hydraulic Experts and the Challenges of Water in Early Modern Times', in *Urbanizing Nature: Actors and Agency (Dis)Connecting Cities and Nature Since 1500*, ed. by Tim Soens, Dieter Schott, Michael Toyka-Seid, and Bert De Munck (London; New York: Routledge, 2019), 193.

8 Leo Hollestelle, 'De zorg voor de zeewering van Walcheren ten tijde van de Republiek, 1574–1795', in *Duizend jaar Walcheren: Over gelanden, heren en geschot over binnen- en buitenbeheer*, ed. by P. A. Henrdickx, J. A. Lantsheer, and A. C. Meijer (Middelberg: Koninklijk Zeeuwsch Genootschap der Wetenschappen, 1996), 104.

9 Middelburg, Zeeuws Archief (ZA), Notulen van de Staten (en Gecommitteerden van de Breede Geërfden), 1511–1812, 23 November 1730, 3000.20.

10 ZA, 3000.20, 28 December 1730.

11 Van de Ven, *Man-made Lowlands*, 65–67, 87–88, 119–21.

12 Erik Thoen, 'Clio Defeating Neptune: A Pyrrhic Victory? Men and their Influence on the Evolution of Coastal Landscapes in the North Sea Area', in *Landscapes or Seascapes: The History of the Coastal Environment in the North Sea Area Reconsidered*, ed. by Erik Thoen et al. (Turnhout: Brepols, 2013), 397–423; Petra J. E. M. Van Dam, 'Eel Fishing in Holland: The Transition to the Early Modern Economy', *International Journal of Maritime History*, 15, 2 (2003), 163–75; Petra J. E. M. van Dam, 'Rabbits Swimming Across Borders: Micro-Environmental Infrastructures and Macro-Environmental Change in Early Modern Holland', in *Ecologies and Economies in Medieval and Early Modern Europe*, ed. by Bruce Scott (Leiden; Boston: Brill, 2010), 62–90.

13 Tim Soens, Greet De Block, and Iason Jongepier, 'Seawalls at Work: Envirotech and Labor on the North Sea Coast before 1800', *Technology and Culture*, 60, 3 (2019), 688–725.

14 Wiebe E. Bijker, 'Dikes and Dams, Thick with Politics', *Isis*, 98, 1 (2007), 109–23; Adam Sundberg, *Natural Disaster at the Closing of the Dutch Golden Age: Floods, Worms, and Cattle Plague* (Cambridge, UK; New York: Cambridge University Press, 2022).

15 Petra J.E.M. van Dam, 'Ecological Challenges, Technological Innovations: The Modernization of Sluice Building in Holland, 1300–1600', *Technology and Culture*, 43, 3 (2002), 500–20.

16 Piet Boon, 'Voorland en inlagen: De Westfriese strijd tegen het buitenwater', *West-Friesland Oud en Nieuw*, 58 (1991), 78–113.

17 Hollestelle, 'De zorg voor de zeewering van Walcheren ten tijde van de Republiek', 106, 125–26.

18 Michiel Bartels, 'Het Bolwerk tegen de woede van de zee', in *Dwars door de dijk: Archeologie en geschiedenis van de Westfriese Omringdijk tussen Hoorn en Enkhuizen*, ed. by Michiel Bartels (Hoorn: Stichting Archeologie West-Friesland, 2016), 134–40.

19 Alfons Fransen, *Dijk onder spanning: De ecologische, politieke en financiële geschiedenis van de Diemerdijk bij Amsterdam, 1591–1864* (Hilversum: Verloren, 2011); C. Baars, 'Nabeschouwing over de paalwormplaag van 1731/32 en de gevolgen daarvan', *Waterschapsbelangen*, 15 (1993), 505.

20 Soens, De Block, and Jongepier, 'Seawalls at Work', 699–703.

21 Peter Paalvast and Gerard van der Velde, 'Distribution, Settlement, and Growth of First-Year Individuals of the Shipworm *Teredo navalis* L. (Bivalvia: Teredinidae) in the Port of Rotterdam Area, the Netherlands', *International Biodeterioration & Biodegradation*, 65, 3 (2011), 379.

22 Albert van Brakel, 'De paalworm in Hollandse zeedijken', *Tijdschrift voor Waterstaatsgeschiedenis*, 24, 2 (2015), 78–79.

23 Cornelis Alard Abbing, *Geschiedenis der stad Hoorn, hoofdstad van West-Vriesland gedurende het grootste gedeelte der 17e en 18e Eeuw, of vervolg op Velius Chronyk*, Vol. 2 (Hoorn: Gebr. Vermande, 1842), 243.

24 Anon., *Beschryvinge, van de schade en raseringe aan de zee-dyken van Noort-Holland en Westfriesland, door de worm in de palen, en de daar op gevolgde storm, en vervolgens* (Hoorn: Jacob Duyn, 1732), 9, 11.

25 Petra J. E. M. van Dam, 'An Amphibious Culture: Coping with Floods in the Netherlands', in *Local Places, Global Processes: Histories of Environmental Change in Britain and Beyond*, ed. by David Moon, Paul Warde, and Peter Coates (Oxford: Windgather Press, 2016), 82.

26 Jan de Bruin and Diederik Aten, 'Een Gemene Dijk? Verwikkelingen rond dijkzorg in West-Friesland, De Watersnood van 1675–1676', *21ᵉ uitgave van de vrienden van de Hondsbossche, Kring voor Noord-Hollandse water-staatsgeschiedenis*, 21 (2004), 29.

27 Hoorn. Westfries Archief (WFA), Ambacht van Westfriesland genaamd De Vier Noorder Koggen, 'Verbaal doen maken bij den Dijkg ... aan de Hoog Edele Heeren', (1732), 1558.1820.

28 Joop W. Koopmans, 'The Early 1730s Shipworm Disaster in Dutch News Media', *Dutch Crossing*, 40, 2 (2016), 139–50.

29 The *Europische Mercurius* was a bi-annual publication divided into two parts. Both Volume One and Two of the 1732 publication included special appendices about the shipworm 'plague'. *Europische Mercurius, Volume 1* (Amsterdam:J. Ratelband en Compagnie, 1732), 296–309; *Europische Mercurius, Volume 2* (Amsterdam:J. Ratelband en Compagnie, 1732), 281–313. Shipworm reportage appeared again in its 1733 volumes. *Europische Mercurius, Volume 1* (Amsterdam:J. Ratelband en Compagnie, 1733), 285–92. *Europische Mercurius, Volume 2* (Amsterdam:J. Ratelband en Compagnie, 1733), 288–300.

30 Jacob Harkenroht, *Worm in Nederlands paalwerk voor de zeedyken, tot een buitengewoone straffe, van den Heere der Heirschaaren beschikt* (Groningen: Laurens Groenewout, 1733), 60–62.

31 Pieter Boddaert, *Stichtelyke gedichten van Pieter Boddaert* (Leiden: Samuel and Johannes Luchtmans, 1735), 48.

32 Sundberg, *Natural Disaster*.

33 Nicholaas Kist, *Neêrland's Bededagen En Biddagsbrieven* (Leiden: S. and J. Luchtmans, 1849), 327.

34 Sundberg, 'Uncommon Threat'.

35 The Hague, Koninklijke Bibliotheek (KB), Wijnandt Nieuwstadt to William Jacob's Gravesande, 128 D3.

36 Sundberg, *Natural Disaster*, Chap. 3.

37 Richard Unger, *Dutch Shipbuilding before 1800: Ships and Guilds* (Amsterdam: Van Gorcum, 1978), 10–11, 47; Johan de Jong, 'Standvastigheid & erwachting: A Historical and Philosophical Inquiry into Standardization and Innovation in Design and Production of the VOC Retourship during the 18th century' (Unpublished master's thesis, University of Twente, 2010), 59–60.

38 Wendy van Duivenvoorde, *Dutch East India Company Shipbuilding: The Archaeological Study of Batavia and other Seventeenth-century VOC Ships* (College Station: Texas A&M University Press, 2015), 20–21.

39 De Jong, 'Standvastigheid & Verwachting', 59–80.

40 James T. Carlton and J. Hodder, 'Biogeography and Dispersal of Coastal Marine Organisms: Experimental Studies on a Replica of a 16th-Century Sailing Vessel', *Marine Biology*, 121, 4 (1995), 722.

41 Richard Barker, 'Careening: Art & Anecdote', *Mare Liberum*, 2 (1991), 177–207.

42 Nicolaes Witsen, *Aeloude en hedendaegsche scheeps-bouw en bestier* (Amsterdam: C. Commelijn & J. Appelaer, 1671), 267.

43 Wendy van Duivenvoorde, 'Use of Pine Sheathing on Dutch East India Company Ships', in *Between Continents: Proceedings of the Twelfth Symposium on Boat and Ship Archaeology* (Istanbul: Yayinlari 2009), 241.

44 Cornelius van Yk, *De Nederlandsche scheepsbouw-konst open gestelt* (Amsterdam: Andries Voorstad, 1697), 91.

45 Van Duivenvoorde, *Dutch East India Company Shipbuilding*, 198–99, 203.

46 Wendy van Duivenvoorde, 'The Use of Copper and Lead Sheathing in VOC Shipbuilding', *International Journal of Nautical Archaeology*, 44, 2 (2015), 350–51.

47 John Harland, 'Piet Heyn and the Early Use of Copper Sheathing', *The Mariner's Mirror*, 62 (1976), 1–2.

48 Van Yk, *De Nederlandsche scheepsbouw-konst open gestelt*, 121.

49 Van Duivenvoorde, 'The Use of Copper', 350–51.

50 Dániel Margócsy, 'Technological Transfer, Ship Design and Urban Policy in the Age of Nicolaes Witsen', in *Knowledge and the Early Modern City: A History of Entanglements*, ed. by Bert de Munck and Antonella Romano (London; New York: Routledge, 2019), 149–70.

51 Richard Unger, 'The Technology and Teaching of Shipbuilding, 1300–1800', in *Technology, Skills and the Pre-Modern Economy in the East and the West*, ed. by Maarten Prak and Jan Luiten van Zanden (Leiden; Boston: Brill, 2013), 161–204; David H. Roberts, *18th-Century Shipbuilding. Remarks on the Navies of the English & the Dutch from Observations Made at their Dockyards in 1737 by Blaise Ollivier* (East Sussex: Jean Boudriot, 1992), 6; Davids, *Rise and Decline*, 390.

52 Stephan R. Epstein, 'Transferring Technical Knowledge and Innovating in Europe, c. 1200 – c. 1800', in *Technology, Skills and the Pre-Modern Economy*, 39–40; Davids, *Rise and Decline*, 262–63.

53 There is evidence of small-scale woodboring activity in Dutch waters as early as 1580, though attribution to *Teredo navalis* is uncertain.

54 Advertisement, 5 July 1732, *Amsterdamse Courant*.

55 *Europische Mercurius*, Volume 2 (1732), 303–5.

56 *Europische Mercurius*, Volume 2 (1732) 304–5.

57 The Hague, Nationaal Archief (NA), Gedeputeerden van Haarlem ter Dagvaart, 'Notitie van missiven, memorien, en requesten ... van middelen tot verdrijving van de Zee wormen', 3.01.09 1238.

58 NA, 3.01.09 1238, no. 8.
59 NA, 3.01.09 1238, 'Isack Levi'.
60 WFA, 1562.1443, 27 Mar 1733, 'Memorie van de proefpalen'; NA, 3.01.09 1238, 'Frederick Duim', 'Gilbertus Verschoor'.
61 Ludovicus Morgenster, *Memorie overgelevert ... inhoudende zeker gereed, duurzaam, en min kostbaar Middel* (Utrecht: Melchior Leonard Charlois, 1732), 8; NA, 'Notitie van memorien, missiven &c: vervattende remedien tegen de wormen', 3.01.09 1238.
62 WFA, 9 September 1732, 'Memorie van de missiven en Persoonen en merken der Proefpalen', 1562.1443.
63 WFA, 'Memorie', 4 December 1732.
64 ZA, 3000.20, 18 January 1731.
65 ZA, 3000.20, 1 February 1731.
66 ZA, 2.1.332, Resolution 6 June 1735.
67 ZA, 3000.20, 9 November 1731.
68 WFA, 1562.1443, 9 February 1732.
69 WFA, 1562.1443, 31 July 1734.
70 ZA, Verzameling Handschriften Gemeentearchief Veere, 'Testament van Bommenee', 2854.59, fol. 213.
71 *Europische Mercurius,* Volume 2 (Amsterdam: J. Ratelband en Compagnie, 1732), 285.
72 Seger Lakenman, *Het wonder Ordeel Godts, ofte een kort verhaal van de ongehoorde bezoeking dezer Provintie door zekere plage van Zee-wormen* (s.n.,s.l., 1732), 15.
73 Johannes Jouke Schilstra, *Wie Water Deert. Hoogheemraadschap van de Uitwaterende Sluizen in Kennemerland en West Friesland, 1544–1969* (Wormerveer, Meijer Pers, 1969), 136–37; Michiel. H. Bartels, Peter Swart, and Harmen de Weerd, 'Wormspijkers in het Medemblikker havenhoofd. Archeologisch en historisch onderzoek naar de maatregelen tegen de paalworm in het havenhoofd van Medemblik, West-Friesland', *West-Friese Archeologische Rapporten,* 80 (2015), 28.
74 Sundberg, 'Uncommon Threat', 129–32.
75 C. Baars 'Het dijkherstel onder leiding van de Staten van Holland', *Waterschapsbelangen,* 6 (1989), 197–201.
76 Pieter Straat and Pieter van der Deure, *Ontwerp Tot Een Minst Kostbaare Zeekerste En Schielykste Herstelling Van De Zorgelyke Toestand Der Westfriesche Zeedyken* (Amsterdam: J. Oosterwyk, 1733), 6.
77 Baars, 'Dijkherstel', 200.
78 Baars, 'Dijkherstel', 202.
79 Fransen, *Dijk onder spanning,* 218.
80 Heleen Kole, *Polderen of niet? Participatie in het bestuur van de water-schappen Bunschoten en Mastenbroek vóór 1800* (Hilversum: Verloren, 2017), 136–40.
81 Fransen, *Dijk onder spanning,* 213–45.
82 Baars, 'Nabeschouwing', 507.
83 Hollestelle, 'De zorg voor de zeewering van Walcheren ten tijde van de Republiek', 121.
84 ZA, 3000.2021, 29 March 1735.
85 ZA, 3000.2021, 26 May 1735, 23 June 1735.
86 Hollestelle, 'De zorg voor de zeewering van Walcheren ten tijde van de Republiek', 122–23, 125–28.

87 ZA, 3000.2021, 29 August 1737.
88 Hollestelle, 'De zorg voor de zeewering van Walcheren ten tijde van de Republiek', 124, 126.
89 ZA, 2854.59, fols 214–215.
90 Hollestelle, 'De zorg voor de zeewering van Walcheren ten tijde van de Republiek', 128.
91 Sundberg, *Natural Disaster.*
92 Peter Paalvast and Gerard van der Velde, 'New Threats of an Old Enemy: The Distribution of the Shipworm *Teredo navalis* L. (Bivalvia: Teredinidae) related to Climate Change in the Port of Rotterdam Area, The Netherlands', *Marine Pollution Bulletin*, 62, 8 (2011), 1822–29.

Contributors

Ovanes Akopyan is a Marie Skłodowska-Curie fellow at Ca' Foscari University of Venice. He is the author of *Giovanni Pico della Mirandola's* Disputationes adversus astrologiam divinatricem *and Its Reception* (Brill, 2021) and the editor or co-editor of three collections of essays. His second monograph, provisionally entitled *Explaining Natural Disasters in Early Modern Europe: Science, Politics, Rhetoric*, is under contract with Johns Hopkins University Press.

David Rosenthal is a historian of early modern Italy, specialising in urban social history and public history with mobile media. He is the author of *Kings of the Street: Power, Community, and Ritual in Renaissance Florence* (Turnhout: Brepols, 2015), co-editor of *Hidden Cities: Urban Space, Geolocated Apps, and Public History in Early Modern Europe* (Routledge, 2022) and has written numerous articles. He is a research fellow at the University of Exeter and co-director of Hidden Cities apps.

Gerrit Jasper Schenk is Professor of Medieval History at Technische Universität Darmstadt (Germany). His research is on historical disaster research, environmental history, history of infrastructure, urban history, rituals and ceremonies. Chronologically, he focuses on the late Middle Ages and early modern period, and spatially, the empire as well as Italy. He has curated smaller and larger historical exhibitions; currently, he is leading research projects on the 'fluvial anthroposphere'.

Lydia Barnett is a historian of science and the environment in early modern Europe. She teaches at Northwestern University in Evanston, IL, where she is an Associate Professor of History. Her first book, *After the Flood: Imagining the Global Environment in Early Modern Europe* (Johns Hopkins University Press, 2019), argued for the importance of religion in the emergence of global environmental consciousness in premodernity and was the winner of the 2019 Morris D. Forkosch

Prize. She is currently working on a book about labour and ecological knowledge in the eighteenth century.

Sara Miglietti is Senior Lecturer in Cultural and Intellectual History at London's Warburg Institute. She specialises in European intellectual history (1500–1700), with a focus on political philosophy, book history, and translation and reception studies. She is the editor of *Reading Publics in Renaissance Europe* (2016), *Climates Past and Present* (2017) and *Governing the Environment in the Early Modern World* (2017) and is currently completing a monograph on the reception and transformation of ancient climate theories in the early modern period.

William M. Barton studied Greek and Latin in Britain and Canada, receiving his Ph.D. in Classics from King's College, London in 2015. Barton's research interests centre around the uses of Ancient Greek and Latin in literary production after the Classical period, in particular the late-antique and early modern periods. His work has focused on the themes of the representation of the natural landscape in both poetry and prose, the place of the Classical languages in the history of science, and more recently autobiography and ego-literature.

Martin Korenjak was born in 1971 in Wels (Austria). He studied Classical Philology and Linguistics at Innsbruck and Heidelberg (1990–1996), worked as research assistant at Innsbruck (1997–2003), and was Professor of Classical Philology at Bern (2003–2009), before returning to Innsbruck in 2009. Since 2011, he is working there partly for the University and partly for the Ludwig Boltzmann Institute for Neo-Latin Studies. His research areas include Greek and Latin poetry, rhetoric, literary theory and criticism, the reception of classical antiquity and Neo-Latin literature.

Milena Viceconte obtained a Ph.D. in History of Art in 2013, a joint degree between the University of Naples Federico II and the Universitat de Barcelona. She took part in several research groups within the University of Barcelona, focused on artistic circulation between Italy and Spain in the early modern period. Since 2018, she has been a postdoctoral fellow at the University of Naples Federico II in the framework of the DisComPoSE project, within which she deals with issues related to the imageries of disasters in the territories of the Spanish Monarchy through the analysis of figurative sources (sixteenth to eighteenth centuries).

Felicia M. Else is Professor of Art History at Gettysburg College, Pennsylvania. She received her Ph.D. in Art History from Washington University in St. Louis and has published on the connections of art, water, cartography and natural history in (sixteenth-century Florence, including the work of Bartolomeo Ammannati. She is the author of *The Politics of Water in the Art and Festivals of Medici Florence: From Neptune Fountain to Naumachia* (Routledge, 2019) which explores the intersection of water, art and festivals in the age of Cosimo I de' Medici and his heirs, Francesco I and Ferdinando I. Her current projects include an article on a representation of Peru in a (sixteenth-century entry into Florence and an extended study of a notorious nickname, 'Biancone', given to Ammannati's famous colossal statue of Neptune and its ties to the traditions of the mock heroic and carnivalesque.

Pamela O. Long is a historian of late medieval and early modern European science and technology, and cultural history. Her fellowships include those at American Academy in Rome, the Guggenheim Foundation, the John D. and Catherine T. MacArthur Foundation, and the Institute for Advanced Study at Princeton. Her books include *Openness, Secrecy, Authorship: Technical Arts and the Culture of Knowledge from Antiquity to the Renaissance* (2001); *Artisan/Practitioners and the Rise of the New Sciences* (2011); and *Engineering the Eternal City: Infrastructure, Topography, and the Culture of Knowledge in Late Sixteenth-Century Rome* (2018).

Emanuela Ferretti is an associate professor at the University of Florence. She received a graduate degree in 'Art History and Archaeology' from the University of Siena (2000), and a Ph.D. in Architecture ('History of Architecture') from the University of Florence (2004). She had a fellowship at Villa I Tatti – The Harvard University Center for Renaissance Studies (2012–2013). Her interests and publications focus on questions of design, style, building materials, and structures, from the fourteenth to the end of the sixteenth century (Leonardo and Michelangelo, in particular). A particular research interest is the modes of communication used in architectural projects – e.g., drawings, models and administrative documents – from the Renaissance to the present. Her more recent publications include 'Con lo sguardo di Leonardo. L'arte edificatoria e il microcosmo del cantiere', Firenze, Giunti, 2023, in the series 'Laboratorio Rinascimento', edited by Paolo Galluzzi.

Monica Azzolini teaches history of science at the University of Bologna. She has published widely on the anatomical studies of Leonardo da Vinci, Renaissance astrology, and the formation and circulation of scientific knowledge, with particular emphasis on the relationship between orality and the written word, and the use of scientific illustration. She is the author of *The Duke and the Stars: Astrology and Politics in Renaissance Milan* (Harvard University Press, 2013). In recent years, her interest has shifted to the underground: she has explored the relationship between natural disasters and cult saints during the

Counter-Reformation, and she is currently working on a history of the underground in early modern Italy.

Adam Sundberg, Ph.D., is an associate professor of history at Creighton University in Omaha, USA. He works primarily at the intersection of environmental history and the history of disasters. His first book, *Natural Disaster at the Closing of the Dutch Golden Age: Floods, Worms, and Cattle Plague,* was published by Cambridge University Press in 2022. He is currently working on a global environmental history of aquatic and marine species introductions.

Index

For Product Safety Concerns and Information please contact our EU
representative GPSR@taylorandfrancis.com
Taylor & Francis Verlag GmbH, Kaufingerstraße 24, 80331 München, Germany